The Farm Security Administration
and Rural Rehabilitation in the South

The
Farm Security Administration
and Rural Rehabilitation
in the South

Charles Kenneth Roberts

The University of Tennessee Press
Knoxville

Copyright © 2015 by The University of Tennessee Press / Knoxville.
All Rights Reserved. Manufactured in the United States of America.
First Edition.

Portions of chapter 7 were first published in *Agricultural History*, volume 87, no. 3 (Summer 2013). Reprinted by permission of the Agricultural History Society.

Portions of chapter 8 were first published in *The Alabama Review*, volume 66, no. 2 (April 2013). Reprinted by permission of the Alabama Historical Association.

The paper in this book meets the requirements of American National Standards Institute / National Information Standards Organization specification Z39.48-1992 (Permanence of Paper). It contains 30 percent post-consumer waste and is certified by the Forest Stewardship Council.

Library of Congress Cataloging-in-Publication Data
Roberts, Charles Kenneth.
The Farm Security Administration and rural rehabilitation in the South / Charles Kenneth Roberts.
    pages cm
ISBN 978-1-62190-160-0 (hardback)
1. United States. Farm Security Administration.
2. Agriculture and state—Southern States—History—20th century.
3. Farms, Small—Government policy—Southern States—History—20th century.
4. Rural development—Southern States—History—20th century.
5. Rural poor—Southern States—History—20th century.
6. Southern States—Economic conditions—20th century. I. Title.

HD1773.A5R63 2015

338.1'80975—dc23   2014040820

Contents

Preface ix
Introduction xvii
List of Abbreviations xxxi

PART I.
Finding Solutions: The Origins and Evolution of New Deal Rural Poverty Policy — 1

Chapter 1.
From Relief to Rehabilitation: The Origins of Rural Rehabilitation — 5

Chapter 2.
Going Back to the Land: The Division of Subsistence Homesteads — 29

Chapter 3.
Unifying Rural Reform: The Resettlement Administration — 49

Chapter 4.
The Farm Security Administration before World War II — 71

PART II.
Fighting Rural Poverty: Farm Security in Practice — 89

Chapter 5.
Regaining a Lost Security: The Rural Rehabilitation Program — 91

Chapter 6.
Creating Family Farms: The Tenant Purchase Program — 109

Chapter 7.
Rehabilitation in Action: Credit and Supervision — 129

Chapter 8.
The Resettlement Communities around Birmingham, Alabama — 157

Chapter 9.
The Farm Security Administration and World War II — 179

Conclusion 201

Notes	207
Bibliography	267
Index	285

Illustrations

Figures

Dunmovyn, One of the Palmerdale Homesteads Near Birmingham, Alabama	x
Nurse Brings Hookworm Medicine to the Lewis Family	xxviii
Making Farm and Home Plans with Rural Rehabilitation Clients	23
Construction of a Community Center at Bankhead Farms, Alabama	35
Alabama Coal Miner, Bankhead Mines, Walker County, Alabama	39
A House at Palmerdale Homesteads	43
Cartoon: "Modern Gulliver Helps the Little People"	50
Rexford G. Tugwell	54
C. B. Baldwin	75
Examining the Mouth and Teeth of a Client's Mare	79
Tenant Purchase Client on a New Hay Rake	112
Nat Williamson and Farm Security Administration Official E. H. Anderson	119
Charlie McGuire with His Mules and Cultivator	124
Interviewing Applicants for Homesteads in Birmingham, Alabama	135
Homesteader with Some Baby Chicks	143
FSA–sponsored "Know your Farmer" Tour Stops	149
Terraced Fields at Greenwood Homesteads	162
Mr. and Mrs. J. A. Britain in Their New Home	168
Plowing a Field at Palmerdale, Alabama	175

Tables

1. FSA Programs by Expenditure, Fiscal Year 1941	73
2. Rural Rehabilitation Average Loan Size	139
3. Distance in Miles to Industrial Centers from Birmingham Homesteads	169

Preface

This book is dedicated to persons who have ideas of what can be done to help suffering, hopeless, unhappy humanity back to brighter, more useful lives; To the courage it takes to get the job done; and To the joyous feeling of happiness when success comes. This book is the answer to the question, "Was the Palmerdale Project a Success?" The answer comes loud and clear, "YES!!!" It was a WONDERFUL success.!![1]

Mattie Cole Stanfield's self-published 1984 history of the resettlement project and community known as Palmerdale begins with a discussion of Daniel Murphree, who in 1816 arrived in the area that would one day be known as Birmingham. The Murphree family ended up giving its name to the valley where the Palmerdale project was located and remained a source of community pride. Doubtless, part of the reason Stanfield included this story was for a little color and to share a proud bit of local history. But it also reflects a sense among many of those who suffered through the Great Depression and benefitted from the New Deal that their story was an integral part of the larger American one: the story of people looking for, and finding, a better life. The Great Depression threatened that effort, but with the help of the federal government and the hard work of those who built lives in communities like Palmerdale, that constant improvement in the lives of all Americans could continue. The Murphree family and those like them made a better society; the Palmerdale project helped maintain that tradition. Whatever the criticisms of the program or of Franklin Roosevelt, people like Mattie Cole Stanfield believed their own lives and communities proved the value of that government help.[2]

The willingness of the federal government to provide help, however, was not so certain. The New Deal as a whole failed to adequately help the rural poor or to create new, permanent government programs that might continue to provide aid to marginalized or struggling small farmers. The iconic New Deal agricultural program, the price supports for agricultural commodities that began in the Agricultural Adjustment Administration (but did not end when the Supreme Court struck down the Agricultural Adjustment Act), benefitted large farmers most of all and disproportionately hurt small farmers, tenants, and sharecroppers. Some New Deal programs had a better track record but lacked the national impact of the AAA and similar programs. The Tennessee Valley Authority, for

DUNMOVYN, ONE OF THE PALMERDALE HOMESTEADS NEAR BIRMINGHAM, ALABAMA. LIBRARY OF CONGRESS.

example, had a more positive impact on the lives of rural Americans than did AAA, but its necessarily limited geographic reach and its ever-closer ties with local power structures limited its impact. The Bureau of Agricultural Economics shared many of the FSA's goals, but as primarily a planning and information agency, its direct impact was necessarily limited.

The New Deal's most effective effort to improve the lives of rural Americans on a mass scale was the collection of programs eventually known as farm security, housed in the Farm Security Administration and its predecessor agencies. The work of the FSA most widely remembered is its Photographic Section, which even today defines the public image of the New Deal and the Great Depression, or the community-building program and the political controversy that surrounded it. Early on, New Dealers looking at rural reform proposed a wide range of programs—rural resettlement, community building, land use reform, various credit programs, and more. By the time the program had reached its peak around 1940 and 1941, however, New Dealers committed to the immediate improvement of the lives of small farmers and the rural poor emphasized rural rehabilitation, and especially supervised credit, as the solution. Despite popular recollections of the New Deal and rural America, rural rehabilitation was the broadest (and most broadly supported) effort to help small farmers.

The neglect of rural rehabilitation is easy to understand. Resettlement and photography, for example, left behind striking reminders of their history, physical manifestations in images and communities that even today shape Americans' understanding of the New Deal; and, ultimately, rural rehabilitation (like the agency that housed it) did not survive the antireform wave during World War II. But the results of rural rehabilitation, while not as apparent, were no less significant for its participants. Rural rehabilitation included tenure improvement, debt adjustment, cooperatives, and more, but the most important part of it, which came to define farm security by the late 1930s, was supervised credit. The federal government, through the FSA, provided tenants and small farmers who could not otherwise receive credit with loans, so long as the client family agreed to accept supervision from farm and home-management supervisors. Most loans were relatively small, but starting in 1937, a select few tenants received larger loans and more supervision to promote farm ownership. These loans and their accompanying supervision meant practical improvement for small farmers and tenants, providing them credit when none could be had or only at ruinous interest rates. But even more important, rural rehabilitation specifically and the FSA in general represented an opportunity. It had the potential to change the government's relationship with small farmers. Over eight hundred thousand families, perhaps even over a million, received loans and supervision before the end of the FSA, with about forty-seven thousand larger loans with more intensive supervision going to tenants for the purpose of turning them into farm owners. Many thousands more received help in the way of grants, tenure improvement, debt adjustment, or other forms of rural rehabilitation.

Many New Dealers considered such efforts absolutely necessary. Belief in the economic importance and moral value of farm life necessitated rehabilitating rural life—recovering the lost sense of security and the chance to contribute to and participate in American life, which farmers in previous generations had enjoyed. This was not really a new idea. The inherent values of farm life had been taken for granted for most of American history, at least going back to Thomas Jefferson's imagining the independent yeoman farmer as the sturdy foundation of a new democratic society. Most of the ideas behind the important New Deal programs addressing farming and rural life had been around for at least a few decades by the time the Great Depression hit. What made the Farm Security

Administration and its predecessor agencies like the Resettlement Administration unique was a willingness to extend government-backed aid and credit to small farmers and tenants, to restructure or at least improve rural communities and rural life on a mass scale, and to include at least some rural Americans whose voices otherwise went unheeded in debates over agricultural policy. No other New Deal agency, so directly or in such large numbers, provided government aid to marginalized and impoverished rural Americans as did the Farm Security Administration; indeed, no other federal program could or would even meaningfully propose to do so. When we remember the New Deal, that memory should include this unprecedented effort to extend the promise of farm security to at least a portion of small farmers and the marginalized rural poor.

That the New Deal failed in its efforts to provide farm security, that the federal government would cast aside rural rehabilitation as a meaningful goal or its programs as valuable tools for improving rural life, can be attributed to a number of factors. Would-be reformers deserve some of the blame. Romantic notions about the nature of rural living and a paternalistic view of poor farmers hampered building an effective relationship with rural clients; bureaucratic ineptitude and (especially in the early years) ideological confusion meant that many efforts were contradictory, expensive, and ineffective. Resettlement projects, for example, generally improved the lives of those who ended up living in them; however, they did so at great cost in terms of time, money, and political hostility. Rural rehabilitation and, more specifically, supervised credit worked, though after a good bit of experimentation and experience, and never perfectly or ideally, but better than the policies that replaced them—or rather, failed to replace them. These kinds of shortcomings in the farm security program had significance in that they made reform efforts more vulnerable to political assault. The FSA's ultimate failure resulted from the attacks of its political and ideological opponents, which could only be successful after former supporters in the federal government had withdrawn their protection.

The FSA, then, represents both a great success and a great failure. The New Deal was a governmental and social effort to end the Great Depression, if necessary by restructuring the economy itself, but it eventually became a guarantee that the federal government should permanently provide at least some measure of economic security to the American people. The history of farm security matches this course almost exactly—from the heady early days when New Dealers believed they could create new communities and new ways of living, to settling on the belief in rehabilitating farmers where they lived and helping some deserving tenants to that most American goal of owning their own farms. And, for millions of Americans, the federal government did all of that: creating new communities, reforming land use, saving marginal farmers from catastrophe, and even helping a few into farm ownership. And yet the FSA's commitment to reform did not survive past World War II; hostile political forces gutted it almost entirely by 1943 and then in 1946 replaced the FSA with the Farmers Home Administration. The kind of New Deal liberalism that had emerged by the late 1940s, whatever its rhetorical commitment to individual security, did not make room for everyone. The government proved unwilling to provide the same kind of help to small farmers and renters that it provided by the billions of dollars to larger producers (much less to bring into the mainstream truly marginalized groups like migrant farm laborers or the poorest sharecroppers).

This failure to consider the most vulnerable farmers, tenants, and small farmers who would stay on the land and protect their farms from the increasing dominance of industrialized agriculture suggests deep flaws in the consensus about the role of the liberal state emerging in postwar America. People were not merely left behind during America's postwar economic growth; the federal government actively encouraged policies that prized efficiency and productivity above all else. Even before Earl Butz phrased it so memorably, "get big or get out" was the reality for many American farmers. Whatever the very real questions about what the Farm Security Administration or similar programs might have done for small farmers and the rural poor after World War II (at a minimum, it would have had to continue to evolve and expand programs like rural rehabilitation, committing to the democratic promise that supervision could be a negotiation, a partnership, between supervisors and farmers), the utter disregard that federal policy had for those who suffered during the government-promoted shift to capital-intensive agribusiness indicates that things could not have been worse.

But the marginalization of small farmers and the rural poor continued, especially in the South. The federal government did not entirely create the trends favoring big farmers, but it did introduce policies that strengthened the position of the already largest and most influential. After World War II, most rural southerners realized that the farm had no place for them and moved to cities. Some who remained prospered through mechanization; others stayed impoverished, on farms that barely broke even or failed altogether. Farming became capital-intensive agribusiness, and rural southerners remained mired in poverty. FSA planners hoped at a minimum to use government-provided credit and expertise so that family-sized farms might continue to operate in an era of big farming. The defeat of the FSA meant that the federal government would not solve the problems that its embrace of ever more efficient, capital-intensive farming created. America has an agricultural system today that manages to be both hyperefficient and deeply dysfunctional. That system has its roots in the failures of the Farm Security Administration.

It would have been impossible to complete this project without a lot of help. Kari Frederickson provided me with the original idea which became this book, and she has been an advisor, mentor, friend, sounding board, teacher, and everything else while I worked. Her Proseminar and Seminar in Southern History classes set me on the course I have taken into the history of the American South. Along with Kari, Andrew Huebner, Lisa Lindquist-Dorr, George Rable, and Pete Daniel read through my original dissertation, and I thank them for their questions, feedback, and for getting through a manuscript that turned out much too long. In addition to their help with the book specifically, they, along with Lawrence F. Kohl, Joshua Rothman, George McClure, Paul Hagenloh, Harold Silesky, and Luke Niiler, taught me how to be a better student, teacher, and writer, and for that I thank them. R. Volney Riser, Kristopher Teters, Jess Gilbert, and reviewers at the *Alabama Review* (where parts of a different version of chapter 8 appeared) and *Agricultural History* (which featured parts of what became chapter 7), looked through portions of this manuscripts and in general put up with my questions, and I appreciate their aid and insight. The University of Tennessee Press has been exceedingly kind and helpful; I would particularly like to thank Thomas Wells, who is doubtless tired of my emails.

The Department of History and the Graduate School at the University of Alabama provided me with the financial support necessary to complete this project. The good people at Interlibrary Loan at Gorgas Library have put up with a number of obscure requests, and I thank them for it. More people than I can name in the University of Alabama History Department, from undergraduates to full professors, have helped or inspired me while I worked on this project: I appreciate them all. I also have to thank the friends I have made in Tuscaloosa, who have helped me to keep going with their support and camaraderie.

I could not have completed this project without the help of researchers and research assistants at archives and libraries across the country. I must express my thanks to the people at the National Archives at College Park, Maryland; the National Archives Southeast Region at Morrow, Georgia; the Alabama Department of Archives and History; and the Mervyn H. Sterne Library in Birmingham, Alabama. I particularly appreciate the help and patience of Robert G. Richards, who helped me the first time I visited the Southeast Region Archives, when I did not really know what I was doing or where I was going with this project. I also thank Elizabeth A. Boyne and Kathryn Coursey, who helped me obtain copies of records in Iowa and New York.

I would like to thank my family for all their support while I have been working on this book. My parents, Ken and Glenda Roberts, and my brother, Andy, have provided me with all the love and encouragement anyone could hope for. Rufus and Madison Roberts have kept me company while writing, and with only minimal distraction. Finally, I have to thank my wife, Lauren, for her love and support during the long, tiring, and poorly compensated years of graduate school and teaching. She has put up with me and my project since its beginning, and for that I cannot thank her enough.

Introduction

In January 1932 New York governor Franklin D. Roosevelt explained to the state legislature his plan for moving the urban unemployed to new semi-rural subsistence homesteads. The plan displayed Roosevelt's romantic view of rural living and his certainty that, on the farm, one could never truly suffer the kind of poverty so apparent in America's cities. "They may secure through the good earth," Roosevelt announced, "the permanent jobs they have lost in overcrowded, industrial cities and towns."[1] As president, Roosevelt would come to understand, if only partially, that the problems of rural America ran much deeper than he had realized. But his response—and that of the New Deal—would reflect these earlier misconceptions about rural life.

By and large, farmers found themselves left out of the prosperity that made the 1920s urban America's "Jazz Age." Rural Americans across the country saw conditions worsen as the decade went on. The concentration of poverty was greatest in the South, and southern poverty had deep historical roots, notably a long history of an impoverished agricultural caste. While small farms, misuse of the land, rural isolation, and depleted resources were common throughout most areas of rural poverty, the South was plagued by a devotion to one-crop farming, an unbalanced credit system, racial antagonism, and high levels of tenancy, especially in the cotton-heavy Southeast. The difficult conditions of the rural South, as in the rest of the nation, intensified when the Great Depression hit. The desperation and economic instability that dominated the lives of the rural poor came to characterize most of American agriculture and the rest of the country.

The New Deal, in general, failed to alleviate or in many cases even to adequately address the problems of the rural poor and American small farmers. One reason was the sheer magnitude of the problem, not just in rural areas but in the United States as a whole; only so much could be done. This did not stop people from trying. From the inner circle of the Roosevelt Administration down to individuals worried about local conditions, Americans came up with any number of ideas, ranging from the hopelessly utopian to the effective and promising, to fix the problems of rural America. But there were also a great many people and organizations opposed—for political, economic, or ideological reasons—to government efforts to improve the lot of small farmers. Missteps by agricultural reformers made their job more difficult, particularly in the case of the legislative and institutional origins of the federal programs and of the related problem of an inability to build sufficient political strength to weather hard times.

To evaluate New Deal efforts to improve rural life, this book traces the political and administrative history of the Farm Security Administration (FSA) and its predecessors, examining the ideological and institutional evolution of New Deal anti-rural poverty and rural reform programs and how they translated into practical, on-the-ground solutions. New Deal liberal thought and action about how to best address rural poverty evolved considerably throughout the 1930s. Early reform efforts varied widely, encompassing rural rehabilitation, land use reform, and back-to-the-land schemes.[2] Many New Dealers knew very little about what rural life was actually like; Franklin Roosevelt, for example, idealized the family farm as a place of refuge, where physical hardship simply could not exist so long as an honest and hardworking family was willing to draw its livelihood from the soil. By 1937 the New Deal's approach to rural problems had settled on the concept of rural rehabilitation, a system of supervised credit and associated ideas that came to profoundly influence the entire FSA program. The New Deal did not consistently or adequately address the needs of poor rural Americans, but to the extent that it did, it did so through rural rehabilitation.

Additionally, this book explores efforts to remedy rural poverty and the marginalization of small farmers, with an emphasis on farm security in the South, by looking at how the FSA's actual operating programs (rural rehabilitation, tenant purchase, and resettlement) functioned, especially at the local level. FSA employees and clients had a massive job in trying to solve the problems of rural poverty, and they did so with fairly limited resources and in an increasingly hostile political environment. Its success, while modest, proves that the FSA had developed an effective program that could have ameliorated the enormous human cost of rural poverty and improved the conditions of small farmers. This is particularly the case for the South, where rural poverty and associated ills (such as harmful tenancy practices) were among the worst.

The primary goals of the New Deal—relieving human distress, promoting economic recovery, and reforming American economic and governmental institutions—resulted in varied and often conflicting methods. Providing relief for those most in need, like migrant farm laborers or the urban unemployed, was often opposed by those economic leaders expected to help revive the economy, such as large landowning farmers or business owners, and New Dealers struggled to maintain a balance. The New Deal had, in the words of Anthony Badger, "little ideological coherence."[3] During the Great Depression, the United States was struck by a monumental disaster made up of many separate, if often related, problems; the overwhelming economic catastrophe pushed aside traditional remedies. New Dealers tried new and untested ways to help those hit by the Depression, to get the economy going again, and to ensure that such an economic downturn would not happen again. This necessarily meant false starts and retreats of policy. The New Deal's breadth and experimental nature also reflected the mind of Franklin D. Roosevelt, who could not only believe two inconsistent ideas at the same time but also proved able to put them into practice in politically effective ways. Roosevelt was not ideological in the sense that he believed that one best solution existed for every problem; he was a pragmatist who was famously willing to try something and, if it failed, to try something else.

This experimental approach to administration was apparent in the general unrest in New Deal agricultural organization, which meant contradictions and reversals in policy and bureaucratic structure (made worse by the need during the New Deal to create an administrative organization to oversee all the new programs that Roosevelt and Congress rapidly created as they searched for a solution to the Depression). The federal government worked so fast and on so many different programs that oftentimes its right hand did not know what the left was doing. New Dealers poured time, money, and effort into improving crop yields and raising efficiency among both large and small farmers, while at the same time paying farmers to take land out of cultivation in order to reduce crop production and raise prices.[4] New Deal credit programs aided farmers in holding on to heavily mortgaged property, and homestead programs encouraged people to return to the farms and increase the farm labor population; at the same time the federal government was injecting capital into the agricultural market to encourage mechanization and efficiency—programs, in other words, that reduced the need for both farmers and farm laborers. The intended recipients of one program's benefits often bore the costs of another's. In almost every case, when conflict arose among the interests of different groups or classes of farmers (real or imagined), the smaller and poorer farmers lost out, ultimately resulting, among other things, in the destruction of the Farm Security Administration in 1946 and the success of an agricultural policy that benefitted big producers.

The convoluted administrative history of rural reform efforts during the New Deal deserves a brief explanation, as it played a large role in the FSA's operation and ultimate fate. The FSA's various predecessors brought political enemies as they were assembled into a single organization; its wide range of activities stepped on a number of powerful toes; and its constant reorganizations made the agency less efficient and effective while robbing it of any sense of permanence or solidity. Between 1933 and 1946, five different organizational units were established to direct the main rural rehabilitation, resettlement, and tenant purchase programs (not to mention other programs like land use or suburban community building). The Division of Subsistence Homesteads operated in the

Department of the Interior between June 1933 and May 1935, and the Rural Rehabilitation Division was part of the Federal Emergency Relief Administration (FERA) between April 1934 and June 1935. The larger units were the Resettlement Administration (originally an independent executive agency before moving to the USDA in early 1937), which absorbed the previously mentioned two units and existed between April 1935 and August 1937; its successor, the Farm Security Administration, which existed between September 1937 and August 1946 in the USDA (though it bounced around in a number of wartime reorganizations of the Department); and the FSA's successor, the Farmers Home Administration.

The FSA and the New Deal in general operated during a transitional period in American agriculture. Farmers in the 1920s and 1930s were less self-sufficient than ever before. Well before the Great Depression hit, it had become clear that the future of American agriculture would be defined by its relationship with the government. The New Deal represented an opportunity to ensure that relationship would include small farmers and the rural poor as well as the large landowners who increasingly dominated American agriculture as it became more and more mechanized and capital-intensive. The FSA and its predecessors (and smaller, related New Deal programs, such as those in the Bureau of Agricultural Economics) tried to ensure that government benefits were shared with small farmers through a variety of different programs, but the most important approach was rural rehabilitation. The FSA made credit and government-trained agricultural experts more available to small farmers who, FSA administrators hoped, could still operate successfully in the changing agricultural economy.

The failure of the FSA was part of a larger marginalization of the rural poor and small farmer, particularly in the South. In Texas, for example, 1938 saw record-breaking decreases in the farm population. "Farmers are leaving the farm not because they want to," FSA Regional Information Advisor John H. Caufield wrote, "but because they are being shoved aside in the rush toward bigger units, more tractors and less men per acre."[5] The process accelerated during and especially after World War II. The majority of rural southerners moved to cities, pulled by postwar urban prosperity and pushed by mechanization and government programs that favored larger farming operations. The rural remainder generally fell into two groups: a more prosperous set took advantage of mechanization and expansion, while the poorer lower classes, lacking the skills and capital necessary to take up the practices of modern agriculture, continued to scratch out a living just as impoverished southerners had done before World War II. The result was that, more than ever, farming was a large-scale operation. High levels of poverty persisted, and the rural South remained the nation's poorest region. With the defeat of the FSA, the federal government gave up its efforts to alleviate the human costs of the changes it had done much to accelerate.[6]

This transition in agriculture matched a similar transition in American politics. The New Deal represented the ascendance of a political liberalism that demanded an expanded federal government involved in the regulation of the economic and social life of Americans to an unprecedented extent. However, the exact goals of that intervention varied. Partly it included the protection of individual rights and the promotion of the consumer economy (as would become dominant in the postwar period), but it also included a class consciousness and social democratic emphasis that American liberalism discarded after World War II. American liberals (and, practically speaking, American politics in the postwar era) focused on individual rights, especially in terms of racial discrimination, while more or less rejecting a consideration of the class and structural elements of exclusion and discrimination. This separation meant that federal programs became tiered-based on economic status: policies aimed at the poor and lower classes remained unpopular and poorly funded, but those that benefitted middle- and upper-class clients increased in popularity and size.[7] Programs aimed at the rural poor and small farmers, like the FSA, would have no place in America's political consciousness until Lyndon Johnson's Great Society, and even then the solutions would be halting, the programs underfunded, and the benefits meager.

What role the FSA might have played in the postwar agricultural revolution is unclear. One might even argue that the FSA and its predecessors were, if anything, too conservative in trying to preserve an older style of agriculture based on small farms owned and operated by a single family. In a consumerist age, Joseph Eaton wrote in 1943, reformers not only had to consider "people who have always been poor or who got poorer during the depression" but also the fact that "their demands and goals have expanded. The advance of technology has whetted their appetite for material, educational, recreational and artistic advantages which make even the best 'good old homesteads' appear to have been a primitive way of life."[8] The rural poor were seemingly less able than ever to afford a comfortable lifestyle, just as the average conception of what that meant was expanding. Possibly the FSA might have been completely ineffective in the face of such enormous economic and cultural change.

Whether or not this was true, many policymakers and government officials did not consider it an important enough question to find out. As FSA Assistant Administrator Robert Hudgens asked about a hypothetical rural rehabilitation client, "Is it desirable to keep him on the farm or is it not?"[9] The answer was apparently not, at least not desirable enough to make a concerted effort to keep farmers on the land or to ease the dispossessed's transition into urban life. Small farmers produced a decreasing portion of American agricultural output, and poor southerners left behind by the shift to commercial agriculture remained on the outside of political, economic, and cultural influence. FSA reformers proved unable to alter the course being set by the South's (and the country's) economic and political elite and by economic and technological changes. Instead, in the South as in the rest of the country, agriculture became dominated by large-scale, industrial agribusiness that allowed little space for the small farmer to survive.

Scholars have done valuable work on the history of the FSA, rural rehabilitation, resettlement, and the New Deal's impact on rural America. The most comprehensive history of the Farm Security Administration to date remains Sidney Baldwin's 1968 work *Poverty and Politics: The Rise and Decline of the Farm Security Administration*. Baldwin uses the agency as a case study of the strengths and weaknesses of the American political system and the difficulties inherent in political efforts to address the poverty. This has the effect of pushing the workings of FSA programs to the sidelines as Baldwin focuses on how administrators made goals, tried to build institutional legitimacy, and handled (or failed to handle) political and bureaucratic assaults.[10]

Few other histories review the FSA in its entirety. Paul Mertz and, most notably in recent years, Jess Gilbert have analyzed the FSA in terms of its ideological and political background.[11] Other scholars have focused on a particular element of the FSA's activity. Recent works by Michael Johnston Grant and Jarod Roll deal with specific difficulties that New Deal rural rehabilitation and resettlement programs faced and created. Their works put FSA programs within a larger context of what it meant to be a small farmer and farm laborer in the face of economic and political pressure toward mechanization and large-scale agribusiness.[12] New Deal resettlement and community-building programs have received a great deal of scholarly attention, especially in works by Brian Q. Cannon, Paul Conkin, and Donald Holley. Concentrating on a small group of resettlement communities allows for a deeper understanding of the ideology behind the programs, how well they functioned, and how the communities were influenced by the residents living in them. In recent years, the New Deal's contribution to conservation and land use planning has been the subject of important works by Sara M. Gregg, Neil M. Maher, and Sarah T. Phillips, which expand our understanding of New Deal history and open up new avenues of investigation through environmental history.[13]

Using the FSA as a case study or focusing on particular aspects of its program (such as land use or resettlement) can obscure or overshadow the rural rehabilitation and farm ownership programs. The early anti-rural poverty program encompassed a wide variety of ideas, but soon most New Dealers, including administrators at the top of the agency and the actual FSA employees who put their programs into action, came to focus on supervised credit as the key to solving rural problems. This book focuses on the solutions FSA administrators, employees, and clients believed to be the most important part of agency's program, why they came to believe that, and how those programs functioned in practice.

Part 1 describes the origins of rural rehabilitation, resettlement, and the tenant purchase program, and the various administrative changes and shifts they went through. The New Deal's initial response to the agricultural crisis, the Agricultural Adjustment Administration, generally worsened conditions for small farmers and particularly for tenants and sharecroppers, who often found themselves at a reduced status or even expelled from the land entirely. Programs aimed at the problems of the rural poor, at least until the passage of the Bankhead-Jones Farm Tenant Act in 1937, were something of an institutional mess, built on a variety of executive orders with only the thinnest veneer of legislative sanction. From 1933 to 1935, the New Deal's anti-rural poverty efforts were a scattershot affair, uncomfortably situated in the Federal Emergency Relief Administration and the Department of the Interior, among other departments. Even when the Resettlement Administration consolidated the programs in 1935, it was as an independent executive agency. The farm security programs only found a measure of stability in 1937, after the inspiring but polarizing first head of the RA, Rexford Tugwell, stepped down and was replaced by Will W. Alexander, and the RA moved to the USDA to become the Farm Security Administration following the passage of the Bankhead-Jones Farm Tenant Act.

Part 2 focuses more closely on the operation of rural rehabilitation, resettlement, and tenant purchase as mature, functioning programs, and concludes their story with a look at the FSA during World War II. The most important of these by far was rural rehabilitation, which took up the largest share of the budget and personnel and which strongly influenced the operations of both the resettlement and tenancy programs. These programs operated primarily at the project or county level, so these chapters look at how farm security operated at the local level. The individuals who lived in resettlement communities or who received rehabilitation and farm ownership loans by no means found the government's programs uniformly ideal, but their living conditions by and large improved both materially and socially. Similarly, FSA employees who oversaw such clients generally considered their work valuable and important. Although by 1942 the FSA had developed effective antipoverty programs aimed at improving the lives of small farmers, it had not built sufficient political strength. C. B. Baldwin, considered by his opponents to be a radical in the vein of Rex Tugwell, had taken over the FSA. An attack came during World War II, as anti-New Dealers and established agricultural interests, embodied by the American Farm Bureau Federation and its congressional allies, marked the FSA for elimination. In 1942 and 1943, the FSA's opponents successfully reduced its appropriations to the point of impotency. The agency dragged on until 1946, when it was replaced by the Farmers Home Administration, essentially a rural credit agency aimed mostly at veterans.

The specific problems of the southern rural poor, especially tenant farmers, deserve special mention, as they would prompt many of the most important reforms that the FSA and other anti-rural poverty efforts aimed to solve. The structure and ideology of southern tenancy made it different from that of other regions; one agricultural economist wrote in 1937 that "consideration of farm tenure in the United States might well be made on the basis of contrasting situations, comparing the South with the rest of the country."[14] Interest rates were high, any gain in profit by a tenant could be wiped out by the store bill, and merchant or landlord credit remained easy to get no matter what; tenants therefore became discontented, produced bad crops, and took a leisurely view of making repayments.[15] Most tenants had little chance of making a consistent and sufficient income. "What is earned at the end of a given year is never to be depended on," James Agee wrote, "and, even late in a season, is never predictable."[16]

There were three major classes of renting in the South, all of which might mingle on the same farm or even within the same family. Under sharecropping, the tenant furnished labor and usually half the fertilizer, while the owner provided everything else; the tenant and landlord then split the crop in half. Share renting meant the landlord provided the land, home, fuel, and a quarter or third of the fertilizer in exchange for a quarter or third of the crop, while the tenant provided labor, work stock, tools, feed, seed, and the rest of the fertilizer. In cash renting, the landlord furnished the land, house, and fuel, taking in return a fixed amount of cash or the crop while the tenant got everything else. In 1930 about 1.1 million southern operators, or about 44 percent of southern farmers, owned their farms, while about 1.3 million were tenants of some kind. About 646,000 of these, or 27 percent of southern farmers and 48 percent of tenants, were sharecroppers.[17] As a practical matter, the difference between a sharecropper and a share renter did not have much real impact except for the method and ease with which a landlord could enforce a claim. In either case, tenants found themselves at a legal disadvantage in any contest with their landlords; local government and landlords operated under the assumption, in attorney A. B. Book's words, that tenants of all kinds were "a class which would default in its obligations if the slightest opportunity existed."[18]

Tenancy knew no color line. The growth of the number of white tenants in the cotton South had actually increased more than had the rate of tenancy overall. Between 1920 and 1930, the number of white tenants in the cotton states increased by two hundred thousand families, about a million people; the number of African-American tenant families decreased by about two thousand because of urbanization.[19] Social scientists went out of their way to point out the interracial nature of tenancy and to overcome the idea that picking cotton was a mostly black job, but in the years leading up to and during the Great Depression, racial conceptions of tenancy remained. This misconception had a practical impact on policy and public opinion, as potential reform efforts ran into hostile defenses of white supremacy and apathetic disregard for black poverty.[20] Even those who had little interest in improving the lives of black tenants pointed this out: Alabama Senator John H. Bankhead assured a correspondent that "our primary interest is in the white tenants, and it may surprise you to know that a majority of the farm tenants in Alabama are white."[21] Black farmers still faced racial as well as class oppression: Ned Cobb remembered that his father "wasn't a slave but he lived like one. Because he had to take what the white people gived to get along."[22] But contrary to popular perception, white farmers could find themselves near the bottom of the agricultural ladder as easily as could black ones.

For the average southern farmer in 1930, farming had not changed very much over the preceding five or so decades. Most farmers still used animal power. Mechanical improvements like combined reapers and threshers introduced during the nineteenth century had reduced the need for labor in some fields, but in 1930 only a small portion of southern farmers used such implements. Tractors, which would rapidly expand farm productivity in the 1930s and after, were neither small nor cheap enough for most southern farmers to consider using until the late 1920s, when the worsening agricultural depression made it harder to afford expensive new equipment.[23] One study of a small southern farming community in the 1930s found only three tractors, ten trucks, and two stationary engines among 189 farm families.[24] While mechanization had begun to increase production in parts of the South, especially in the Southwest, tradition and poverty meant that most southern tenant farmers in, say, 1925 used the same farming methods as had their parents and grandparents.[25]

Despite having to raise their children, cook, wash, sew, clean, tend the garden, and fulfill other housekeeping obligations, farm women also had to work in the fields, chopping and picking on cotton farms or priming, stringing, and curing tobacco. Women often preferred field work to housework: social interaction was more common and one could actually see the end of a day's work (as opposed to the never-ending demands for cooking, cleaning, and such). Age, poor health, and child care tended to change women's minds about field work or force them to stay at home. Farm life for women, no matter what the crop, was, in historian Melissa Walker's words, "dominated by ceaseless labor."[26]

The South had among the lowest farm incomes in the country, especially for those who did not own their farms. In one study of medium and large southern plantations, the average yearly net income of laborers was $180, for croppers $312, and for tenants $417, for a total average of $309.[27] Landlords and merchants used a combination of economic and social means to keep incomes so low, which were necessary for the southern tenancy system to survive. Short-term debts were treated as essentially advances on the season's income. Even in good years, a large number of tenants failed to pay back their advances. More debt resulted from the "furnish"—additional food, feed, and other necessities, provided by the landlord and to be repaid from the tenant's share of the crops. Annual interest was high—10 percent on cash advances, 20-25 percent on goods—in part because the average duration of all loans was for the growing season, only about six months. In practice, most merchants provided farmers with as much credit at as high an interest rate as possible without completely crushing the debtor.[28] As a result, possible gains were minimal; possible losses were catastrophic. Alabama farmer Ned Cobb described seeing his father twice "stripped of everything he had had . . . and he never did prosper none after that."[29] Even defenders of sharecropping, like William Percy, conceded that it offered "an unusual opportunity to rob without detection."[30]

In part, landlords could maintain this system because relations between southern landlords and tenants were almost entirely informal. Detailed agreements were not usually worked out, and contracts were almost always of the oral or handshake variety. The landlord usually kept any records of advances and repayments. Sharecropper Arnold Berry, for example, described in the 1930s how, as far back as he could remember, his bill at the plantation store was the same every year: $234, the exact amount of his maximum allowed yearly credit. Berry reasonably decided, after realizing this, to use all the credit his landlord made available up to its limits, as he would be in debt for the same amount no matter what. Berry considered himself fairly lucky, as he was at least getting some cash income; friends on adjoining plantations received payment in scrip.[31] Even if a sharecropper had maintained records, few would have been willing to challenge a landlord. Tenants did not even have to threaten their landlords directly; if they did manage to save any money, for example, tenants might find that local landlords would refuse to sign a lease the next season.[32] For black tenants and sharecroppers, challenging the landlord-tenant relationship edged dangerously close to threatening the color line; southern racial mores meant that black tenants (and owners) had to put up with considerably worse treatment than did whites.[33]

Just as the system encouraged many of the supposed class sins of tenants, so did it create and exacerbate the worst tendencies of creditors. Big land owners also claimed that they "hated bankers" and "hated merchants" because "they had robbed us."[34] Hardships for the landlords and merchants generally trickled down, meaning hardships for their tenants and debtors as well. H. L. Mitchell, one of the founders of the Southern Tenant Farmers Union (STFU), dealt with a landlord in Cross County, Arkansas, named C. H. Dibble, who had evicted twenty-one sharecropping families after word of his negotiations with the STFU had reached the local bankers. If Dibble had refused to evict the tenants whom bankers saw as troublemakers, then no ginner would accept his cotton; the bank even threatened to foreclose on his mortgage. Nor did merchant or landlord status protect a grower from economic downturns; a Resettlement Administration administrator discovered in one southern community he visited that all of the merchants had gone bankrupt, despite charging 30–50 percent interest to tenants. The Depression ruined even creditors who charged outrageous fees and wielded great power.[35]

However much planters and merchants complained, the situation was the worst for the already impoverished farmers of the South. Bad housing conditions were perhaps the most obvious sign, particularly as the Depression wore on. The President's Special Committee on Tenancy reported that many southern homes were "of poor construction, out of alinement [sic], weather-beaten, and unsightly.... Often the roofs are leaky. The surroundings of such houses are bleak and unattractive." Charles Johnson was blunter: "The drab ugliness of tenant houses might be condoned if they were comfortable."[36]

The situation inside the home was sometimes even worse. One contemporary description of tenant living conditions in tobacco country found homes that outwardly looked modest but livable, with a "respectable mien." Inside one, however, the floor sloped so much that one side of the bed had to be propped six inches higher than the other. Newspaper tacked onto the wall failed to prevent mud from crumbling from exposed logs. Repairs to the house made it impossible to close the window.[37] In no southern state, and in only four states nationwide, did more than half of farms receive central-station electricity in 1935; in Mississippi the number was less than 1 percent of all farms. Compared to a national average of 34 percent, only 13.7 percent of southern homes had a telephone in 1930.[38] In 1934 more than two-thirds of southern farmers had unimproved outhouses. One in five white farm owners (economically the best off) had no sanitary facilities at all; the number was almost one in three for black tenants. Only 4.6 percent of white owners, 1 percent of white tenants, 0.2 percent of black owners, and 0.1 percent of black tenants had indoor flush toilets.[39] There was little incentive (when there was the ability) to improve housing because tenants moved so much, especially sharecroppers and African Americans, either to other farms or (increasingly after World War I) to northern cities.[40]

Since southerners were so poor, schools and other public services were inadequate. In one study, a third of adults in a thousand southern farm families were functionally illiterate. Rural southern children, especially black ones, received lower-quality education, with fewer going to school, and for less time out of the year, than children elsewhere in the country.[41] In Gorgas, Alabama, white residents were proud of their high school, perhaps the most substantial structure in the entire community. But it was poorly laid out and lacked teaching materials such as maps, globes, and dictionaries. The high school science lab failed to meet even the low requirements set by the Alabama Department of Education. African American students in the community had only a poorly built one-room wooden schoolhouse, heated by an unjacketed wood stove. That school offered courses only for the first four grades.[42]

Despite living in generally good growing areas with abundant land, southern tenants and sharecroppers in the first two or three decades of the twentieth century had the worst diets of probably any large group in America. One 1941 FSA report asserted that even among rural clients not on relief, more than half had inadequate diets.[43] Strikingly, tenants had to spend as much money on food in the summer, when a garden would be most productive, as they did throughout the rest of the year. Some poor southerners apparently ate virtually no vegetables at all in the summer and winter. They instead purchased the "three M's": meat

A NURSE BRINGS HOOKWORM MEDICINE TO THE LEWIS FAMILY, RURAL REHABILITATION CLIENTS, COFFEE COUNTY, ALABAMA. LIBRARY OF CONGRESS.

(fat salt pork), meal (usually corn but sometimes wheat), and molasses. This may still overestimate the variety of a poor farmer's diet throughout the year; for example, pork is difficult to store, and it may have been eaten almost all at once, during slaughter in the late fall and early winter.[44] Rural southerners, especially tenants, generally had enough to eat, but they lacked the variety necessary for healthy living; one woman recalled that her average breakfast was cornbread and syrup, and her average lunch was cornbread pancakes.[45] Poor landowners outside the cash-crop areas faced different kinds of nutritional problems; residents of the hill and mountain country, for example, endured shortages of food in the winter and early spring, even as their diets the rest of the year were more balanced.[46]

Unsurprisingly, bad diets, bad sanitation, and a low standard of living led to bad health. The South as a whole had inferior medical care compared to that of the rest of the country. Rural southerners had access to fewer, and less qualified, doctors, nurses, and dentists than did city dwellers; rural hospitals, medical schools, and other medical facilities were both inferior and relatively less numerous than in urban areas. Sociologists Arthur Raper and Ira de Reid noted that it was easier for a tenant to find a veterinarian in the South than a doctor.[47] Pellagra and other nutritional diseases were common, the result of an inability to buy or failure to produce enough vegetables for home consumption. Poverty encouraged the spread of disease; for example, southerners were too poor for screens to stop mosquitoes or the medicine to treat malaria. In 1938 the South had about two million cases of malaria, reducing the economic output of the South by an estimated one-third.[48] The conditions of childbirth were often dreadful. In some cases, dirty or old quilts and cloths were collected because they could be thrown away or used to avoid staining new clothes or mattresses. Only boiling water, or in a few cases Lysol, was available as an immediate antiseptic. Doctors and midwives were rare. Unsurprisingly, attitudes toward childbearing were ambivalent; in general, rural mothers were proud of the children they had had but hoped to avoid future pregnancies. Farm women, at least, approved of contraceptives, even if most did not use them regularly or use the more effective methods. Poor families often turned to patent medicines, home herbal remedies, or folk magic. Turpentine and castor oil were used to treat almost any malady. In many cases the rural poor simply had to live with painful and often debilitating diseases. Chronic illness was common, as were serious handicaps.[49] At any given time, perhaps one in ten rural southerners in, say, 1930 was permanently disabled or too acutely ill to get about normally. Another large portion of the population had some chronic or manageable affliction.

Poor and rural southerners lacked the clout to find political solutions to their condition, as one study of Gorgas, Alabama, in the mid-1930s indicates.

Elections were often a source of excitement in the small rural community: the long campaigns, with their speeches and rallies, greatly interested the often bored citizens. But of the 330 eligible white householders, only 37 percent qualified to vote, and 81 percent of these were either landowners or their related tenants. No black voters qualified. The yearly poll tax of $1.50 per person was too much for many residents, and the tax accumulated so that one could not vote unless the entire back tax was paid. For those who could vote, the process itself was confusing and intimidating. In Gorgas, there were twenty-two candidates for the nine county offices and overall seventy-one candidates seeking thirty positions. The ballot with all the names and offices printed was over two feet long.[50]

This was the backbone of the "Solid South": counties where almost no blacks and few whites voted. But even where white tenants formed a majority, they had little influence on local government. The courts, relief agencies, and other government functions were controlled by planters: the tendency of relief agencies to quit giving relief during cotton-picking season did not originate during the Great Depression. The result, as V. O. Key wrote in 1949, was that "electoral apathy in the South generally assumes colossal proportions."[51]

List of Abbreviations

AAA	Agricultural Adjustment Administration
AFBF	American Farm Bureau Federation
BJFTA	Bankhead-Jones Farm Tenant Act
CWA	Civil Works Administration
FCA	Farm Credit Administration
FERA	Federal Emergency Relief Administration
FmHA	Farmers Home Administration
FPA	Food Production Administration
FPHA	Federal Public Housing Authority
FSA	Farm Security Administration
NIRA	National Industrial Recovery Act
PHA	Public Housing Administration
PWA	Public Works Administration
RA	Resettlement Administration
SCS	Soil Conservation Service
STFU	Southern Tenant Farmers Union
USDA	United States Department of Agriculture
WFA	War Food Administration
WPA	Works Progress Administration

The Farm Security Administration
and Rural Rehabilitation in the South

PART ONE

Finding Solutions
The Origins and Evolution
of New Deal Rural Poverty Policy

The tough conditions in the rural South were not caused by the Great Depression, though that economic catastrophe made things worse. Rural poverty and farm failures were old problems in 1929, particularly in those regions characterized by tenancy and bad farming practices. The Great Depression worsened these problems but also brought them out into the open, creating political conditions that offered the hope for some improvement. But the New Deal response left much to be desired. The chief program for agriculture was the Agricultural Adjustment Administration. At best, the AAA provided little aid to the rural poor; at worst, it made things more difficult for small farmers, particularly tenants and laborers.

But many New Dealers saw an opportunity to improve the lives of rural Americans, in both the short and the long term. An idealized picture of rural life motivated many of their efforts. Relief administrators, in both state and federal governments, imagined the farm as inherently self-sustaining and that, with just a little help, farmers could get back to taking care of themselves. The Federal Emergency Relief Administration created the Rural Rehabilitation Division in the spring of 1934 to replace relief to struggling farmers with a program of expert help and government loans based on a home and farm management plan. Administrators hoped this would be not only more effective than relief, but also less expensive, as loans would be for a limited duration and repaid. At around the same time, the Department of the Interior was at work on a series of back-to-the-land projects through its Division of Subsistence Homesteads. Motivated by people like Alabama Senator John H. Bankhead Jr., the subsistence homesteads program was founded on the notion of the farm as a refuge against the modern industrial economy. More than simply rehabilitating existing farmers, as the rural rehabilitation program was hoped to do, the Division of Subsistence Homesteads was intended to improve the lot of small farmers and the masses of unemployed or underemployed industrial workers by demonstrating a new way to live. Uncertain when, if ever, the industrial economy would recover, proponents of subsistence homesteads wanted to combine modern farming techniques with part-time industrial employment at new, government-created communities around the country. Supporters hoped these communities would become models for a new way of industrial-subsistence living.

Other federal agencies and departments during the early New Deal took a variety of different approaches to rural poverty (as in, for example, the Farm Credit Administration's debt adjustment program). All these efforts faced similar handicaps. Created in a period when the back-to-the-land movement was at its peak, early programs operated under the belief that rural life was essentially self-sufficient and could provide, if not a life of luxury, then at least a guarantee of food, shelter, and a comfortable existence for any family willing to work hard for it. This led to unrealistic ideas about how successful, and how easily implemented, an antipoverty program could be. A related problem was that the federal anti-rural poverty effort lacked political, institutional, and ideological focus. This was true among the various programs (for example, a number of different agencies operated resettlement and community-building programs of some type) and within them (the success of rural rehabilitation varied greatly from state to state and even county to county, depending on local policy and personnel).

By 1935 that setup was no longer viable, and New Dealers increasingly believed that a new approach was necessary. Reorganization resulted in a new agency, the Resettlement Administration; much of its work (and that of its successor, the Farm Security Administration) involved solving problems created by those earliest flaws in the New Deal's anti-rural poverty program. Rexford Tugwell, who took over the new RA, lacked the illusions of the back-to-the-land movement and its ideal of moving millions of industrial workers to subsistence homesteads. In their place he had a different set of illusions about the ability of the federal government to effect rapid, mass social change and, more important, about his own ability to oversee that change in a contentious bureaucratic and political framework.

Tugwell's self-assurance and vision helped turn a collection of widely spread and often ineffective federal programs into a massive and effective national agency. It was far from the all-powerful and unified behemoth that its critics imagined, but Tugwell and those he brought to work with him in the agency successfully made the RA into a coherent, organized bureaucracy by the time he decided to resign in late 1936. By that point, the RA had a strong sense of its purpose, responsibility, and general administrative shape.

The RA lacked, however, a firm political foundation. Its reputation for bureaucracy and for skirting the dictates of Congress, not to mentions its ambitious reform goals, made enemies. As an independent executive agency, the RA lacked permanency; its ability to withstand the attacks of those enemies was uncertain. Three major changes in 1937 promised to improve its political position. First, Will W. Alexander, who was more effective as a politician and less of a lightning rod for criticism, replaced Tugwell as RA Administrator. Second, the RA moved into the United States Department of Agriculture. And, most important, in 1937 the Bankhead-Jones Farm Tenant Act created a new set of responsibilities for the federal anti-rural poverty program: a program of federal loans for tenant farmers to enable them to become farm owners. For RA administrators, the tenant purchase program held out the hope that its disparate clump of emergency programs might become part of a long-term, comprehensive program by the federal government to address rural poverty.

In the fall of 1937, the RA became the Farm Security Administration, the product of years of institutional and ideological struggle. It represented what the New Deal approach to rural poverty, which had been evolving for years, would be. The FSA conducted programs of rural rehabilitation, tenant purchase, and resettlement (in that order of importance). In the prewar years, the FSA put into place a relatively effective set of programs to address rural poverty, centered on rural rehabilitation and more specifically supervised credit, as it made its way through the political terrain of the prewar USDA. If the New Deal solution to rural poverty was going to undergo any further changes in its policy or institutional shape, as it had dramatically done over the previous years, those changes were cut short when World War II led to dramatic changes in the rural economy and the political climate.

CHAPTER ONE

From Relief to Rehabilitation
The Origins of Rural Rehabilitation

Measuring intangibles like interest in a farm life or farming initiative proved difficult for the county case workers trying to put together a rural rehabilitation program in the Federal Emergency Relief Administration, the beginnings of the program that would become the most influential part of the farm security effort. Usually they attempted to do so by in-person interviews, recommendations, and similar methods, but some relief workers adopted less orthodox approaches. In Alabama, county relief workers used oxen to determine who was serious about becoming a rehabilitation client. Originally, Alabama administrators wanted the state's rehabilitation clients to use oxen because they were relatively cheaper than mules (the traditional form of animal power on a southern farm), which cost between fifty and seventy-five dollars a head, versus twenty to twenty-five dollars a head for steers; additionally, oxen were cheaper to keep, as they could graze and needed only about two dollars a month for food, as compared to ten dollars for mules.[1] But more than that, from the relief workers' perspective, the use of a steer provided, in the words of an Alabama Relief Administration report, an "acid test" for "an applicant's sincerity and purpose. Acceptance of a steer by an applicant was proof to the Relief Administration that the farmer wanted to earn his own living."[2] Mules were considered so integral to southern agricultural that a common way of describing the size of the farm was by indicating how many mules it took to operate it.[3] Clients who could overcome that traditional attachment to using mules, relief workers believed, were truly committed to the program. Not all farmers appreciated the importance of using oxen; one landowner reportedly exclaimed, "Hell! This ain't no New Deal if we-all got to go back to plowin' steers!" But most small farmers and tenants had few problems actually using oxen, seemingly justifying the hopes of rural rehabilitation planners.[4]

Relief workers had to find such novel solutions because they were in unusual circumstances. The conditions of the Great Depression were historically difficult, but even more than that, these relief workers represented an unprecedented effort by government agents to improve the lives of the rural poor. Agricultural policymakers had previously been slow to implement programs that benefitted the poorest farmers, tenants, and laborers. Things did not change immediately when Franklin Roosevelt became president; the New Deal's earliest responses to the rural Great Depression, particularly the Agricultural Adjustment Administration (AAA), reflected the traditional emphasis on helping big producers. During the early New Deal, when the Roosevelt Administration was first grappling with the problem of rural poverty, the United States Department of Agriculture (USDA), one of the largest departments in the federal government, played little part in providing rural relief (despite great need among impoverished farmers). The USDA created only one of the three most important federal programs dealing with poor farmers in the early New Deal, the AAA, and it had the worst record of helping them. The Division of Subsistence Homesteads, created in 1933, had its home in the Department of the Interior, and the rural rehabilitation program was part of the Federal Emergency Relief Administration.[5] Those latter two were in part responses to not only the catastrophic proportions of rural poverty in the 1930s, but also to the federal policies and programs that made that conditions of the rural poor even worse.

Aid of some kind was obviously necessary, particularly in the South. Poor farmers there had the largest families, the least education, the smallest farms, and the least operating capital and income.[6] A Works Progress Administration (WPA) official described the typical cotton-belt farmer on relief in the summer of 1935. He was an unemployed but able-bodied sharecropper or farm laborer, about forty years old. He, his wife, and two or three children had to get by on relief grants of ten dollars a month if he was white and seven dollars if he was black. Essentially illiterate, the farmer owned no livestock or machinery, and his family lived in an unpainted frame shack. The family had been on relief for at least five consecutive months and out of work in four of the previous six months before going on relief.[7] Reformers believed that this farmer, and the millions like him throughout the South and across the country, needed help and that some sort of new federal agency was the best vehicle for that help.

The USDA response (the AAA) suggested that the agency and its supporters did not understand the situation facing struggling farmers; nor did the reformers in the Roosevelt Administration. They introduced programs that had less to do with the realities of rural poverty and more to do with their own idealized, even sentimental, notions of life on the farm. FERA's Rural Rehabilitation Division and the Department of the Interior's Division of Subsistence Homesteads reflected the same misconceptions, in both the arguments for their creation and in their operations. Administrators simply could not imagine large-scale, long-term poverty among the nation's farmers. For FERA, this misunderstanding about the nature of rural poverty took the form of repeated assurances that not only could small farmers, with a little help from the federal government, get back on their feet and out of poverty, but that it could be done at a much reduced cost, perhaps even a profit in some cases, for the federal government. Ideally, a tax burden would be turned into a tax base, and the government would find most of its expenditures repaid before the program ended. This would be accomplished via a program of rural rehabilitation: targeted loans and expert supervision for the rural poor who proved themselves to be the most capable of taking advantage of it. In shaping how the new federal rural rehabilitation program would work, misconceptions about the relative ease of alleviating rural poverty would shape the long-term future of rural rehabilitation and the larger federal program to reduce rural poverty.

Another long-term impact of the Rural Rehabilitation Division's early days involved its legislative and political underpinning. The Rural Rehabilitation Division was, like the Division of Subsistence Homesteads, something of a bureaucratic orphan. Its most natural home, the USDA, remained dominated by hostile influences. Instead, since it was considered a replacement for rural relief, the Rural Rehabilitation Division found itself oddly placed in FERA. The result was a scattershot program whose effectiveness varied greatly on a state-to-state, even county-to-county, basis as local social workers, relief agents, and rural rehabilitation experts grappled with how best to implement the program. As they did so, the answers they found would become the most important part of the larger federal anti-rural poverty program.

The term "agricultural ladder" came into popularity in the 1910s and 1920s, although it reflected a much older conception of agricultural life as a series of steps whereby a farmer moves, through hard work and native intelligence, from the bottom of the farming hierarchy to the top. A 1919 description identified four rungs. In the first, the unpaid laborer, usually on a family farm, learns the basics of farm life. On the second rung was the hired laborer; on the third, the tenant farmer. Finally, the owner stood at the top on the fourth rung. Straightforward in theory, tenancy in the South had unlimited permutations of farming on thirds or fourths or cash, based on race, crop, region, time period, and factors like the relationship between the landlord and renter.[8] With the decline in southern agricultural fortunes during the previous decades, the agricultural ladder had become an agricultural slide—60 percent of southern farm laborers reported that their current status as laborer was somewhere below their highest previous status attainment. Many more had fallen off the ladder altogether. In 1934 some 1.7 million rural families were on some form of relief.[9]

Government aid to agriculture did not necessarily improve things for tenants and small farmers. In fact, it took a long time for most Americans to even recognize its need. In the early twentieth century, most Americans (even as they moved in large numbers to the growing industrial cities) would have agreed that a rural life was an inherently better life. This was the inspiration behind the Country Life Movement, which got its start with Theodore Roosevelt's Country Life Commission in 1908. The burst of proagrarian sentiment in the late nineteenth and early twentieth centuries reflected a belief that, while farm life had its problems, scientific expertise could make farming "modern" and return farmers to their rightful place in society. In 1919 the American Country Life Association formed on the idea that the benefits of rural life could be maintained while still providing the economic, cultural, and technological advantages of modernization. Experts could apply the new technical and social sciences, which were transforming the urban economy and culture, to rural life.[10]

Similarly, agriculturalists in the United States Department of Agriculture wanted the federal government to improve the lives of farmers. Prior to the 1920s, the most important avenue for the government to improve farm life was the State Extension Agencies. The Hatch Act of 1887 funded agricultural experiment stations in each state to reach farmers who lived far from agricultural colleges. The Smith-Lever Act in 1914 institutionalized the Extension Service; under the Act, the USDA worked with agricultural colleges to channel information from agricultural research agencies to farmers. Agricultural extension agents taught farmers better ways to farm, including the latest advances in soil conservation, hybrid seeds, fertilizers, and new equipment. Home demonstration agents, usually women trained in home economics, taught housewives how to garden, provide a good diet for their families, preserve food, and make clothes or other home accessories. Supporters of the program pointed to increased production and the general expansion of agricultural knowledge as proof of its value.[11]

These benefits were hardly shared equally. The problem, as extension agents admitted, was that the educational nature of the Extension Service meant that it only really aided those who were capable of using its advice. In other words, it focused primarily on more prosperous farmers.[12] Soon after he became head of the rural rehabilitation program in the Federal Emergency Relief Administration, Lawrence Westbrook described the fundamental problem that the federal government's approach to agricultural problems simply could not fix. A "county agent can make a farmer want a cow, sow, and hen," Westbrook wrote, "but he cannot always ferret out the means by which the want may be fulfilled."[13] At times the

Extension Service seemed not even to recognize that smaller producers existed. One state assistant extension director, for example, advised in 1930 that, for visits by county agents, sending a letter or making a phone call beforehand made a visit go more smoothly, and that the county agent should expect to call upon the same farmers year after year.[14] The obvious expectation was that county agents would be visiting better-off farmers, as one could not in the 1930s count on tenant farmers or small operators having telephones or being literate, and tenant farmers moved too frequently to be available for visits over multiple years. Even if some low-income farmers *could* use its help, the Extension Service generally served those who already had higher incomes, since large producers could most easily implement the kind of changes or farming practices that county agents recommended. As the agricultural economy worsened in the 1920s, soon the Extension Service could not even meet the demands of just relatively better-off farmers. And in the South, racial politics came into play. White southerners, who controlled the state Extension Services, had no interest in allowing a full share for black farmers, who could be economic competitors and threats to the racial status quo. As a result, African American farmers could look for help only to underfunded, segregated black extension divisions or, just as often, no help at all.[15]

The Extension Service was not alone, however, in occupying itself mostly with larger producers; it was part of a web of federal agencies and legislation that primarily provided assistance and aid to those farmers who, relatively speaking, needed it the least. Policymakers were instead primarily concerned with prices, which were constantly dropping. The basic problem with American agriculture was also its greatest strength: enormous productivity. American farmers produced more than the American market could absorb, which dropped prices; this encouraged farmers to plant more, further pushing down prices. Agricultural policymakers and lobbyists called for parity, a fair exchange of value between agricultural goods and manufactured items. Farmers would have to receive higher payments for their crops, equal to prices in the five years before World War I (popularly held to be the last time when industry and agriculture were in balance). George N. Peek introduced the idea in 1921. He wanted the government to subsidize foreign sales, even if it meant selling at a loss, to raise farm income. Eventually a different approach, the idea of limiting production, came to dominate agricultural circles. By growing less, farmers would reduce supply and thereby raise agricultural prices.[16]

As the agricultural situation worsened with the Great Depression, the conviction that something had to be done grew stronger. Many New Dealers came to believe that agricultural recovery would be the basis for a general American recovery, a belief heavily encouraged by agricultural spokesmen. The first and most influential piece of New Deal legislation aimed at agricultural recovery was the Agricultural Adjustment Act of 1933.[17] It created the Agricultural Adjustment Administration (AAA), replacing the old Farm Board. The AAA began a long process of trying to increase agricultural income by controlling production. Government production and distribution controls were an old idea, pushed in various forms by the Grange, Farmers' Alliance (better known as the Populists), Farmers' Union, and other agrarian and populist groups; it was popular enough by 1932 to get at least some support by the major political parties.[18]

The AAA rested on a single essential premise: American farmers must match their production to the American market. The AAA would raise farm prices (and thereby create parity) through marketing agreements and a voluntary domestic allotment system for large commodities (in 1933 those were cotton, wheat, and hogs; tobacco and corn were added in 1934). The allotment program was the most controversial aspect. Farmers cut production of certain crops by assigned percentages, with the idea that reducing supply would increase price, and received compensation from the AAA. Participating farmers elected committees to determine quotas for each farmer and the production for a local area.[19]

It took several weeks of votes and hearings to get the Agricultural Adjustment Act passed. The agency, created in May 1933, had to act immediately to have any impact on cotton prices for the 1933 market. It would have to destroy already-planted cotton (and over six millions hogs and piglets) to bring down the yield of those commodities that had already begun production. Only one existing organization was capable of overseeing such a project: the Extension Service. County agents set up the local machinery responsible for the administration of the allotment system. The AAA's ties to the Extension Services and its allies meant that it would necessarily be strongly influenced by the conservative agricultural elite. The American Farm Bureau Federation, for example, took on much of the administration of New Deal agricultural programs, and its membership more than doubled between 1933 and 1938.[20]

While many social scientists and liberals initially supported the program, the AAA from the beginning was primarily concerned with what its administrators saw as practical farm economics and politics, not improving the lot of small farmers. This is evident in the selection of George Peek to head the new agency. Peek had experience working with agricultural leaders and was known as a friend of agricultural interests, and his selection signaled that pragmatic concern for production and income, not theory and idealism, would animate the new agency.[21] Only the AAA's rapidly expanding legal division had any significant liberal or leftist influence (for lack of a better term, they were often called the urban liberal group). Peek and the leader of the urban liberals in the AAA, Jerome Frank, were longtime rivals whose dislike and distrust of each other went back a decade.[22]

The AAA cotton contracts for 1933 were of two types. The first was a straight rental: the government leased up to 40 percent of an owner's cotton land, to be taken out of production in exchange for a rental based on the average yield of the leased land. The other plan combined rental with the granting of options on government-owned cotton. Only landowners were allowed to sign the contracts. The AAA's Cotton Section had essentially the same view of tenants and sharecroppers that the landlords did, that they were incapable of handling their own affairs and that landlords would best take care of them. And from a practical standpoint, getting the approval of the landlords was most important. The program could function without the voluntary cooperation of renters, but without the owners' approval, it would fail.[23]

For the 1934–35 cotton contracts, benefits changed and were divided into rental payments and parity payments. Landlords got the entire rental payment, and the parity payment was split in proportion to the previous division of the crop between landlords and tenants. This meant that the payments overwhelmingly favored even the landlords who dealt entirely fairly with their tenants.[24] For a sample farm that had a fifteen-acre base of cotton, with six acres rented to the government and nine acres planted, resulting in two hundred pounds of cotton, a sharecropper working on halves would receive six dollars in government payment compared to $48.00 for the landlord under the 1934–35 cotton contract.[25]

The cotton production code was almost entirely designed for the landlord's benefit; the actual administration of it took care of any other problems the landlord might have had. Many landlords simply kept the money for themselves. Others required tenants to have payments credited to old debts, with the tenants receiving nothing. Landlords also got to decide the status of their tenants. A "managing tenant" as defined by the AAA received rental as well as parity payments, so landlords lied about or changed the station of those renting their land. Renters became sharecroppers, and sharecroppers became laborers. Some landlords evicted their tenants entirely, so they did not have to provide the furnish, split the government benefit checks, or let the tenants make any decision on the use of the rest of the acreage.[26]

The apparatus for addressing abuse left much to be desired. Even when not entirely made up of landlords, the committees and investigators dedicated to handle complaints about mistreatment and breaking contracts were still largely sympathetic to them. Disputes at the local level went to committees dominated by big planters and their allies. In many cases, complaints sent to the USDA were sent to AAA administrators, then to county agents, and then to the landlord against whom the complaint was made.[27] As of June 25, 1934, landlord-tenant field representatives investigated 617 complaints and found 419 to not be justified—conclusions that, even if technically correct, only underlined how tilted the system was toward landlords.[28] Historian David Conrad estimates that there is "no way to determine the extent of landlord chiseling in 1933," because "the machinery for processing complaints in 1933 was inadequate and many tenants did not understand their rights or were afraid to assert them."[29] Tenants had little ability or will to protest. Any resistance resulted in immediate and often violent counterattack, especially for black tenants and sharecroppers. But some fought back, joining the Southern Tenant Farmers Union. This was risky; tenants who joined the STFU often found themselves evicted or threatened with arrest and violence.[30]

The AAA had problems. Most notably it proved to be more or less ineffective. Cotton prices rose temporarily after the plow-up but soon declined again. Instead, cotton farmers found relief in government loans, primarily from the Commodity Credit Corporation, established under the Reconstruction Finance Corporation. And it was in the tobacco and cotton producing sections of the country that, relatively speaking, the AAA did the best—it was an even bigger failure nationally.[31]

There were also, at least until 1935, considerable divisions within the AAA. Conservatives in the AAA often recognized that the program would favor landlords, but they also saw the entire program as a risky and uncertain business. From their perspective, to introduce social change into an economic program would invite disaster. Jerome Frank and other liberals in the agency believed AAA leadership to be indifferent to the harmful impact that the AAA had on tenants. The conflict came to a head in early 1935. Frank, following the recommendations of his subordinate Alger Hiss, issued a legal opinion calling for landlords to retain their tenants for the life of their AAA contracts. Angry over the opinion, Chester Davis (George Peek's successor as director) removed numerous liberals from the agency, including Frank.[32]

The AAA was also unpopular with many segments of the American farm population. Kenneth Bindas found, reviewing six hundred interviews with rural southerners discussing the impact of the New Deal, that the AAA was one of the least frequently mentioned programs; at the time, and in their memories, most rural southerners felt that the AAA did little for them.[33] But it was the Supreme Court in January 1936 that finally killed the AAA. In *United States v. Butler,* also known as the Hoosac Mills case, the court struck down the Agricultural Adjustment Act. This apparent defeat actually benefited the price-supports system, giving the program a new start after essentially three years of testing. The Soil Conservation and Domestic Allotment Act of 1936 and the Agricultural Adjustment Act of 1938 continued the practice of offering price supports in return for crop reduction (under the guise of soil conservation) and introduced new features like federal crop insurance.[34]

The AAA did not help impoverished farmers in the South or across the country, driving tenants from their farms and pushing them down the agricultural ladder. But this was only part of a larger process in the South, as the old system was collapsing. It was obvious that the AAA was not a solution for the rural poor living through this transformation. Instead, other government programs would have to try to preserve the well-being and place on the land of impoverished southerners. "Even as the old tenure system deteriorated," Pete Daniel writes, "liberal New Dealers tried to patch it up."[35] Some believed that they had found these patches in government efforts to rehabilitate and resettle tenant farmers, sharecroppers, and the rural poor.

In rural areas, most of the factors that influenced the development of government relief in cities were lacking. As rural sociologist E. L. Morgan aptly put it, rural relief had not become institutionalized. Instead, Americans expected immediate emotional appeals, neighborly good will, and kinship to cover the plight of the rural poor, which was considered to be less difficult than the needs of their urban counterparts. As a result, no welfare institution with trained social workers developed in rural areas.[36] Misconceptions about rural poverty in turn influenced the planning of federal anti-rural poverty programs. Despite years of hardship, most people did not think of farmers as being poor at all. To many, the very existence of a farm and thus the means to sustain oneself meant, if not luxury, then at least safety and security for the farmer.

The Federal Emergency Relief Appropriation Act of 1933 easily passed Congress early in Franklin Roosevelt's first hundred days. It authorized the creation of the Federal Emergency Relief Administration to make direct relief grants to the states. To deal with the problem of some 12 million unemployed Americans, FERA had half a billion dollars to spend in direct aid and another $3.3 billion to be spent for public works.[37] It provided money to state Emergency Relief Administrations, which controlled their own programs with a bit of federal oversight. Federal funds were intended to supplement state funds, but in the cash-strapped South, with its tightfisted state governments, the federal government contributed most of the money spent to alleviate poverty. In November 1933, FERA formally started the Civil Works Administration (CWA), which operated a program of work relief instead of providing direct relief.

At first, FERA did not make any special effort to improve the lot of farmers in general, and government aid did not differentiate between unemployed workers and poor farmers. Farmers had long been considered autonomous, self-sustaining independents who needed less help than the rest of the population. Well after the rural rehabilitation program had begun, this conception remained. "Rural families can be more easily rehabilitated," a pamphlet describing FERA's "Purpose and Activities" vaguely claimed, "because, with the proper help, they can be put in a position to supply themselves with food and other necessities of life."[38]

New Dealers explicitly contrasted this supposed notion of the farm as a bastion of security with the declining position of industrial workers, who could not fend for themselves. David R. Williams, chief of the Planning Section of the Rural Rehabilitation Division, argued in 1934 that even in their depressed state, America's farms could be a refuge for those displaced by the declining industrial sector.[39] Irish poet George W. Russell told Rural Rehabilitation staff members that a "race dies out within two or three generation of city life," comparing the physical deterioration of the contemporary Londoner with the health and vitality of his rural forebears.[40] This idea of rural life as self-sustaining, immune to the vagaries of the economy, was also evident in the justification for the Division of Subsistence Homesteads, which was already engaged in the construction of resettlement communities based on the idea of subsistence farming.

In addition to this belief in the stability and inherent safety of farm life, practical concerns limited how useful FERA's earliest efforts were for the rural needy, particularly for the worst-off rural poor in the South. As a system designed to help the unemployed, it did not really fit the agricultural South, with its traditional slack seasons. Further, many landlords believed that their tenants were always employed, no matter how much pay or support the landlord provided. And with living conditions already so bad, it was difficult to define a standard of living to be met by relief. For the first few months, relief was broadly aimed, encompassing a variety of groups in dissimilar situations: the urban unemployed, the rural poor, and the simply unemployable. For farmers, this meant that relief was in a practical sense nonexistent: local relief was organized in towns and cities, often at a distance that made participation impossible. Those who did manage to get on relief found that the depressed agricultural economy meant that there were few options when local relief ran out.[41]

There was also the fear that rural relief could prove counterproductive. Harry L. Hopkins, a former New York social welfare worker and state relief director and FERA's first administrator, was hesitant, for example, to take responsibility for migrant agricultural workers, who he feared would be taken advantage of by farm owners foisting support for their employees onto the federal government.[42] Similarly, landlords heavily influenced (or even operated) some relief programs, with the result that tenants had to lose virtually everything to get relief. One study of farm families in December 1933 showed that about

30 percent of the relief families in Alabama were helped by landlords to get on relief in order to ease the landlords' own financial responsibilities. In addition to using relief as a way to save themselves money, landowners also influenced relief agencies to maintain a pliable supply of labor; for example, some rural relief offices closed during the two months of cotton picking. This helped both the relief officials, by reducing their expenditures, and the landlords, by not disturbing the low local rates for cotton picking. Furthermore, rural relief might not even reach those who needed it most. Black farmers, for example, were kept off relief rolls because of discrimination and because landlords kept them around longer. Local relief was even used as a weapon against groups that challenged the power of big planters. The Alabama Relief Administration, for example, undermined organization efforts by sharecroppers and tenants. By the summer of 1934, the Share Croppers' Union had attracted the interest of many sharecroppers and had begun to prepare a large-scale strike in Lee and Tallapoosa Counties for all cotton pickers. As soon as the union announced its plan, however, relief director Thad Holt declared that all able-bodied workers who did not voluntarily pick cotton for wages would be dropped from the relief rolls.[43]

Even taking into consideration these risks, it was evident that a wide spectrum of the rural poor relied heavily on relief and needed additional help. Farmers who technically qualified as landowners often labored under a heavy burden of tax liens and mortgages. On the basis of the actual realization value of their property, an owner's net equity was often nonexistent. These owners generally lacked the necessary work animals and farm implements for the depleted soils of their small acreages. Farm tenants depended on the production of staple crops, but the production and prices of these had dropped so much that their landlords no longer maintained a normal work force. Families who had abandoned farm work drifted to nearby towns and villages and lived in substandard conditions that effectively constituted rural slums. Other families had formerly worked in rural industries like timber or mining until these enterprises collapsed with the Depression, leaving them without any means of moving or making an income. All of these groups had been under economic strain for several years, and even the minimal resources available to impoverished city populations were unavailable to them. To them, relief felt like a godsend. Many tenants would have rather stayed on relief than risk going back to farming; it offered the highest standard of living they had ever known.[44]

FERA officials discussed whether the agency should add raising the standard of living or curing poverty to its mandate of providing unemployment relief. The expense of relief and the enormous, chronic poverty in rural areas strengthened the argument that something further had to be done. Federal administrators looked to the states for solutions. The emerging idea of rural rehabilitation, which appeared to be successful in certain southern states, seemed to be the key. Starting with Alabama and Texas, cotton states began to provide tools and livestock to relief clients. They were cheaper than relief and encouraged farmers to continue to produce. FERA officials saw this shift from relief to rehabilitation as the grounds for a new long-term agricultural policy.[45] The similarities between the rural rehabilitation programs in these states and the eventual national program indicate something of their influence. The Alabama Relief Administration's rural rehabilitation program, among the closest to what the national FERA rural rehabilitation model would be, was designed as a mass program to provide loans and seeds in conjunction with farm management expertise via "farm foremen," who traveled across the state teaching clients how to budget, harvest, and perform other necessary agricultural tasks.[46]

The goal of rural rehabilitation was to find a more cost-effective and long-term solution for thousands of rural families in need of immediate relief. To use only direct relief payments would be ruinously expensive and probably impossible. The solution lay in making the poor self-sustaining by putting them back into a position to grow enough food to support themselves, providing credit to purchase necessities, and giving supervision to ensure that the money was well spent. In a larger sense, FERA was the beginning of a new national program designed to deal with the problems of rural poverty. It would take farm families off the relief rolls, get them farming again, and create self-supporting rural citizens. The imminent expiration of the Civil Works Administration, at any rate, made some kind of replacement program necessary. The time and ideas were right, and federal officials decided on a transition to the new program. Along with a new work program for the urban poor, relief efforts would include rural rehabilitation and, for stranded rural populations, relocation and community building.[47]

Hopkins began making moves toward formally creating what became the Rural Rehabilitation Division of the Federal Emergency Relief Administration in the spring of 1934. It had, he said, three aims: to make farmers self-supporting again, to move farmers from bad land to good, and to establish self-sustaining communities of stranded industrial workers.[48] He gave responsibility for the program to Lawrence Westbrook, relief administrator for Texas. Hopkins intended the division as a federal-state project in which the states would administer programs based on federal policies and recommendations. State relief administrations created Rural Rehabilitation Corporations to act as the financial agencies of these agencies. The only direct control that FERA would have would be control of funds and possession of all capital stock of the various corporations. Work and direct relief clients from rural areas moved to the rural rehabilitation program on April 1, 1934.[49]

This new rural program differed from urban relief in two significant ways. First, rural rehabilitation was conceived as a solution or even a preventative to at least some of the problems of rural poverty. While most urban unemployment relief was designed to help those already in need of relief, the rural FERA program ideally took on borderline cases to keep them from going on relief. Second, work relief for urban and industrial workers was a cash wage, a payment for work done, while the money given to the rural needy was a loan to be repaid. (That was the theory, at least; in practice, a number of loans had to be written off as unrecoverable).[50]

FERA officials promoted their program in part as a way to save money, a general emphasis in all FERA self-promotions; Harry Hopkins, for example, titled his 1936 account *Spending to Save: The Complete Story of Relief*. When describing its work in late December 1934, Westbrook noted that rural relief differed significantly from regular relief and that "there existed in rural areas opportunities for rehabilitation which could be taken advantage of at relatively small cost." He went on, claiming that "many thousands of destitute farmers ... could be made self-sustaining by an investment in equipment and supplies which would not exceed the cost of extending relief over a period of several months."[51] As Paul V. Maris, supervisor of field service, wrote, rural rehabilitation meant not simply "getting people off the relief roll" but rather "making it possible for them to get off the relief roll." He went on to assert, "Rural Rehabilitation is not a program of public charity" and that every expenditure was to be repaid.[52] The program, supporters claimed, might even make money. FERA's *Rural Rehabilitation* magazine, for example, described an "actual instance" of a rural rehabilitation client moving from the ranks of the "tax-eaters" to the "taxpayers" thanks to supervision and a farm plan, debt adjustment mediation, and a small loan.[53] The program also reflected the widely held view that government handouts contributed to the demoralization or decline of its recipients. "The government is providing them with the necessities of life," Westbrook wrote of stranded rural workers on relief, "but not with the necessities of living."[54]

The national Division of Rural Rehabilitation did not make the loans directly to clients; instead FERA made loans via the State Emergency Administrations. The state relief agencies organized their own rehabilitation divisions, in most cases reallocating funds to county relief administrators; by March 1935, FERA consolidated these into forty-six state Rural Rehabilitation Corporations. The broad, undefined nature of the decentralized, state-based programs was evident from the beginning. Rural projects developed in three different directions, embracing rural rehabilitation via loans and supervision, community creation and resettlement, and land reform.[55]

The use of home and farm management plans, which would become the centerpiece of the overall farm-security effort, began in the Rural Rehabilitation Division of FERA. Paul V. Maris wrote, "Rural Rehabilitation seems to be very correctly described as a form of 'Supervised Credit.'"[56] Early FERA loans were generally restricted to promote subsistence-level farming. Inexperienced and determined to get the program into place in a hurry, FERA's local agents tended to underestimate the size of loan needed. As a result, loan advances were both small and frequent: the 397,000 family cases received almost 1.27 million individual advances.[57] Many states required that all loans be paid back in one year, so loans were often made with that goal in mind rather than in consideration of what the most effective amount would be. A family might get, for example, a two-hundred-dollar loan, that being the largest amount it could pay back. But that almost certainly necessitated a loan the next year for almost as much money, because the initial loan was insufficient to get a family-sized farm working profitably. And especially in the first few months, loans were made with neither the clients nor the agents entirely sure of what they would be used for, a practice that created problems and encouraged more careful planning when the program moved to the Resettlement Administration.[58]

In most states, the funds for loans and grants were reallocated to the counties to do whatever rehabilitation activities that county relief officials felt best suited their caseload. In Arkansas, for example, canned goods were the most common source of repayment after twelve hundred community canning centers had been set up, producing almost three hundred thousand cans of fruits and vegetables weekly. In September alone, the government took in repayments of about four million cans, which were then given to other needy families.[59] Particularly in the South, credit took the form of groceries, feed, fertilizer, work stock, and other goods, building on the southern tradition of the furnish. County supervisors were aided by assistant county supervisors, sometimes called "farm foremen" and imagined as part-timers selected among the best farmers in the community, and home rehabilitation assistants, generally women trained in home economics and whose primary responsibility was periodic home visits to help plan budgets, food production, meals, and similar concerns. One measure of the program's success in Alabama, as Lorena Hickock reported, was that these supervisors often became "so interested that they [were] working way overtime, without pay."[60]

At first, the rural rehabilitation program was primarily a southern affair. Many states had made a total transfer of rural cases from general relief to rural rehabilitation, and several southern states already used a loan basis for their relief programs. The South also already had a strong credit tradition: rural rehabilitation loans easily slid into the place of landlords and merchants in the furnish system (sometimes with the previous creditors' approval, sometimes not), especially as the Depression had resulted in a decrease in credit available from these traditional sources. Finally, the generally small-scale nature of southern agriculture made the problems of rural poverty in the South appear relatively straightforward: give money to these poor farmers already on the land they cultivate.[61] The overall southern emphasis of the program is evident in that half of the total loan amount expended went to only six southern states (Alabama, Arkansas, Georgia, Louisiana, Mississippi, and Texas). In February 1935, more than half the advances went to cases in just Alabama and Louisiana.[62] Even as the program developed a more national scope, many of the rules and procedures developed to address southern problems remained—for example, the requirement that landlords sign a waiver to the effect that they would not seize any of the year's profit or crops for tenants' debts.[63]

Although preparation of applications was not quite as rigorous as it would become under the Resettlement Administration, the Rural Rehabilitation Division did require a good bit of information before considering a poor farmer. Applicants had to list the size, type, condition, and value of the farm they currently worked or their history of farm labor, along with their possessions (both agricultural and personal) and outstanding debts. The entire family was considered in the application. Potential clients described their families, particularly in terms of capacity to work: children were grouped by those over or under sixteen years of age, a common dividing line for a full day's worker, and applicants had to list any family member at least "16 years of age who are unable to work," family members "especially qualified for supplemental industrial employment," and any military disability by the family head. Recognizing the credit situation of small farmers, the rural rehabilitation application asked about seven possible debts—first and second mortgages, chattel mortgages, past-due interest on other debts, delinquent taxes, store bills, and borrowings against life insurance. [64]

Most state relief administrators assigned selection to their social service divisions, since caseworkers were already familiar with the farm families on relief rolls in the state and with the techniques of family investigation. Beyond the few minimum standards of age, health, experience, and intelligence, caseworkers generally had freedom to select families for each project. This was often a long process. Caseworkers told families about the projects during their interviews; based on their responses to the interview questions, families were rejected or recommended. Eligible families were told about the possibility and given time to consider, then interviewed further for information about less quantifiable factors like initiative or farming ambition. Desire for life in a rural environment was the most important consideration; other qualifications included farm experience and health. Recommended families were informed of their eligibility and reinvestigated. Caseworkers met with interested families, who then went into conference with farm experts, social workers, and personnel experts. A five-person local committee then reviewed the entire record and decided. Approved families were notified and got physicals.[65]

FERA rural rehabilitation administrators found themselves with a problem that would face their successors for the next decade in evaluating clients and, more specifically, in deciding for whom their program was actually intended. Providing loans for the neediest farmers meant burdening them with loans that might never be repaid, risking the Rural Rehabilitation Division's reputation by making bad loans and wasting a good bit of money. Raising the eligibility level for loans and grants meant ignoring those who needed them most and increasing the number of farmers on the relief rolls, the reduction of which had been part of the initial justification for the program. In general, FERA tended to select farm families in a better economic position more often than sharecroppers and day laborers. One study found that only about a fifth of clients were wage hands and a third of the rest were sharecroppers.[66]

The rural rehabilitation program considered both a family's farm income and its personal well-being, things that had previously been discrete elements of American rural life. In the South, many small farmers saw the credit system as not really a part of farm life but as an interruption or danger to it. As Robert W. Hudgens, who had long been involved with rural rehabilitation and would be until after World War II, put it, when rural southerners had thought about their expenses, "The banker was in competition with the family living." Credit threatened a family's well-being the same way locusts or a heat wave might do. The Rural Rehabilitation Division was the first agency to put the farm budget and family living together.[67] Bringing together and integrating unrelated activities led to some early mistakes, but it changed the relationship between creditor and client. It also made rural rehabilitation, at least in theory, much more effective—unlike banks or credit agencies that would lend money to a bad farmer with good credit, or an extension service giving good advice to farmers who could not afford to implement it, the rural rehabilitation program could consider the whole farm family and thus better tailor its approach.

The Rural Rehabilitation Division was quite popular with clients, social scientists, administrators, and the press. Indeed, some early reports were almost hyperbolic: an *Atlanta Constitution* headline praised the "pioneer spirit" awoken by the "amazing rural rehabilitation plan."[68] While the program was not large enough to put a significant dent in rural poverty, one observer claimed that "the results have been gratifying" in that the "condition of many individual families has been materially improved," which raised morale not only for those who took part but also those who saw the program in action.[69] Farmers who prided themselves on their independence found that loans and even grants had less social stigma than did direct relief, and loans had the added benefit of costing the government less than direct or work relief. One observer claimed that to the relief workers, "the rural rehabilitation program offers more hope than anything else these days."[70]

The general public favored the principle and goals of the program, although they sometimes questioned the methods. Part of the reason for its popularity may have been how many people it helped; in its busiest month (April 1935), FERA had just short of 210,000 cases, involving over a million people.[71] FERA advanced rehabilitation funds to 397,130 cases between April 1934 and June 1935, most of which would have been considered nonstandard loans or grants by its successors—that is, loans unlikely to be paid back. Rural rehabilitation money was particularly valuable for the poorest farmers. The program was heavily concentrated in the cotton South, where the furnish had provided funds to tenants and sharecroppers and, to a lesser extent, in the drought-stricken Plains and Great Lake states—in other words, in those areas of the country where farmers were in the worst shape.[72] In the tenant-heavy cotton areas, 97.9 percent of southern clients received "subsistence goods" loans—for food, clothing, medical care, and similar needs. In contrast, less than a third of clients needed subsistence loans in the northern "Hay and Dairy Region."[73] Rural rehabilitation did not help all, or even most, families in need, but it was easy for observers to imagine the possible impact if the program could manage on a large scale what individual projects did for the families involved. The Cumberland Project (later renamed Skyline Farms) twelve miles north of Scottsboro, Alabama, for example, only helped 250 people while administered by the Alabama Relief Administration, but each farmer moved, in one reporter's words, "from servile sharecropper to proud land-owner."[74] This kind of success, many supporters hoped, could provide a foundation for the future.

Despite this popularity and apparent success, observers recognized that a loan program alone could not solve the problem. In fact, in some cases, it made things worse. FERA could give farmers the chance to borrow money much more easily than it could the capacity to repay it. A farmer often borrowed against the farm and, unable to repay the loans, risked losing it altogether. Nor was it always clear to farmers that they had even incurred a debt; many had viewed the rehabilitation loan as a grant or relief and only gradually realized after a year or two that they now had debts to the federal government. All of a sudden a family found itself with yet another debt and with little to show for it, a situation most reminiscent of the old-fashioned furnish system.[75]

Intended in part to remedy some of the shortcomings of the Rural Rehabilitation loans, FERA also engaged in various community-building and resettlement programs along the same line as those developed by the Division of Subsistence

Homesteads. This aspect of the program was aimed at two different kinds of rural citizens. First, a number of rural people (perhaps a quarter of the nation's population) found themselves stranded in areas lacking employment prospects for half or more of the potential workers living in the area. Many, like lumber and mine workers, had their old jobs eliminated through mechanization, exhaustion of resources, or the Depression (or all three). Natural disasters, loss of markets, and acreage reduction across the country reduced agricultural production and thus demand for agricultural labor. In these cases, FERA tried both to better the situation of local residents by improving their communities and to cut down on relief costs by paying locals for the construction of such improvements. In most cases, this meant working at the level of individual farms, but in others, FERA aimed to reconstruct entire rural communities.[76]

One pressing need was for stranded families who were left behind by the economy and had little chance of finding any kind of meaningful employment or advancement, much less self-sufficiency, in their current locations. For those families, FERA administrators hoped to create new rural-industrial communities or revitalize existing towns or villages to such an extent that they became entirely new communities. These new communities would combine, ideally, the production of food and raw materials for home use with industrial activities aimed at producing products for outside sale. Residents would live on small tracts where they could grow food for their families, surrounded by and intermingled with social and economic centers. Between 100 and 150 such communities, composed of perhaps up to one thousand families but no less than two hundred each, would dot the American landscape. Part-time farming and decentralization of industry, two trends that supporters were certain to continue in light of the Depression, were the backbone ideas of the plan.[77]

Rural Rehabilitation director Lawrence Westbrook intended these communities to operate in much the same way that the rural rehabilitation program did. The states could run their own programs. Residents could get small loans for things like mules, chickens, or equipment. Westbrook hoped that industry would move to the projects, but he made no effort to directly encourage it. In all cases, keeping the government out of the way was paramount; FERA administrators wanted only to provide an opportunity. "We are not trying to do all the thinking for these unfortunate people," Westbrook said about a project carried out by the Texas Relief Administration. "We are trying to give them an opportunity. They'll accomplish more for themselves than we could in the long run. I think that we who plan, often plan too much."[78]

The new communities selected families much as the rural rehabilitation program did, with local decision making, usually carried out by social workers who best knew the families and the needs of the communities. In the resettlement

RURAL REHABILITATION AGENTS MAKING FARM AND HOME PLANS WITH CLIENTS MR. AND MRS. E. H. WISE, COFFEE COUNTY, ALABAMA. LIBRARY OF CONGRESS.

projects, however, there were additional decision-making layers. The 105-unit Irwinville, Georgia, project is a good example. There, the Georgia Rehabilitation Corporation put W. P. Bryan, a member of the corporation's board of directors, almost entirely in charge of organizing the project. He selected the land for the project, handled purchases, and chose the families who would start the project. These families all carried with them case reports from their time on relief, and Bryan used these and personal interviews to select the residents. This personal influence curried some benefits; Bryan even loaned the residents his own money until they got settled. But this sort of entirely local, individual-driven approach also created problems, most notably in how slowly the program got started. It was not until two years after the project began, and under a new and more centralized agency (the Resettlement Administration), that new houses and farm buildings were finally built, farm and home management plans were created, and a medical program was begun.[79]

The resettlement projects had a much worse reputation than did the rural rehabilitation program. Many were badly planned. The Coffee County Farms Project, scattered throughout Coffee County and five other Alabama counties, was begun in 1935 by the Alabama Rural Rehabilitation Corporation to aid families on the sixty thousand acres facing foreclosure in the area and suffering from bad crop years, eroded soil, and low produce prices. The state government bought most of the foreclosed farms and turned them over to the RA. After surveys, submarginal lands were planted with pine trees and the remainder were divided into family-sized farms.[80] The goal, in the words of one agricultural economist, was "to raise the standard of living of rural people whose standard of living has been almost unbelievably low."[81] But FERA's good intentions were not enough. Robert W. Hudgens asserted that "there was no sound economic basis for the project as it was originally conceived." Ernie Pyle described the project's start as "not by design, especially, but because government people and the local agencies got enthusiastic, and it just grew up under them."[82] The Coffee County Project appears to have been started in large part because the Alabama Rural Rehabilitation Corporation was able to get so much land at such a low cost. The land was poor and prone to erosion, and the proposed farm plan would only make the problem of erosion worse. Nor was construction of the project handled much better; many of the men working on the construction phase of the project mistakenly believed that their homes and farms would be provided for free, as an additional payment for their work at the Alabama Conservation Camp.[83]

Skyline Farms, in northern Alabama, similarly created headaches for administrators. It was developed to ease the relief burden in northeastern Alabama and to provide homes for families forced to move by the Tennessee Valley Authority. It and similar projects prompted fears that the federal government would compete with private industry or would promote one business at the expense of another. Critics worried (correctly, as turned out to be the case) that the government would get involved in promoting opportunities for industrial employment if local industry failed to revive on its own and if the project failed to attract new employment opportunities. Less remarked upon at the time was that Skyline Farms reflected a frankly racial conception of who made a good candidate for resettlement. The families living there, a report explained, were "some of the purest Anglo-Saxon strains in the country" but had been "handicapped for generations by low incomes and lack of normal opportunities." Being white was considered proof enough that clients could make a project work.[84]

Business leaders in particular resented the resettlement program, especially the implication that private businesses had any sort of responsibility to ensure the success of new communities. Henry I. Harriman, president of the United States Chamber of Commerce, claimed that such changes should develop gradually because business was "in a pretty hot spot." He argued that removing factories to the fringes of a subsistence community was unrealistic but that conditions could change in the future. "As their plants wear out and replacements are needed," he said, industry might consider a transition to the kind of development FERA had in mind, but it was unrealistic to expect such expansion while the economy was so bad.[85]

Besides operating the rehabilitation and resettlement programs, FERA was part of a larger effort toward general land reform, which was actually an effort to address a series of related problems. FERA, the Soil Erosion Service, the AAA, the National Park Service, the Forest Service, the Division of Subsistence Homesteads, and a variety of other government agencies and divisions had a role in creating and undertaking new land use policies. The retirement of submarginal land, the improvement of land, promulgation of soil erosion techniques, reforestation, and other programs had been an interest of Roosevelt's since his time as governor of New York. Rexford Tugwell thought a long-term federal land program, in which the government took possession of unprofitable or marginally profitable land and put it to other uses like parks, forests, wild life preserves, and so on, to be the only realistic option for lasting stability. Nor was it just the liberal New Dealers and other reformists who saw the need for land use reform; even the *Extension Service Review* dedicated articles to promoting it. Some planners, like L. C. Gray, saw sound land use planning as the most important and valuable thing the government could do for rural America, and he praised the government's willingness to implement a better land use system during the New Deal.[86]

The need for such a program certainly appeared great. For example, one reason for the South's poverty was its misused soil. Cotton and tobacco rapidly took nutrients out of the soil; most southern farmers neither knew nor cared about this—nor could they afford to replace soil fertility. Erosion not only ruined the soil, but it also silted reservoirs, filled up navigation channels, and made flood control more difficult. Erosion was a problem across the country—the dust storms that cut across the Great Plains being a good example—but the problem was widely associated by the 1930s with the problems of tenancy. Run-off cut gullies throughout the South. The enormous Providence Gully in Georgia was in spots two hundred feet deep and six hundred feet wide. It meandered along for several miles and had by 1934 swallowed a barn, a church, a schoolhouse, a graveyard, some homes, and several farms.[87]

The problem of land use was inseparably tied up with rural poverty. The most soil-damaging crop, cotton, was most likely to be grown by those who least cared about soil damage, tenant farmers. As of 1930, tenants operated 73 percent of farms that relied primarily on cotton for income.[88] Tenants had no incentive to improve the land; on the contrary, it made the most economic sense for them to take as much as they could. Likewise, landlords had little reason to repair or improve housing, barns, fences, or wells on land they rented, as these had little direct impact on their own well-being. Furthermore, even if the tenant did want to improve the land, most could only afford to plant cash crops, not soil-improving cover crops (much less let the land lie fallow); if they purchased fertilizer, it was only for the purposes of a short-term increase in production.[89]

The South's treatment of its other natural resources followed a similar pattern. Urban growth increased the demand for lumber, and the South, like the rest of the country, responded with enormously wasteful destruction. Timber companies crossing the country left deserted mills, stranded part-time farmers, and completely cutover forests. This not only encouraged erosion and reduced future sources of timber from local farmers, but it also contributed to the problems of low-income, submarginal farmers: a number of the rehabilitation families in the 1930s had been trying to make a living on farms originally intended for part-time farming combined with timber and other rural industry.[90]

A number of New Dealers saw the land use program, despite its halting start, as the key to ending such abuses and restoring agricultural prosperity. Rexford Tugwell wanted a long-term federal land program in which the government took possession of unprofitable or marginally profitable land and put it to other uses like parks, forests, wildlife preserves, and so on. The most efficient farmers would remain on the best lands, continuing to increase their efficiency without overwhelming the markets. This was, in Tugwell's view, the only realistic option: wholly private ownership of agricultural land had proved itself inadequate, while large-scale restriction of land use to reduce agricultural output only created more expense without addressing the fundamental problems. A land program, combined with programs to raise standards of living (and thus consumption), would provide for long-term stability.[91]

∽:∽

Despite high hopes, FERA's overall rural program was essentially a holding effort, designed to keep farmers on their farms at a subsistence level during troubled times. That work relief was still available for many of them to repay their loans makes it clear that early administrators did not think of these farmers as entirely self-sufficient or as being able to support themselves solely from agricultural pursuits. Further, many FERA administrators saw rural poverty as a temporary problem, brought on by the Depression or natural disaster, and correspondingly believed that their work would be temporary. It was not a surprise, then, that so much of their early supervision was simply not very good.[92] Many of the principles set down in this early period, however, would prove to be enduring, including the connection between credit and supervision, along with related or complementary programs like resettlement, land reform, and debt adjustment to ensure that clients remained good candidates for agricultural success and the repayment of loans.

FERA efforts also prompted a wave of criticism and opposition from conservatives who opposed it on ideological grounds, from business and agricultural leaders who resented the intrusion and regulations, and from within the federal government as new agricultural programs stepped on the toes of established departments and agencies. For example, many in the Extension Service believed, as the West Virginia state director wrote in 1930, that if Congress introduced a land-retirement program, "extension work will undoubtedly have to shoulder much of the responsibility of interpreting it to the rural people and enlisting their cooperation." The creation of a new agency outside the Department of Agriculture did not fit into this vision. This criticism was made more dangerous by the fact that FERA as a whole lacked permanency: it was part of no major executive department, with few institutional protections and defenders. Despite this opposition, FERA did have successes. The rural relief rate in the Southeast began to decline in the early months of 1934 and continued to do so irregularly during the life of the FERA programs, as opposed to the generally increasing relief rolls in the rest of the country. Beyond any quantitative success, the rural rehabilitation program also seemed to point toward a new way of approaching rural poverty. As the New Deal got underway with a second round of reform and agency making in 1935, liberals considering the problems of agriculture and the rural poor sought to incorporate the program into a new effort at national planning and reform.[93]

CHAPTER TWO

Going Back to the Land
The Division of Subsistence Homesteads

In early 1934, one reporter described a new way of life that had caught the imagination of many Americans during the early years of the Great Depression. "Midway between factory town and typical farm," Henry Edison Williams wrote, "lies the Promised Land—a land of comfortable subsistence for everyone willing to work."[1] This was the appeal of subsistence homesteads, an idea that at times captivated Americans from the lowest farm laborers up to senators and presidents. Various colonization efforts had been proposed over the years, but subsistence homesteads had considerable appeal, drawing on long-held cultural notions of self-supporting individualist farmers. This was a far cry from the most influential part of the early New Deal's agricultural program, the Agricultural Adjustment Administration, which had been planned almost exclusively for the benefit of larger producers and generally worsened the position of small farmers and pushed tenants off the land. John H. Bankhead, Democratic senator from Alabama, was one of the earliest influential supporters of subsistence homesteads. Like President Franklin D. Roosevelt and millions of other Americans both in and out of government (including the planners who were at the same time getting the FERA rural rehabilitation program underway), Bankhead believed in the notion of the farm as a refuge against the uncertainties of the industrial economy. He hoped that a new federal program could take advantage of the natural advantages of rural life to better the condition of those trapped in poverty and unemployment.

This conception drew on a curious mix of extreme pessimism about the economy in general and extreme optimism about the restorative powers of rural living. Planners and supporters had an almost panicked feeling that a return to industrial prosperity would not happen for years or decades, maybe never. Some alternative had to be found for the mass of unemployed who had, from this perspective, no future in their current occupations. The solution would be a new form of living, perhaps even a new form of community, that would reflect the budding back-to-the-land movement capturing imaginations across the country. These new subsistence homesteads would provide everything that a rural or semirural citizen needed for survival. Some would be for farmers moved from submarginal lands or for former industrial employees like stranded coal miners. But the most hope lay with communities that would combine the best of rural and urban life. These would be a modern combination of mechanized farming techniques and small subsistence farms that still provided the opportunity for seasonal or part-time industrial employment.[2]

Generally speaking, nothing about the subsistence homesteads worked out as hoped during the year-and-a-half existence of the Division of Subsistence Homesteads. From the very beginning—even before the beginning, when subsistence homesteads were still just a proposal—unexpected delays and obstacles made it impossible to legislate, plan, construct, and operate the homesteads as anticipated. As the homestead projects around Birmingham, Alabama, illustrate, the Division of Subsistence Homesteads would not even begin construction on most of the projects it planned. That would be left to its successor agency, the Resettlement Administration, which struggled under the burden of the mistakes made and obstacles not overcome by the Division of Subsistence Homesteads.

The back-to-the-land ideal undergirding the early subsistence homesteads program was not new; many in the early 1900s joined the back-to-the-land movement in arguing that the best solution for both rural and urban problems would be to move city folks back to the country. This settled uncomfortably in the Country Life Movement, which by and large favored modern intensive farming as the solution for agricultural ills and therefore frequently had little use for amateur urban farmers. By the late 1920s, a back-to-the-land movement seemed to have apparent basis in observable action. Early in the Great Depression, the trend toward rural migration to the city reversed itself. Jobless workers left urban areas to go back to the supposed security of the farm to avoid paying rent, to live with relatives, to squat, to grow food, and to become independent and self-sufficient. By 1930 eight states and numerous localities had started back-to-the-land programs to harness this new movement. The Depression seemed as never before to strengthen the agrarian, back-to-the-land belief.[3]

Subsistence homesteads looked like a good middle ground between rural and urban life. Land utilization experts argued that perhaps a third of the country's farms should be removed from cultivation because of poor soil and instead be converted to woodland or pasture. This meant moving farmers off the land, a process that would only accelerate with improvements to farm machinery.

Yet unemployment in the cities, proponents of subsistence homesteads argued, indicated that industrial centers were overpopulated. Farmers could not go to the cities; in fact, urban workers needed to move out. Subsistence homesteads would be the place to send these excess workers, making them neither farmers nor factory workers, but something in between.[4]

On these plots, marginally employed workers or farmers would produce a substantial amount of food needed for their family's consumption. Cash income would come from home production and industrial or other outside employment. The acreage would be fairly small—bigger than just a yard and a vegetable garden, but considerably smaller than that required for a commercial farm. In part this was because the would-be homesteader, with outside employment, would not have time to cultivate a large plot, and in part because it takes a fairly small bit of land—between one and five acres, depending on how much cash income is available for supplementary food—to produce enough food to supply a single family. Administrators expected some small surplus farm produce to be sold, but this would only constitute a minor part of the family's overall budget.[5]

Two relatively conservative back-to-the-landers who gave the movement political strength were the governor of New York, Franklin D. Roosevelt (who was soon to be elected president), and agricultural expert and United States Department of Agriculture administrator M. L. Wilson. Roosevelt at times displayed an inherent distrust for urban living and efforts toward urban reform, and he had a strong sense, as did many back-to-the-land proponents, that it was simply impossible to encounter real physical hardship in rural areas. "I think it is fair to say," he claimed in a radio address, "that with certain exceptions most of these people in the country are not faced with actual starvation or actual eviction . . . actual distress and starvation and lack of fuel and lack of clothing exist primarily in the cities of the nation."[6]

In his inaugural address, Roosevelt noted the "overbalance of population in our industrial centres," calling for "a national scale in redistribution" as part of his "endeavor to provide a better use of the land for those best fitted for the land."[7] Technological changes, he believed, would make the family farm more efficient than it had ever been. Thus some kind of homestead program could be a solution to the problems of urban employment as well. Those who lived in the city were "in the very places where it is most inconvenient and expensive for society to help them." Urbanites moving to the country would live both cheaply and more healthily—getting more milk and fresh vegetables, at a lower cost, than was possible in the city.[8]

Roosevelt introduced a subsistence-farming program while governor of New York, and he told the state legislature in January 1932 that it was more than a mere "back to the farm" plan. The "distribution of population during recent years has got out of balance," FDR explained, and "a readjustment must take place to restore the economic and sociological balance."[9] He presented the idea as the next step beyond the immediate, urgent needs of short-term relief, crafting an argument that combined rational economics with notions of the superiority of rural life. Roosevelt went on to claim that unemployed relief families could find "a position at least to partially support themselves in healthy surroundings, of which many of them are now deprived."[10] This would be between the extremes of rural and urban life; as Roosevelt said, "there is a definite place for an intermediate type [of living] between the urban and the rural, namely, a rural-industrial group."[11] If FERA's Rural Rehabilitation Division was aimed at those farmers who had gotten too far away from the market, then the subsistence homesteads would help those who had become too enmeshed in industrial capitalism.

M. L. Wilson was active behind the scenes promoting subsistence homesteads long before the idea reached the point of legislation. He was one of the few leading homestead advocates who was actually a farmer; he had been involved in developing and promoting farm improvements. After graduating from the State Agricultural College at Ames, Iowa, and studying postgraduate agricultural economics at Wisconsin, Wilson had become a widely recognized expert on wheat production (he had been one of the chief authors of the AAA domestic allotment plan), as well as on subsistence farming. He had seen in person the possibilities for very small plots: he had organized just such a development in Montana based on the success of small farmers in the Cache Valley, Utah, working on farms barely larger than two acres. Wilson was a proponent, as were Bankhead and others, of a mixed farming-industrial homestead. As late as 1940, he said that to those who saw the need for a cash income among subsistence or small-time farmers, "it seems that in a national economy there would be room for a combination of part-time industrial employment with part-time farming."[12]

Wilson and Roosevelt shared the belief that American cultural and economic changes made the moment ideal for the government to create a back-to-the-land program, as did many of the administrators under them. Oscar M. Dugger, a regional supervisor in the Division of Subsistence Homesteads, believed that a move back to the land was a result of both the new Depression-era environment and something inherent in humanity. After the crass materialism of the Jazz Age, he claimed, it was only natural that the yearning for home and security should come to the forefront, a reflection of mankind's true values. "Sentiment for home is deeply rooted in the heart of the human creature," Dugger claimed. Reflecting the sentimentalized view of the family farm and home, he also said, "A man, no matter how strong he may be in professional or business qualifications is always a man-child who wants to go at the end of the day to a home and a woman to mother him." A subsistence homestead, Dugger and other supporters insisted, would be the perfect way to nurture this inherent family instinct.[13]

Not everyone agreed with the back-to-the-land assumptions: since the turn of the century, many proponents of rural reform had found little value in agrarian moralizers or proponents of rural self-sufficiency, and that division continued into the New Deal. Like most of the other members of FDR's Brains Trust, Rexford Tugwell put a great deal of emphasis on the importance of agricultural recovery. The Depression had begun in agriculture, he believed, and it would be solved there. But Tugwell found Roosevelt entirely too sentimental about rural life.[14] As he later wrote of his early discussions with FDR, "I differed sharply with him about remedies," even as both agreed about the basic problem.[15] Tugwell had little use for "common platitudes" about "the advantages of fresh country air and sunshine," which "never could fully hide the fact of cruelly long hours and short pay," as he wrote in 1935.[16] With this hard life in mind, he opposed the popular subsistence homesteads idea.

The fact that FDR's particular back-to-the-land ideal ran contrary to such advice was not especially important. It was a politically potent idea, and it followed Roosevelt from Albany to Washington, especially as the national mood seemed to favor such a program. Across the country, a back-to-the-land movement had support. For example, Columbus, Georgia, began a back-to-the-land project in January 1932, moving more than a thousand people to farms. Organized by the local welfare department, the project cooperated with local community groups and organizations to furnish provisions and market produce.[17] Some who called subsistence homesteads "one of the most Utopian of the Roosevelt economic ventures" still favored the program. "Business men like it," a *Dallas Morning News* editorial reported of subsistence homesteads, "because with all its idealism, it is fundamentally conservative." Even if the program had "not much chance of succeeding in practical application," the editor continued, something had to be done. "The age of specialization has brought evils that should be corrected."[18]

In March 1933 Bankhead first introduced a bill to move the supposed surplus industrial population from urban centers to subsistence farms. It would have provided loans, via a $400 million grant to the Reconstruction Finance Corporation, for unemployed industrial workers with farming experience who had recently worked in agriculture or wanted to return to a farm. This bill died in committee, as did a second, much smaller bill, which called for the slightly different "subsistence homestead" instead of "subsistence farms," which might include colony or communal settlements as well as individual farms.[19] The considerable opposition to the bill came with claims that it was "too socialistic," despite (or because of) the fact that it had the support of many social scientists like T. J. Woofter, who called it "one of the most common sense approaches to actual needs of tenants which has been proposed."[20]

Bankhead, with the support of Roosevelt and like-minded New Dealers, then opted for a less direct approach. With little fanfare, a subsistence homesteads program was practically sneaked into the National Industrial Recovery Act (NIRA), as Section 208, to "provide for aiding the redistribution of the overbalance of population in industrial centers" by "making loans for and otherwise aiding in the purchase of subsistence homesteads."[21] The 1933 act empowered the president to create an agency to make loans for the purchase of subsistence homesteads with an allocation of $25 million to finance the program, with repayments from the homesteaders intended to finance its continuing operation. On July 12, 1933, Roosevelt authorized the secretary of the interior to implement Section 208.[22]

Bankhead's program was hardly connected to the rest of the NIRA, and going about creating the subsistence homesteads program in such a way had considerable impact on its administration. It was in Title II, which created the Public Works Administration (PWA), and the subsistence homesteads program received $25 million of the $3.3 billion allocated for fulfilling the purposes of the title. Labor regulations laid down for relief labor had to be used, which caused a number of problems in the operation of the homesteads.[23] The rather tricky legal underpinning required administrative maneuvers that contributed to the reputation of the Division of Subsistence Homesteads for violating legislative intent (which would follow as it became part of the Resettlement Administration and the Farm Security Administration). For example, at the Bankhead Homesteads in Jasper, Alabama, planners had to take creative steps to build a school. The law funding construction of the project established a community center but had no provisions for a public school. Thus the public school building was named the Bankhead Farmsteads Community Center. Except for the auditorium that could host other community functions, it was primarily used as a school, but the name and official purpose alone seemed to have satisfied the letter of the law.[24]

The lack of care in crafting the program created long-term problems. As a relief program, the subsistence homesteads projects had to be put together in a hurry. There simply was not time for careful reflection or study, as rural sociologists noted at the time.[25] The haste also obscured the fact that the purposes of the subsistence homesteads remained unsettled. The discussions over the initial homestead bills indicated that some supporters imagined subsistence homesteads as relief for the industrially unemployed, who needed immediate food and housing. Others saw a more long-term project that would harken back to an older tradition of agricultural workers with part-time employment. In any

CONSTRUCTION OF COMMUNITY CENTER AT BANKHEAD FARMS, ALABAMA.
LIBRARY OF CONGRESS.

case, no clear ideas emerged on what the program should actually look like. Bankhead (like many supporters) straddled these two main ideas, writing, "the only form of permanent relief" is getting "people out of the industrial centers and get[ting] them on small farms where they can dig out of the ground a living for themselves and their families."[26] There were no hearings or debates over the way subsistence homesteads got into NIRA in 1933; the result was that the differing, contradictory, vague ideas of the movement went on unresolved. In the words of rural sociologist Carl C. Taylor, even discussing the planning was difficult, as "no one has yet come forward with a satisfactory definition of what a subsistence homestead is or should be."[27]

Had further reflection taken place, homestead advocates might have recognized their own negativity about the possibilities of their program. High and rising unemployment and growing relief rolls, along with over a decade of agricultural stagnation, meant that some New Dealers worked in what they considered a rather hopeless situation. In 1932 Bankhead asked, "Can we reasonably expect an early return of . . . reemployment by industry of the great army of the unemployed? . . . How many years lie ahead of us before industry can again employ all the idle workers?" M. L. Wilson asserted simply, "Industry in the United States [planned or not] will never again be able to absorb the total available manpower at full time." J. Blaine Gwin of the American Red Cross noted in 1934 the belief "that we have a permanent condition of poverty and must move in the direction of security at the expense of a richer living." Subsistence homesteads advocates believed that full industrial employment would never return.[28]

Roosevelt brought Wilson out of the Department of Agriculture, where he was head of the Wheat Section in the Agricultural Adjustment Administration, to become director of the Division of Subsistence Homesteads when the program was officially established on August 23, 1933.[29] In a way similar to Roosevelt's general approach to New Deal planning—the idea of saving capitalism from its own excesses—Wilson saw the subsistence homesteads as a way to chart a middle course. In a world "where natural resources have been largely appropriated . . . where labor power is increasingly abundant and even harder to sell," pure capitalism was out of date. A completely planned economy, however, would not suit the "individualistic traditions of the American people." Carefully developed, government-instituted national planning would be the solution.[30] The early 1930s, he believed, was the ideal moment for such planning: social, economic, and technological changes meant that the old idea of subsistence farming had a new value. In fact, such programs were already underway—the Tennessee Valley Authority being a key example—which embraced the shifting of populations, regional planning, and the decentralization of industry.[31]

Wilson's goal was a decentralized, grassroots sort of organization. The best structure seemed to be a government-owned corporation; to facilitate decentralization, each project was to have a separate corporation whose stock was owned by the parent corporation. Wilson and most of his staff believed the projects could only be successful if supported, planned, and run locally, so they wanted strong, autonomous local corporations.[32] Philip Glick, who first worked with him at Subsistence Homesteads, called Wilson "a great decentralist" and "a thorough-going democrat of the small 'd' as well as with a capital 'D.'" "Democracy," historian Paul Conkin writes, "was the nearest thing to an absolute to Wilson."[33] If planning did not start at the bottom, on a solid foundation of democratic opinion, it would necessarily fail. This meant that Wilson wanted both gradual change (because any change in democratic opinion would be gradual) and a locally oriented program.

Because the appropriation for the program was so small, administrators decided from the beginning to focus on creating homestead communities rather than providing aid and loans to individual families. The project was so small in comparison to the problem at hand that, as the *Christian Science Monitor* put it, the program could provide "scarcely more than a demonstration of the soundness of this type of population decentralization."[34] A successful community would be a much better demonstration than would individual homesteaders. Further, many administrators believed that a social organization, which could promote a feeling of local responsibility and group sentiment, might be more effective. This explicitly experimental approach did raise criticisms; Dr. William A. Wirt, a school superintendent from Gary, Indiana, testified before a House investigating committee that Wilson's experimental approach was a violation of the government's main goals of relief and recovery.[35]

But such complaints did not deter Wilson, who envisioned three different kinds of communities. First, and most common, would be part-time farming communities near already-existing industrial employment. A second group would be rural colonies for farmers removed from submarginal land. A third, heterogeneous group included the most experimental and controversial programs, such as the communities for stranded coal miners.[36]

The secretary of the interior created the Federal Subsistence Homesteads Corporation on December 2, 1933, which then set up subsidiary corporations for each project. At the first meeting of the directors of the new corporation, eleven subsidiary corporations were set up, each issuing one thousand dollars of common stock, all owned by the Federal Subsistence Homesteads Corporation (so that it could influence the policies of each subsidiary corporation). The parent corporation acted as the division's administrative branch in Washington. The stock of the Federal Subsistence Homesteads Corporation was to be held in trust by Harold Ickes, but the purpose of such an arrangement was local control. Each local company did have some responsibility to the parent Federal Subsistence Homesteads Corporation: it had to expend advances on loans as laid out in the fairly general "Project Book" created for each project to roughly lay out its purposes, planned construction, and organization. But the subsidiary corporation did most of the local work. It owned the title to the land acquired for each project, retained attorneys, created contracts, selected homesteaders, and supervised the communities. The original directors were to be local citizens with representation of influential local groups: agricultural and business leaders, government officials, labor groups, and eventually the homesteaders themselves.[37]

Sponsors of several projects, like the ones planned around Birmingham and Jasper, Alabama, were optimistic about the possibilities for immediate results; John Bankhead, for example, wrote to the director of the subsistence homesteads in the summer of 1934 with plans to start producing tomatoes for the fall.[38] Birmingham in particular seemed like the ideal place for subsistence homesteads. Its metro population of four hundred thousand was among the most industrialized in the South. The large numbers of families on relief living in the apparently overpopulated city center (some 17 percent of Jefferson County families were on relief, about 18,700 cases) could move a few miles away, where they could farm. The loamy soil could be worked easily for intensive, small-scale cultivation. Planners hoped that the land would need very little clearing, as most of the land had been at some point cultivated in the previous few years. Supporters pointed to the existing trend in subsistence homesteading and community gardening in Birmingham; they would only be building on what industrial leaders in Jefferson County already encouraged.[39] The need was obviously great, and sponsors assumed that since rapid growth meant that many of industrial workers were recently from rural areas, the unemployed had enough farming experience to produce food both for themselves and for those on relief.[40]

But the projects quickly ran into problems. One of the most difficult was client selection. The division's methods for approval were influenced by the decentralist Wilson. The division in Washington prepared applications, which it sent out to locals, and Wilson made recommendations about the structure, requirements, and procedures for family selection based on health, family background, stability, attitude, financial status, and similar qualifications. But decision making happened at the local level, and methods thus varied from project to project in the early months.[41] A significant problem was that planners never really defined the ideal candidate. In many cases, in part a result of the hurried creation of the Division of Subsistence Homesteads, planners for local projects had competing ideas about qualifications. The delayed construction of the homesteads around Birmingham, for example, reflected a fundamental division among planners over the exact nature of the Birmingham homesteads.

The National Industrial Recovery Act implied that families had to be from industrial areas and with low incomes, and administrators at first hoped that low-income people with monthly incomes of $50–$100 would be the average occupants, with a monthly income of about $125 as the upper limit.[42] The description of the merits of the Palmerdale project in 1934 neatly encapsulates the thinking of many of those planning the Birmingham homesteads. The project, planners wrote, would "relieve the concentration of population in the industrial area of Jefferson County, Alabama, by shifting a portion of the industrial employees and 'white collar' workers" away from their homes, which had "unattractive surroundings, where housing conditions are not satisfactory." Instead, workers would move to new, attractive homesteads, "located reasonably near existing industries." Those living in these homesteads would gain income from "part time industrial employment and employment of 'white collar' workers with part time agriculture." This living situation provided "a combination of the best elements of both city and country living."[43] Residents would not produce for the market:

"The homesteaders," planners asserted, "will get their subsistence vegetables, berries, fruits, poultry, meats, and milk which they themselves produce. The individual homesteads will be too small for any type of commercial farming and they will not be encouraged to produce farm products for the market." Instead, homesteaders would earn money (expected to be something like $1,000 in annual income) from part-time industrial employment.[44] This small income would allow them to pay cash expenses and to repay their mortgages. The properties were expected to cost some $2,500, to be repaid over a period of twenty years.[45]

But many observers, early on, recognized the possibility that residents would be middle class and white collar rather than poor, semiemployed industrial workers. Erskine Ramsay wrote in April 1934 that, considering the type of housing to be built for the project, "a large proportion of the houses we build will be taken over by 'white collar' people, who will require a good house and they will be able to pay more for a homestead than the poorer class."[46] And initial planning encouraged such a result. The division planned to run the program as a revolving fund for loans to the local corporations. This would have created more stringent demands on purchase price, and the middle-man local corporations likely would have raised interest rates: the Division of Subsistence Homesteads was going to loan money to the corporations at 4 percent, which means that homesteaders could not have had rates much lower than 4.5 percent, which would have made

ALABAMA COAL MINER, BANKHEAD MINES, WALKER COUNTY, ALABAMA.
LIBRARY OF CONGRESS.

it difficult for poorer clients to afford payments. The division planned to offer no subsidies, but it would have been virtually impossible to have such policies and still reach the poor industrial workers who were the intended clientele. At any rate, the RA took over before the division could deal with that predicament.[47]

The conception of the projects often changed over time. The planners of the Bankhead Project at first had in mind a slightly different clientele from the homesteads closer to Birmingham. American coal production, like American agricultural production, had produced more than domestic demand for years. The result around Birmingham and Jasper was stranded miners with farming backgrounds. Initially, planners intended for the Bankhead homesteaders to have the opportunity to grow marketable crops, perhaps even cotton, for cash.[48] However, they increasingly emphasized self-sufficiency as planning moved along. After only a year, planning documents used almost identical terminology to that used for the other Birmingham projects to describe the purposes of the Bankhead Homesteads (to "demonstrate the advisability of decentralizing over-balanced population groups") and its intended beneficiaries (part time industrial workers, in this case coal miners).[49]

Perhaps the best example of how these homesteads were begun without clear objectives could be found in the Cahaba project. The land for the project was purchased without careful study; it was originally intended to be a subsistence homestead like the other communities around Birmingham. Once the land was purchased, it became clear that it was not suited to such a purpose, and for three years it went unused. The project changed into a planned suburban community almost by accident. Rexford Tugwell was visiting in 1936 and casually heard of plans developed by the property manager. He was so impressed with the idea that the project was underway two weeks later.[50]

Such confusion had particular importance for deciding acreage allotments for each community. Low-income industrial workers were the ideal occupants for the projects near Birmingham; how many acres could this kind of worker operate? How much could a part-time worker with an irregular schedule devote to farming? What about a full-time but low-income worker? Decisions had to be made before answers could be found. Agricultural experts claimed that a full-time subsistence farm in Alabama needed ten acres; after more discussion, it was decided that the Palmerdale, Mount Olive, and Greenwood projects should get three to four acres for each homesteader. This was based on the thinking that even full-time industrial employees could farm that small an area, perhaps divided into an orchard, hayfield, pasture, or garden. Bankhead Farms was planned for irregularly employed miners, so its occupants were expected to have a well-rounded program of subsistence farming on units of nine to twenty-five acres, the average being twenty acres.[51] But as it turned out, these kinds of residents failed to materialize, and many of the homesteads became middle-class housing projects.

In part this resulted from local confusion over the nature of the projects and their construction. The Division of Subsistence Homesteads believed that local corporations could carry out the construction and administration. But local leaders had no precedent to follow, and there was no clear understanding of the purposes at the beginning when critical decisions were made. In this period of haste and confusion, the project sites were chosen far away from local industry and mills, selected to be close to the city water supply, sewage systems, and cheap transportation facilities. Some local sponsors, looking to when Birmingham would be a larger city and favoring the English garden-home ideal, wanted the projects to be part of a ring of subsistence homes all around the city, to be completed through zoning or an expansion of the subsistence homesteads projects. The local board responsible for picking the Birmingham sites also had little idea of what type of projects would be undertaken. Individual board members had their own ideas: one anticipated the project to be for destitute relief charges living in simple Civilian Conservation Corps–type shacks. Another assumed that residents would find cash jobs near the projects with no need to travel far, while others expected factories to be located nearby.[52]

Trying to overcome such differences in outlook, Subsistence Homesteads administrators also had to keep in mind the intense scrutiny it was under as an experimental program. Critics who claimed that new residents would use bathtubs as coal bins were eagerly watching the program. On the one hand, that scrutiny meant that administrators looked for those families most likely to make a successful go of homesteading; on the other, it was the poorest families who really needed such a program. There "was some conflict," Clarence Pickett, M. L. Wilson's assistant in the Division of Subsistence Homesteads, admitted, "between these two justifiable desires."[53]

This tension was never really resolved because the division went through so much change so quickly in its brief lifespan. The Federal Subsistence Homesteads Corporation's real strength was that it was able to cut through red tape and adapt quickly because it was not under close control by the General Accounting Office. But in January 1934, Roosevelt issued an order that all agencies and corporations of government had to follow GAO accounting procedures. It was not entirely clear that the program would have to be completely centralized (Wilson believed that some method of local planning could be devised) until March 1934, when word spread that the various local corporations could no longer continue. The official decision came in May 1934. Ickes found the decentralized administration unwieldy and feared the waste or scandal that could result from uncertain local behavior, so he abolished the subsidiary corporations (and with them, local control) and centralized authority in his own department.[54] As he wrote in his diary, local corporations allowed too much local control by those who had no financial interest in success or failure; he would instead insist "on having authority if there is to be responsibility."[55] Ickes placed all projects into a single Division of Subsistence Homesteads. To replace the local boards of directors of the subsidiary corporations, local boards of sponsors and advisers were established, sometimes but not always with the same personnel as the earlier boards of directors: many were offended by the government's intrusive administrative changes and refused to continue serving.[56]

This upset Wilson, who saw it as an end to grassroots democracy. Ickes did not trust or have confidence in the abilities of Wilson or his local appointees, believing him to be a poor judge of personnel. Wilson naturally disagreed, but he also put much less emphasis on efficiency: democracy and self-determination were more important. No matter how wasteful his methods appeared, Wilson believed that no centralized planning could properly administer and direct the various communities. Planning and management had to be local, or they would at some point fail because the interests of a distant, centralized bureaucracy would diverge from the interests of those actually living in the communities.[57]

Wilson was not alone in his opposition to the move. John H. Bankhead, who received a number of complaints regarding the Subsistence Homesteads, wrote in February 1935 that he was "dreadfully disappointed with the administration of the Subsistence Homesteads. It has been made bureaucratic and been so centralized that it is practically broken down from my view point."[58] As it turned out, although Wilson had been more concerned with local control than with efficient operation, the one had encouraged the other. Administrators reported that federalization of the program simply did not work. The local corporations could not function with so much oversight.[59] For example, Robert Hudgens, who joined the Division of Subsistence Homesteads after the centralization, was put in charge of the creation of three South Carolina subsistence projects that were never built—hamstrung by red tape that magnified small obstacles, like a $250 cost overrun or plans for an unnecessary well, into project-killing problems.[60] Smaller opportunities, such as working with a good local tomato producer for the Jasper homesteads, similarly slipped away.[61] That anything happened at all came as a surprise to some. "I am profoundly grateful that I did not know how many things one cannot do under government regulations," Clarence E. Pickett later wrote. "If I had known all the restrictions, I doubt whether the four homesteads for stranded miners would ever have been built."[62]

In June 1934, Wilson left, soon followed by much of the staff. With them went most of the more idealistic goals of the Division of Subsistence Homesteads. Charles E. Pynchon, a former steel and business-equipment executive from

Chicago, replaced Wilson as administrator.⁶³ By mid-September, 117 families occupied homesteads, and another 47 were living temporarily in barns and other outbuildings while their homes were completed. Another 149 homes were almost ready for occupation, and 163 were under construction.⁶⁴ The ambitious hopes and plans for new projects of Wilson and his associates were replaced by an effort to create a more effective administrative organization and simply complete construction. More and more, administrators thought and acted just to get projects finished—the kinds of concessions that would characterize resettlement and related projects throughout their lives.

Part of this shift in emphasis was practical. The situation became increasingly difficult for the Division of Subsistence Homesteads and Pynchon. The factories that supporters had hoped would provide part-time work for homesteaders never materialized near most of the communities. One solution was government-owned factories, but both business groups and congressmen already argued that the government was unfairly promoting industrial activity at the expense of private efforts. Also, FERA and the AAA began to organize similar programs in 1934, prompting the Division of Subsistence Homesteads to stop creating new rural projects.⁶⁵

Another problem arose from the division's NIRA origins. To get relief to the unemployed more quickly, projects began before full plans could be completed,

ONE OF THE HOUSES AT PALMERDALE HOMESTEADS NEAR BIRMINGHAM, ALABAMA. LIBRARY OF CONGRESS.

raising costs. Work relief wage-rate and working-hour provisions applied to the use of labor on subsistence homesteads projects. Delays resulted from not being able to hire the kind of workmen needed and raised building costs by about a third, meaning subsidies or something else had to be added to sell the homes at rates the homesteaders could afford. Will Alexander estimated that these sorts of burdens raised the cost of projects overall by one-third or one-half in most cases.[66]

For example, construction on the units at Palmer Station, the first of the Birmingham-area homesteads, fell behind schedule. Bids for construction were not called for until August 1934.[67] Locals in Birmingham worried about Palmerdale, as a *Birmingham Post* reporter noted: "The reason for the delay is not quite clear here." Moreover, the paper continued, "The idea seemed to be a good one," hinting that perhaps locals had already changed their perceptions of the project.[68] Construction did not begin until February 1935. Planners expected it to take six months, but officials announced that families could begin moving in on November 15. The actual selection of Palmerdale applications was not finished until May 1935.[69] As the *Birmingham Post* reported, "For nearly a year and a half it seemed only a nebulous idea as it found itself enmeshed in red tape. A lot was talked, but there was nothing on the surface to show for it."[70]

From the beginning, using relief labor (as required by NIRA) caused problems: in the first month of construction, requisitioning of workers was held up because an announcement that the Civil Works Administration (CWA) would not be hiring any new workers in Alabama was mistakenly applied to the federal project. Architect D. H. Greer explained to the division's chief of construction that planning the location of the houses could not commence until the surveys were completed, and surveying two thousand acres of land with a small, CWA-adjusted force meant it would take longer. When construction began, administrators found it difficult to hire the kinds of contractors and skilled workers they needed, a problem made worse by the difficulty in matching specifications set by the Construction Section of the Division of Subsistence Homesteads.[71] And the Birmingham homesteads did relatively better than did many other projects. The Casa Grande project in Arizona, for example, included homes built out of adobe bricks. The 120,000 bricks could have been produced by machine, but administrators decided instead to use handmade bricks to employ WPA labor and create jobs. Overall such requirements added about $100,000, or some 15 percent of the final cost.[72] For the Birmingham homesteads, it became necessary to write the value of the houses down to well below their cost in order to make up for the use of relief labor; this raised the concerns of the editors of the *Birmingham Age-Herald* that the federal government was involved in "an unfair competition with the sale of private property."[73]

As the Division of Subsistence Homesteads struggled to get the projects started, its supporters considered improvements. FERA, the Division of Subsistence Homesteads, and other agencies were all doing the same kinds of things in 1934, a sign of the organizational confusion that had resulted from uncertain goals. Tugwell, Wilson, and Secretary of Agriculture Henry Wallace considered moving the program to the USDA to make it more agricultural; there it might find common cause with the submarginal land program. Wallace, for his part, favored pulling the rural resettlement program out of the hodgepodge and placing it in its natural home, the USDA (a place, also, where Ickes would have less influence). Tugwell also agreed that a back-to-the-land program for unemployed industrial workers only hindered any effort to ameliorate rural poverty. In November 1934, FDR considered moving the Subsistence Homesteads Division to FERA, where it could join related projects (a plan supported by several influential homesteads allies, including John H. Bankhead, who appreciated the work Hopkins's Rural Rehabilitation Division had done), and where it would be happily out of Ickes's hair, but Harry Hopkins and other FERA officials did not want it.[74]

In part FERA administrators objected because the subsistence homesteads program simply had not been effective: in spring of 1935, as reorganization became increasingly likely, the division had only spent $8 million of its $25 million appropriation.[75] As a result of such delays, it was losing support. Its poor performance, in the words of one former section chief, was "sickening to those who labored night and day within the Division" and "engendered bitterness in a public that seemed" to have had such high hopes for it.[76] The comparison with the similar work being done by FERA was painful. By October 1934, the *Washington Post* reported, the division had spent only $4 million (of $18 million allotted) and developed sixty-three planned communities. "Unrestricted by requirements imposed on subsistence homesteads," FERA, in contrast, was "now caring for 80,000 families" in forty-two states.[77] What money the division did spend was often wasted: Cumberland Homesteads in Tennessee tried a sorghum plant that went under after two years, a coal mine that mined very little coal, a $55,000 cannery that failed, and an unsuccessful cooperative association that required a $550,000 government payment to avoid collapsing only seven months after its creation.[78]

The Division of Subsistence Homesteads faced other problems. Local support eroded as it seemed like the homestead residents were not paying their share. Federally owned land could not be taxed, so residents of the homesteads were not subject to property taxes. Removing enormous tracts of land from local tax rolls while bringing in dozens or hundreds of new constituents caused budget problems for local communities.[79] Furthermore, the emotional appeal of the back-to-the-land movement and of small homesteads began to dry up by 1935, and the sense of immediate catastrophe eroded after two years of the New Deal. Within the Department of the Interior, the initial objective of removing the surplus industrial population was seen as less worthy.[80]

The Arthurdale project near Reedsville, West Virginia, exemplified the failures of the Subsistence Homesteads for many critics. The need for some help to the impoverished coal miners was plain to everyone from Eleanor Roosevelt on down. Planners had high hopes for it: Wilson described stranded workers "eagerly taking part in the planning and construction" of the project that would mean "gaining of a new conception of a genuinely more satisfactory way of living."[81] Almost from the beginning, however, administrators recognized there might be problems. Ickes wrote presciently on December 2, 1933, "that we are due for some criticism for our work there."[82]

Intended for stranded miners and glass workers, Arthurdale was the first project started by the Division of Subsistence Homesteads. Houses, originally budgeted at twenty-five hundred dollars apiece, ran to an actual average cost of seven thousand dollars, with the government paying two to three thousand dollars of the price. Plans changed repeatedly, and fifty houses were built, torn down, and rebuilt because they could not handle the cold West Virginia winter. Prefabricated houses were chosen, being quicker to build, but then construction was delayed for six months. Chimneys were not planned, so houses had to be rebuilt or expanded to where a poorly planned chimney had been built ten feet away. Occupants were carefully chosen for their capacity as subsistence farmers, but the increased costs and twenty-year mortgages meant that they needed a cash income some fifteen hundred dollars above subsistence. Various proposals and efforts to create local industry—manufacturing post office boxes, shirts, and such—failed. Claims from supporters that the project was a "laboratory" did little to assuage critics: the government, they might ask, had to experiment to discover basic, fundamental construction principles?[83]

Legal problems for the Division also began to mount in late 1934 and into 1935. The solicitor of the Department of the Interior ruled in November 1934 that Section 208 of NIRA specifically provided aid to industrial workers, not resettled farmers; this essentially forbade the rural colonies. In February the comptroller general challenged the legality of the expenditures made by the local corporations, asserting that no legal authority existed for the creation of such corporations or for their actions. In May a further ruling held that the Division of Subsistence Homesteads, only a temporary part of the NIRA, would automatically go out of existence in June 1935, with all remaining funds returned to the Treasury.[84] The division appeared to be at the end of its rope.

Worst of all, at the end of 1934, after nearly two years of experimentation and effort of the New Deal, the Depression had not ended. At first the Democrats benefited from the sense that doing something, anything, would help. But on both his left and his right, Roosevelt found that frustration and disillusionment threatened his programs. Republicans and some unhappy Democrats believed FDR had gone too far. This group, however, was for the time being politically unable to do much; Roosevelt continued to move reform legislation through Congress throughout 1935 and 1936. But, more threatening to his position as leader of the liberal movement in America, many progressives and liberals had become disaffected, threatening to fragment the New Deal coalition or at least reduce Roosevelt's ability to control events. Public demand for increased federal spending escalated throughout the early years of the New Deal. Overwhelmed state and local governments, organizations for the unemployed, labor and agricultural leaders, economists, and others demanded more federal action.[85]

Roosevelt ably handled pressure from the left: he pointed to his program as a commonsense middle ground between intransigent conservatives and dangerous radicals. In what would become known as the Second New Deal, starting in 1935, Roosevelt went beyond a focus on recovery to an emphasis on reform. The New Deal would not just return society to its previous condition, but make it something better. There was more to this than ideological change; some of it was simply seizing an opportunity. The political timing was right, and the groundwork for a push for long-term reform had been laid in the period of experimentation and institution-building since 1933. At the same time, a new group of younger New Dealers came to the fore as Roosevelt was less willing to make concessions to his opposition.[86]

The need for a rural program of some kind was obvious. Poverty in rural areas remained as crushing as it had been in 1930, or generations before that. "Relief" to sharecroppers and the rural poor, wrote relief worker Lillian Perrine Davis, "has meant not a pittance to drag them through till they might be restored to the normal standards of a few years back, but a godsend of plenty such as in all their lives for generations back they have never known before." There was also evidence that more and more landlords were shifting their responsibility toward tenants to relief agencies. Harold Hoffsommer found in a December 1933 study that about a third of tenants had been put onto relief rolls with the aid of their landlords. And the agencies handling rural poverty certainly needed to continue: the Division of Subsistence Homesteads had a place for another half-billion dollars, especially for erosion control, which Ickes claimed was costing the nation $400 million a year.[87]

It was simply impossible, both economically and politically, for direct federal relief to support the unemployed, and the federal government was rapidly decreasing its direct relief efforts in late 1934 and 1935. Roosevelt repeatedly insisted that he wanted the government to move away from "this business of relief."[88] At the same time, local and state relief would not suffice; as late as 1937, only Alabama had a permanent state relief organization. Some New Deal officials argued that the rehabilitation program needed to go further, to change the entire social and economic system of the South; at any rate, for the short term, an enormous number of rural southerners—between one- and two-fifths—were neither self-sufficient nor able to find permanent employment.[89] To handle these problems facing rural Americans, many New Dealers came to believe that a reorganization of the current efforts was in order.

CHAPTER THREE

Unifying Rural Reform
The Resettlement Administration

What supporters praised about the Resettlement Administration was often the same sort of thing that made critics uneasy, as a 1936 *Dallas Morning News* editorial cartoon indicated. It depicted the RA with the title "Modern Gulliver Helps the Little People." An enormous man (looking suspiciously like FDR in a farmer's hat) with the words "Resettlement Administration" across his back picks up in one giant left hand a seven-member family labeled "impoverished families." He crouches down to plant them on "productive soil" around "better homes."[1] This idea of a powerful, effective behemoth moving people to better lands may have been appealing to the rural poor or RA supporters. It suggested at least the possibility of the same kind of relationship between the federal government and impoverished small farmers that the New Deal was creating with other marginalized or distressed groups. But the same cartoon (with slightly different labeling) could have served exactly the same purpose for the RA's opponents, as it demonstrated just what they feared: an uncontrollable entity too large to be harnessed by normal means.

Reformers were still active in rural policy making in the mid- to late 1930s. The purge of the liberals from the Agricultural Adjustment Administration in early 1935 appeared to conclude the story of liberalism in New Deal agricultural policy, just as the defeat of FDR's court-packing plan in early 1937 indicated the waning influence of New Deal liberalism more generally. This is only half true. Reform-minded liberals may have been ousted from the Department of Agriculture, but New Deal policy making continued to include debates over how much the federal government should promote agricultural reform.[2] This was evident in several efforts, like the Bureau of Agricultural Economics, but it was most obvious in the Resettlement Administration and its successor, the Farm Security Administration (FSA)—which, tellingly, originated outside the USDA and the influence of traditional agricultural interests.

"MODERN GULLIVER HELPS THE LITTLE PEOPLE." DALLAS MORNING NEWS, NOVEMBER 17, 1936. REPRINTED WITH PERMISSION OF THE DALLAS MORNING NEWS.

In late 1934 and 1935, the confusion over how the federal government would handle relief encouraged FDR to reorganize rural antipoverty efforts. The enormous new Emergency Relief Appropriation Bill asked for the largest peacetime appropriation in American history. The government would be getting out of the relief business. Instead, FDR emphasized work relief, handled by the new Works Progress Administration (WPA). For the approximately 1.24 million rural Americans on relief, this meant a shift to several other forms of support. The vast majority moved into the WPA or found some sort of private employment; a substantial percentage, perhaps around 10–15 percent in the South, moved to the new RA.[3]

A variety of factors encouraged Roosevelt to create the new agency. For Rexford Tugwell and other liberals, the liberal purge of the Agricultural Adjustment Administration had poisoned the atmosphere, a reminder that USDA politics would be a considerable obstacle for any reforms.[4] But something had to be done; sharecropper and tenant unrest demanded a federal response. Poor farmers in 1935 appeared to be facing two disasters, one natural (the drought conditions throughout the Midwest and Plains states), and one manmade (the expulsion of tenants, croppers, and laborers from the land as a result of the AAA's acreage reduction and payments). Administratively, Roosevelt and Tugwell also recognized that programs addressing rural poverty had been isolated from one another, resulting in all sorts of administrative problems. Some sort of organizational cooperation was necessary. The obvious solution was to create a separate agency in which all current activities could be united, and one outside the USDA to avoid conflict with established USDA policies and personalities.[5]

The Resettlement Administration consolidated several rural antipoverty programs into a single agency, but it still reflected the ad hoc nature of the New Deal's efforts thus far. The RA has been described as representing "a more universal, holistic approach to conservation, recreation, and social programs," which was true to an extent.[6] The creation of the RA was intended in large measure to streamline and improve federal relief of rural poverty and its associated ills, in part through the improvement of administrative and ideological harmony within the Roosevelt administration and among the agricultural agencies of the federal government. But it was still an executive creation without a solid bureaucratic or legislative foundation, taking over the work of similarly confused programs; uncertainty over how it all fit together lasted for almost the entire life of the agency. (In contrast, the more popular and longer-lived Tennessee Valley Authority was a geographically and politically cohesive unit from its beginning.) Exacerbating that problem was political hostility, some aimed at its controversial administrator, Rexford Tugwell, and some springing from established agricultural interests who resented the threat to their power base. There was also small but growing criticism from anti-New Dealers who opposed the creation of (in their view) yet another massive, unnecessary, big-budgeted bureaucracy.

Despite these problems, Tugwell laid the foundation for a long-term government anti-rural poverty effort. The national structure he created outlasted his tenure as administrator and the RA itself, and the RA and FSA's institutional vision was largely Tugwell's own. He left to his successors the equally difficult process of creating a long-term political and bureaucratic foundation.

On April 30, 1935, Franklin Roosevelt created the Resettlement Administration. Citing the Emergency Relief Appropriation Act of 1935, Roosevelt established the agency with Tugwell, at the time undersecretary of agriculture, as its administrator. FDR tasked the new agency with administering the resettlement projects, overseeing land use reform, and making loans "to finance, in whole or in part, the purchase of farm lands and necessary equipment" as authorized under the Emergency Relief Act. Illustrating the still uncertain direction of the new agency, the first order establishing the RA did not include the phrase "rural rehabilitation," though that was amended by executive order on September 26, 1935.[7] The RA was to be a new approach to rural poverty, but what exactly that meant remained up in the air.

The bounds of the RA's responsibilities were broad and vague, encompassing a variety of problems that a *Washington Post* editorial described as "A Staggering Task."[8] Getting organized took time and effort. One former employee recalled that those involved nicknamed the agency the "Unsettlement Administration" during its early days.[9] Laurence Hewes, Tugwell's assistant, wrote, "For months we had no regular life; we ate and slept as we could. Office hours were a bedlam of telephones, visitors, hourly crises; evenings and weekends were devoted to accumulated paper work spewed forth by our infantile field organization."[10] Just figuring out what it had was a big job: the RA inherited something like forty-six separate programs, and they integrated into the new agency unevenly.[11]

Initially the RA had three major divisions, all adopted from previous efforts: Land Utilization, Rural Resettlement, and Rural Rehabilitation. Land Utilization and Resettlement had (relatively) smooth transitions, both transferred directly in by executive order by the middle of May. The Rural Rehabilitation Division's move from FERA to the Resettlement Administration at the end of June proved tougher. The comptroller general informally ruled against the use of state rehabilitation corporations. Grants to the states, or agencies thereof, were forbidden. Instead, the program had to be carried out as an entirely federal activity.[12]

Federalization would not have posed much of a problem had the program relied on direct relief, in which case the state projects could have simply continued, and eventually the federal government would have stepped in to take over operations. But Rural Rehabilitation involved complicated constructions projects, many in the middle of planning or building; loans had been promised and projects begun based on the assumption that grants to the state corporations would provide funds. The new agency could not simply start anew without ignoring loan commitments already made to clients and abandoning partially constructed projects.[13] The various state rehabilitation corporations had operated in a variety of ways. Some were essentially rural relief programs; others operated solely as credit agencies; some already had rural rehabilitation programs up and running. All told, the RA operated at some point through five different kinds of corporations: the state corporations inherited from FERA, the thirty-one Subsistence Homesteads corporations, twelve community resettlement projects operated by twelve homestead corporations, twelve other resettlement projects administered by cooperative corporations, and fifty-eight cooperative corporations that received federal loans as part of the rural rehabilitation program.[14]

Eventually, the Washington office developed a small corporation trust staff, which oversaw the work of a corporation custodian in each office.[15] The process went smoothly enough in most cases, but there were exceptions. In North Carolina, for instance, complications involving personality struggles in the board of the state rehabilitation corporation made it impossible to accomplish the transfer quickly, and it continued to operate in parallel to the RA (and eventually the FSA, as it took over two years to resolve).[16] Other problems were more significant as public relations disasters. For example, late in the summer of 1936, it came out that the RA had accidentally been paying some employees two paychecks at a time, leading to an overpayment of approximately $181,000. This took thousands of hours of work to recover, further fostering the image of the agency as a bureaucratic mess, and the RA's effort to quietly solve the problem made it look like administrators had something to hide.[17]

Despite difficulty in finding qualified staff, the RA expanded rapidly. On May 1, 1935, the RA employed 12 people; by the end of the year that number jumped to 16,386. Only the major executive departments matched or exceeded the RA's size and budget within the federal government. In 1936 it had more employees than the Farm Credit Administration or the Social Security Board (two other new "independent agencies of departmental rank," similar creatures to the independent RA) and the Departments of State, Justice, Commerce, and Labor.[18] Despite efforts to reduce its numbers, at the end of May 1937 the RA still employed about 13,700 people.[19]

These new employees were busy, but the exact scope and institutional order of the resettlement and rehabilitation efforts remained unclear. FERA, the Division of Subsistence Homesteads, and the AAA's Land Policy sections had all handled loans and administration in different ways, and this variety of approaches had to be reconciled. Created from bits and pieces of different agencies, the RA inherited personnel who had worked in various departments and units and who often had little interest in resettlement, rehabilitation, or rural poverty. As a result, many of the RA's new employees had little understanding of the goals, methods, and leadership of the new agency. That Tugwell was technically subordinate to the secretary of agriculture but heading his own agency only encouraged disputes among the transplants from other agricultural agencies.[20]

To oversee the RA, Tugwell put together a national staff with a varied background. He selected Will W. Alexander, a respected authority on southern agricultural and social problems, as his deputy administrator, a position comparable to that of assistant secretary in an executive department. Tugwell also hired a broad range of assistant administrators. The most important of these was Calvin B. "Beanie" Baldwin, who took charge of program operations and administrative work. Paul V. Maris, the future head of the tenant purchase program, was one

REXFORD G. TUGWELL, ADMINISTRATOR, RESETTLEMENT ADMINISTRATION. LIBRARY OF CONGRESS.

of the few professional agriculturalists in the RA's upper levels. Maris assisted the head of the Rural Resettlement Division, Carl C. Taylor, a rural sociologist who had worked in both the Division of Subsistence Homesteads and the Land Policy Section of the AAA.[21]

Unlike the Division of Subsistence Homesteads, which attempted to operate at a remove through state agencies, or FERA's reliance on state corporations, the RA was a national organization. Tugwell had personally been involved in federal programs addressing rural poverty, and he was well aware of how confused the government's responses to date had been. The overlap of work between the Forest Service and the Soil Erosion Service, the abuses created and encouraged by the AAA, and the irregular availability of funds for FERA's Land Program, among other problems, signaled to Tugwell that a single, coordinating force was necessary.[22]

The new RA was thus hierarchical, directed from the national headquarters. Divisions in Washington dealt with finance, procedures, planning, legal issues, and similar obligations. Rural Rehabilitation, Resettlement, and Suburban Town Divisions oversaw the agency's field activity. Below the central Washington office, the RA divided the country into twelve regional offices, determined roughly according to agricultural production and generally along state lines. The state offices were further divided into districts, which usually included between six and twelve county offices as well as the various project offices. These project offices were directly under regional offices, while each county office was part of the district, state, and regional hierarchy. The major division staffs in Washington had no direct command over the functional staffs in the various regions; instead the regional directors unified and coordinated the programs.[23]

A centralized organization had both good and bad results. Most obviously, it avoided the threat that the comptroller general would rule its structure improper, as had happened to FERA and the Division of Subsistence Homesteads.[24] Second, the setup helped avoid political pressure. The state director was not particularly important, acting more as a field operator between the regional and district supervisors. This made it difficult for state politics to influence the agency's local operations. A state supervisor could say to a troublesome politician that he did not make policy but only carried out what came from his supervisors, either in Washington or in the regional office. Since regions covered multiple states, no single state politician could hold absolute sway over regional appointments.[25] These advantages did not, to many, outweigh the downsides; many critics (including within the agency) objected that the RA was overorganized. Instead of simply having four divisions for its four main tasks of rural relief, rural resettlement, land utilization, and suburban resettlement, Tugwell created twelve divisions. This led to overlapping responsibilities, unclear lines of authority, and higher administrative costs.[26]

Field responsibilities eventually settled into four "program" or "operating" divisions: Rural Rehabilitation, Land Utilization, Resettlement, and Suburban Resettlement (also important were the Construction and Management Divisions, which did just what their titles suggest—constructing and managing projects). The Rural Rehabilitation Division came to the RA mostly from FERA and expanded rapidly in the new agency. The division had five separate but closely related operations: a loan program built on farm and home management, an emergency grant program, feed and seed loans, a farm debt adjustment program, and cooperative loans. At first a Rural Resettlement Division handled both rural rehabilitation and resettlement, but in December 1935 the two functions were separated into different divisions.[27]

The most important part of rural rehabilitation was the loan and grant program for low-income farmers, farm tenants, and sharecroppers who could not secure credit elsewhere. Clients received loans for the purchase of feed, seed, fertilizer, animals, equipment, and tools. With these loans came a supervised home and farm plan; thus these recipients were known as "farm plan" clients. Grants were used for emergency cases ("emergency" clients); sometimes loans and grants arrived together, and a family often received grants to get into shape to become borrowers. The loan and grant program was by far the largest aspect of the RA program: within the first year of its existence, about 780,000 farm families received aid.[28]

While the Rural Rehabilitation program was the largest of the RA's operating divisions, Land Utilization was probably the program in which Tugwell had the most confidence as a long-term solution to rural poverty. This program involved the purchase of millions of acres of land and resettling the tens of thousands of displaced farmers. By the time the RA moved to the USDA, it had 206 land utilization projects going, although the term is a bit imprecise, as the projects varied in size between a few hundred acres and millions.[29] In Tugwell's view, rural rehabilitation and resettlement were temporary solutions. But land use reform could be the key to long-term agricultural improvement, finally curing the disease instead of simply treating the symptoms. With a national land program in mind, Tugwell foresaw "a different future, a planned rather than a muddled one, to which we can look forward with new hope of prosperity for agriculture and of protection for the greatest of all our resources—the land."[30]

Despite the importance Tugwell placed on it, however, land use reform efforts remained scattered, and even a definition of the concept remained unclear.[31] The land use program faced practical and administrative problems. The process of retiring land was lengthy.[32] Perhaps more than any other RA program, land utilization was unpopular with many of those it was supposed to help, and it was easy to see such a program as an effort to reduce the number of farmers, or even to depopulate rural areas.[33] Moreover, the federal land program was at times very much an empty-the-ocean-with-a-spoon operation. One estimate found that for every family the government moved off of submarginal land (at great effort and expense), five more moved onto poor lands. Even worse, many of the families moved off of submarginal land did not find a better economic situation when they moved; some became homeless, went on relief, or returned to farming on even worse land than they had left.[34]

To deal with just such displaced farmers, the RA operated the Resettlement and Suburban Resettlement Divisions, taking over approximately sixty programs from the Department of the Interior and from FERA. Tugwell, who did not think highly of the subsistence homesteads program, apparently agreed to accept them into his new agency in large part to keep them from tangling up his efforts from the outside.[35] The first duty of the resettlement program was to complete the subsistence homesteads already under way. Only $7 million of the $25 million allocated for the construction of the homesteads had been expended or committed when the RA took over.[36] The RA also began construction on migratory labor camps in California, the creation of which raised debate within the agency. Some planners worried either that migratory labor camps would turn into a weapon for local producers to keep labor in line, or they would become centers of radical labor organization that would make impractical demands. But most administrators had become so concerned that they would risk the RA being unable to protect laborers from the big producers. The situation was simply so bad, the argument went, that any change would be an improvement.[37]

The RA's suburban resettlement program on the surface resembled rural resettlement, but it served different residents, functions, and goals. Suburban resettlement was more modest than rural resettlement; initially, eight projects were planned, compared to the roughly 151 rural resettlement projects, and even this small number was cut down to four "greenbelt" communities. Because of a court decision and shortage of funds, the planned community near Cincinnati was scrapped, and only three suburban projects were actually constructed: Greenbelt, Maryland; Greenhills, Ohio; and Greendale, Wisconsin. These projects were designed as complete communities, not housing projects; they had stores, post offices, schools, parks, and community centers. The suburban projects were separate communities from the nearby cities, as the very name "greenbelt" suggests: a protective ring of farm- and woodland surrounded each project to prevent further development near the communities.[38]

The greenbelt communities never really fit well with the rest of the RA. They were suburban solutions to urban problems, while the large majority of the RA's work involved rural issues. They overlapped with other government programs at both the federal and local level, and prompted considerable criticism from a variety of directions. Baltimore leaders, for example, had expected the RA to begin a low-cost housing project in the city, and its refusal to do so on the basis that the city had reached its maximum growth led to defensive protests by civic leaders.[39]

The greenbelt program prompted criticism even from allies. John H. Bankhead, still interested in what happened to the projects in Birmingham after they transferred to the RA from the Division of Subsistence Homesteads, felt left out. He had heard reports, he wrote to Tugwell, "that the Suburban Resettlement Division is trying to hog the whole program and spending practically all of the money in four or five cities in the East" while the projects in Birmingham and Jasper went neglected. Bankhead felt personally mistreated. "I have also been embarrassed from the fact that much local publicity was given to the location of these projects and I was given some credit as author of the original program for securing the establishment of Homestead Projects," he wrote. "Nearly two years have passed since the original announcement of these projects, and now I hear that some people in your organization are resisting their implementation."[40]

By the summer of the 1936, the RA had become a more or less mature organization. While it did not always run efficiently, it had faced and overcome the greatest difficulties of reorganization. The administrative structure and regional hierarchy was in place and would survive the move to the USDA and change into the FSA. Administration leadership, at least, had a firm idea of their goals and responsibilities.[41]

However, the RA still had some important, related problems. First, Tugwell and other leaders in the RA knew they faced considerable political difficulties. From the beginning the RA suffered from an image problem: its emphasis on central planning and organized agriculture reminded critics of Soviet farming. Although the RA did cooperate with the Extension Service—for example, to develop veterinary services for local farmers—it was not popular with the administrators of the Extension Service or established agricultural interests. As RA administrators recognized, the agency was both broad and intensive; its wide variety of programs could each be considered objectionable to some large constituency.[42]

The criticism that the RA had too much power and too little oversight was perhaps best summed up in reporter (and longtime RA critic) Felix Bruner's four-part series "Utopia Unlimited," published in the *Washington Post* in February 1936.[43] Calling it "one of the most far-flung experiments in paternalistic government ever attempted in the United States," Bruner drew attention to the RA's independent, almost rogue status, free from congressional oversight. The RA, "really a government within the Government," operated by "executive order" without "Congressional sanction or approval" as it "directly affects, virtually rules over, the lives of hundreds of thousands of people who are told how much they can spend for food, for clothing, for rent . . . how they shall conduct the most minute details of their lives." Tugwell, with "every power possessed by both [the President and Congress] except the power to declare war," administered this "huge experiment" along with an enormous work force of 13,045 people.[44] Bruner pointed out that the agency had thus spent about $46 million, but almost $20 million of that went toward administration—meaning that some 40 percent of the money actually spent, and at least 20 percent of the total amount the RA was obligated to spend, would go toward "supplies and desks and traveling expenses and all the innumerable items that go to set up Utopia." If this was an experiment, between its high overhead and the number of wasted loans and grants given to bad-risk clients, it was an experiment that was failing.[45]

The RA was popular enough among farmers, at least with the ones who knew about its activities. W. R. Altstaetter, an RA official taking a trip through Tennessee and Kentucky in the fall of 1936, estimated that "85 to 95 per cent of persons in the small towns did not have the vaguest idea what Resettlement is trying to do." When Altstaetter asked where the local RA office was, on two separate occasions the response was to ask in return if he "meant that 'land thing.'" However, where projects had been undertaken and were in fairly good order, opinion was fairly positive.[46] Those farmers who both knew about and approved of the RA's work often read more into the agency than it did or could actually accomplish; sharecroppers in the Missouri Bootheel staged demonstrations in part motivated by a belief that they could force the federal government into action, but their demonstrations had only mixed results.[47] One report found that somewhere between two-thirds and three-fourths of all white American farmers favored the RA, but only about half of black farmers did. Part of the problem was, again, publicity: a large number (about half of respondents) did not have enough knowledge to answer one way or the other. "Negro farmers," the report explained, "on the whole know too little about it to express either their approval or disapproval."[48]

Black farmers who knew about it still had good reasons to be displeased with the RA. Joseph H. B. Evans, an RA administrator, spoke at length before the Joint Committee on National Recovery in 1935 on the general history of southern agriculture and on the federal government's progress in farm security activities over the last two years. But when it came to specific benefits for African Americans, he could point only to the general advantages for tenants, programs that would primarily benefit already all-black towns and regions, four projects that would "include Negro participation," and the few "qualified Negroes" employed by the RA. But even that only underlined the limited role of African Americans in the agency. Most held low-level or service jobs: thirty custodians, thirty-six messengers, five chauffeurs, and a garage foreman far outnumbered the twelve clerks, nine typists, eight "professional employees" like engineers and architects, and the one administrative employee.[49]

The RA, along with the rest of the federal agencies dedicated to relief, had a bad reputation for too much bureaucracy. Critics made much of the agency's reputation for delay and paper pushing; particularly harmful was the announcement in late 1935 of the size of the RA's staff. The *Washington Post* editorialized, "If there is any doubt that bureaucracy has reached a new pinnacle in this country that doubt is likely to be dissipated by a survey of the Resettlement Administration's pay roll." While commending Tugwell for hiring relief workers when needed, the *Post* mostly criticized him for creating a vast administrative organization, with a fleet of cars and huge payroll.[50] Even supporters saw problems. The relatively friendly *New York Times* published a front-page story on the RA's 12,089 person staff that only created 5,012 relief jobs (ignoring, Tugwell responded, the 354,000 rural rehabilitation cases and the fact that many of those loans would be repaid). This article, as Tugwell's response hints, indicated a common misunderstanding of the purposes of the RA. In all this bureaucracy, *Times* writer Frank Kluckhold claimed, the RA failed to make its contributions to the work relief program, with "administrative force numbers more than twice as many persons as the relief workers hired to date." He, like many others, thought of the RA as a jobs program, not as a rehabilitative agency. As such, it is no surprise that it seemed to be so wasteful, even to a sympathetic observer.[51]

The criticism that the RA made loans that were in danger of being lost because they were unsound, made to borrowers who did not need or deserve the loans or to borrowers who would somehow be made dependent on government largess, even edged over into a literal concern about the safety of the physical loan notes. Before the 1936 elections, the Republican National Convention attacked RA regional headquarters as "fire traps" where millions of dollars' worth of notes could be destroyed or stolen. Further, the RNC charged, the RA suppressed a report detailing this situation. C. B. Baldwin rejected both charges, stating that multiple copies of both proof of the loans and the allegedly suppressed report were made, but this was almost beside the point. Critics believed the whole RA approach to be fundamentally unsound and unsafe; reassurances about bookkeeping procedures did not resolve this problem.[52]

The RA's reputation for bureaucracy, delay, and unnecessary activity also prompted parody and humor. An imagined series of letters and memorandums between the fictional Mr. Torkad Sill and various officials of a State Emergency Relief Administration, County Emergency Relief Administration, and State Resettlement Administration in a 1936 edition of *Survey Graphic* revolved around a three-month struggle over Mr. Sill's efforts to return "ajitater cows" to the government. Various comedic bureaucratic spats over jurisdiction, case histories of the animals involved, and discussions of how to convince Mr. Sill to lead the cow to an "accidental" death followed. The story ended with the impossibly honest Mr. Sill unintentionally shooting the cow and giving the meat to local "reliefers" in Cooper County.[53]

In partial response to such negative publicity, Tugwell organized the Information Division. Most notably, the Historical Section, also called the Photographic Section, recorded the lives of tenants for both artistic and propagandist purposes. Tugwell hired Roy Striker as the first head of the Historical Section. From within and outside the government, Stryker put together a talented and diverse group of photographers, including Walker Evans, Marion Post-Walcott, Russell Lee, John Vachon, and Dorothea Lange. From 1935 until 1943, Stryker's group took thousands of photographs across the country. While many of these were official portraits of successful government projects and the like, many (including the most famous photographs) sprang from a commitment on the part of Stryker and his photographers to documenting the lives and times around them. The Historical Section's work became for many the image of the Great Depression.[54]

But attempts to portray the agency in a good light sometimes backfired. RA photographs that repositioned the same bleached cow skull when cataloging the western drought created another miniscandal involving the agency. While the RA was vindicated in that case, the conclusion that the photographer had only repositioned a skull, and did not carry one around for pictures, did not help the agency's image.[55]

Had the institutional and political support for the RA been stronger, these image problems would have been less consequential. However, the RA's major bureaucratic problem was that it had no real home. As an independent agency, the RA had no natural allies; even after it moved to the USDA, Congress never regarded it as a full-fledged equal to other departments and agencies in the federal government. As Tugwell wrote later, the RA "never attained the status of a respectable and permanent addition to the family of Federal organizations."[56] For example, the RA still depended on the president to make allocations from the Emergency Relief Appropriations Acts, funds which were uncertain and irregular. Throughout early 1936 Tugwell and his staff had to struggle with the White House to keep loans from being retransferred back to the Treasury and to obtain funds already promised but not yet allocated.[57] Much of this was Tugwell's fault. Operating, as he saw it, outside or above politics, and with little interest on his part in building alliances, Tugwell's RA lacked support. As journalist Marquis Childs wrote, "the fact was inescapable that the planners had been planning in a political vacuum."[58] There was even the possibility that the RA, an independent agency created by decree, could be replaced the same way. Rumors spread in the spring of 1936, for example, that a proposed secret bill would absorb or replace the entire RA with a "Farmers' Home Corporation" funded by $50 million in government capital.[59]

Never one to hide his disdain for those with whom he disagreed, Tugwell had become a lightning rod for criticism. Unlike Harry Hopkins, who was doing a similarly large and experimental job, Tugwell simply did not know how to be discreet.[60] He proved unable to tone down his rhetoric or to build a politically savvy group around him. His successors Baldwin and Alexander became effective (in very different ways) at some of the administrative and political aspects of the job, but this was apparently an afterthought on Tugwell's part in hiring them. More representative was Laurence Hewes, who wrote that his "ignorance of American government and politics was appalling" when he was hired as Tugwell's assistant. Hewes had, for example, voted for Herbert Hoover in 1932 only because he was (like Hewes) a Californian and some of his friends knew Hoover. Yet Hewes had been hired in large part to take over some of the political burden from Tugwell, to handle as much of the "uncongenial Congressional and political duties as [Hewes] should prove capable of managing."[61] Such decisions gave Tugwell (and his RA) a bad reputation even among supporters. As Paul W. Ward wrote in *The Nation* in November, "As a man of action he proved to be a fumbler, and for more than a year the RA was a scene of administrative turmoil that defied the descriptive powers of pen, brush, or camera."[62]

During the 1936 presidential campaign, Roosevelt and his political adviser, James A. Farley, did their best to keep Tugwell out of the papers. Despite the Democrats' easy victories in 1936, Tugwell increasingly understood that he had no place in the administration. At best, he was seen as a valuable administrator but a political liability, and the constant criticism made it impossible for the RA to function. Electoral returns apparently vindicated arguments for a greater government role in the economy, but Tugwell resigned soon after the 1936 election, succeeded by Will Alexander—tellingly, on the same day that FDR announced the creation of a committee to study the problem of farm tenancy.[63]

With his departure imminent, Tugwell recommended directly to Secretary of Agriculture Henry Wallace that the RA be moved into the USDA, where, presumably, he believed it would be safer. But this was not a universal opinion. Many in the USDA feared that the RA would bring controversy that the department did not need. Similarly, some RA employees thought that the USDA would shut down their most ambitious programs and force them to work on behalf of the planters, commercial farmers, and farm organizations that had the most influence in the USDA.[64]

Henry Wallace had become convinced of the need for the USDA to do more to address rural poverty. In late 1936, he was trying to decide what to do with the RA. At the suggestion of Tugwell and Alexander, Wallace spent time in the South observing poverty firsthand. He was shaken by what he saw: poverty he could not imagine existing in the United States. Wallace went out and spoke anonymously with impoverished rural southerners and was shocked; he compared their situation, unfavorably, to that of European peasants. More important, from the perspective of RA supporters, Wallace saw people whose lives had measurably changed for the better thanks to the RA (even if they remained in grueling poverty) and many more who wanted to become RA clients. Tugwell met with Wallace while he toured, part of a successful effort after his resignation to convince Wallace that the RA belonged in the USDA.[65]

On January 1, 1937, the RA transferred to the Department of Agriculture, carrying on much as it had before the move. Director Will Alexander maintained essentially the same authority in the USDA as he had when the agency was independent: issuing instructions for routine procedural matters, authorizing budgets, and holding ultimate authority (in theory, though not always exercised in practice) over approval of loans.[66] Despite the move, the agency still lacked a sense of permanency: it did not get appropriations in the 1938 agricultural appropriation bill because it remained unclear whether the RA would even exist in the next year or be replaced by some other agency, like the proposed Farm Security Administration in the Bankhead-Jones Farm Tenant Act.⁶⁷

In early 1937, during the move to the USDA, the agency worked out routine bureaucratic and procedural details. Baldwin, then assistant administrator of the RA, and Paul H. Appleby, then special assistant to the secretary of agriculture, arranged most of it.⁶⁸ Other than relatively minor personnel moves, the RA had matured into the agency that would carry out the heavy lifting of farm security. The RA's basic administrative shape was more or less set by the time it moved to the USDA and would undergo only minor evolutions before World War II.

The full scope of RA responsibilities, however, remained uncertain. Though the RA had been intended as a consolidation of rural antipoverty efforts, it was increasingly becoming clear that the current setup of the agency was only one stop in a continuing bureaucratic and ideological transformation. Rural rehabilitation remained an important part of the RA's program, but approaches to rural poverty like land reform and resettlement were increasingly pushed to the side. Replacing them was a new concern about the problems of tenancy and its connections to rural poverty. RA administrators took a great interest in this issue and in the slowly developing legislative response, which held out of the hope of boosting their antipoverty efforts and strengthening the RA's political position.

As the Resettlement Administration reorganized and expanded, Senator John H. Bankhead Jr. of Alabama and Congressman Marvin Jones of Texas sought to enact legislation specifically to address chronic rural poverty and particularly poverty associated with farm tenancy. The most important single figure behind the legislative push was Bankhead, the son of Alabama congressman and senator John H. Bankhead Sr. The younger Bankhead had been elected to the state legislature and ran one of the largest law practices in Alabama before going to the Senate. Winning office in 1930, he became an influential voice for cotton interests. While Bankhead was undoubtedly close to the agricultural elite, as his Bankhead Cotton Control Act indicated, he was genuinely, if inconsistently, concerned with the plight of the poor.⁶⁹ His conception of a tenant purchase program was more than simply a favor to American agriculture. As he later wrote, his proposal involved "moral and social elements, as well as financial ones," and "the heart of that program goes far beyond any banking loan program."⁷⁰

J. Marvin Jones, a Texas congressman and chair of the House Committee on Agriculture since 1931, was similarly convinced of the need for legislation to help farm tenants become owners, but he took a more conservative approach. Jones was interested primarily in the more competent farm tenants, and he opposed the idea of getting the federal government involved in actually purchasing and reselling land. Jones also felt more political pressure in the House, where there was less interest in (and some hostility toward) any kind of tenancy legislation.[71] He was slower to act, less active in the early creation of the act, and generally not quite as widely associated with tenancy legislation as Bankhead.

Bankhead met with a variety of USDA officials, agriculturalists, economists, and lawyers looking for an immediate solution to some of the problems of tenancy. M. L. Wilson, who had been in contact with a group of thinkers concerned about the problems of rural poverty and tenancy, asked Bankhead to sponsor a bill. Wilson and other USDA officials wanted the act apparently to come from elsewhere: such a potentially controversial measure would have a better shot if it were connected to an esteemed politician, and it would lessen the amount of criticism the USDA might expect.[72]

But getting the act passed proved difficult and time consuming. Bankhead tried first in February 1935 with a Senate bill known as "The Farm Tenant Homes Act." A couple of weeks later, Jones introduced the "Agricultural Bank Note Act" into the House. The titles of each bill reflected early differences in approach to the problems of farm tenancy. Bankhead proposed a billion-dollar program to purchase land and resell it to tenants, sharecroppers, and laborers. Jones's less ambitious bill would offer lower interest rates for mortgage loans to small farm operators, essentially a broad farm credit bill. The eventual compromise bill, which quickly became known as the Bankhead-Jones Farm Tenant Act (BJFTA), seemed to have a good chance of passing, with apparent support from the Roosevelt Administration and the backing of activist and social groups across the country.[73] The bill had support at the grassroots level; the Alabama Farm Bureau Federation endorsed it to make "it possible for worthy tenant farmers on good land eventually to become land owners."[74]

But many obstacles remained. An anti-lynching bill shook up the Senate, and Jones was reluctant to push a likely losing effort in the House, especially when he had so many other bills to worry about. When Congress adjourned in late August, the bill was defeated for the time being. In early 1936, a new version made practically no progress.[75] And the RA (now closely associated with the bill in some circles) had made enemies. For example, Oregon congressman Walter Pierce caused one of the delays in the bill's passage. He told RA officials that he had done so because, some months before, the RA had refused to move a county office to punish some of his political foes. Pierce had taken his revenge by helping defeat the bill.[76]

While politicians delayed, tenancy remained a problem, manifesting in explosive ways. Many assumed that the purpose of the bill was to address what New Hampshire congressman Charles Tobey called "the situation that has developed among the sharecroppers in certain sections of the South, almost bordering on civil war."[77] Evicted tenants (often forced out by the AAA) migrated slowly across the highways. Thousands sent letters to practically every agency, department, and official of the federal government. Mob violence, organized by planters and landlords, met efforts to create sharecropper unions. The Southern Tenant Farmers Union (STFU) organized in July 1934 to unite sharecroppers into a rural labor union. Within a few months, the STFU had over fourteen hundred members (mostly in Arkansas, Missouri, Oklahoma, and Texas), and locals formed faster than the union leadership could keep up with.[78] Allegations of socialist and even communist influence, while harming the organization's public image, did not reduce the STFU's appeal to many tenant farmers, who by the time of the Great Depression had become more concerned about the effectiveness of an organization than they were with its ideological orthodoxy.[79]

Mainstream voices began to demand action. The national concern about tenancy had begun in the early years of the Depression as tenant conditions worsened. The violence of the mid-1930s reinforced this sense that tenancy was a problem demanding a solution, involving the federal government if necessary.[80] By the mid- to late 1930s, the American public was probably better informed about tenancy than it ever had been before, and the belief that the federal government must do something to alleviate the problems of tenancy in American had become widespread. The USDA played a part in raising interest, in particular in its series of public hearings on farm tenancy. These had, as one scholar puts it, "a contagious effect" as state governors or their representatives, Extension Service state directors, field specialists, and federal officials met to iron out plans for what needed to be done.[81] Even those who disagreed in general with the government's intervention in agriculture acknowledged a problem. For those who opposed federal action, if something had to be done, ownership was the best option. The other possibilities were government protection of and aid to sharecroppers and tenants, which would make them more difficult to control, or to resettle them elsewhere, which would tighten the labor market. It was better from the planter perspective, then, to move a small number into ownership, where they might become market competitors but also political allies.[82]

In the South in particular, government action on tenancy had wide support, even if it was not always due to a concern for tenants themselves. The *Birmingham Post* asserted that the striking tenants, who "see no hope out of their present situation," had "turned in large numbers to the Marxian doctrine" and that the "best way to fight Communism is to remove the conditions which are fertile soil for its growth." Similarly, the *Birmingham News* claimed, the bill would be crucial to "preventing the spread of Communism among Negroes in the South." Henry Gee asserted in the *Southern Economic Journal* in 1936 that the "surest antidote to such revolutionary agitations" would be "making it possible for every worthy tenant farmer to step into the ultimate ownership of the farm."[83]

The evidence of the need for action culminated in the report of the President's Committee on Farm Tenancy, written in large part by James Maddox.[84] The committee called for action to open ownership up to tenants who had experience and ability but who lacked capital and needed credit and advice, and to prevent more farmers from losing ground by providing loans for small owners, tenants, sharecroppers, and laborers. It would also address more general agricultural problems by retiring unsuitable or submarginal farm land and improving the lease system. Being so closely related, these actions should be undertaken by a single agency. The committee, then, essentially recommended long-term legislative support for activities very similar to what the RA was already doing and even asserted that the RA "may well serve as a nucleus" for the organization, although it offered the title of "Farm Security Administration" to better describe its activities. Critics on the left argued that the recommendations failed to go far enough, while conservatives thought the program risky and a threat to the principles of hard work. But the report helped lay the groundwork for a federal program aimed at the vague and contradictory concept of farm security.[85]

In late 1936, with elections out of the way, an apparently improving political atmosphere in Washington, and growing public pressure, the Roosevelt Administration signaled its willingness to work for a tenancy bill. The public demand to do something had increased. Representatives of tenant farmers had been working on the administration for years. STFU agitation was at a peak at the same time that Roosevelt was changing his public message to strongly support some kind of government action addressing tenancy. Henry Wallace, the front man on the issue, had support from farm and labor groups, lending strength to the committee's recommendations. And Roosevelt became more directly involved with both Bankhead and Marvin Jones.[86]

The opening of the Seventy-Fifth Congress in 1937 saw a number of bills intended to promote farm ownership, help the RA, and eliminate tenancy. However, only Marvin Jones's House Resolution 8 and John Bankhead's Senate Resolution 106, two nearly identical bills both titled "The Farmers Home Act," had hope of being passed. The bills envisioned a program that was much more limited than the one proposed by Bankhead's 1935 bill or the recommendations of the Farm Tenancy Committee. The appropriation was half of what Bankhead wanted in 1935, and it had few safeguards to protect local farmers or migrant laborers.[87] And sponsors struggled to keep the program even that large. By May 1937 Bankhead was insisting, in response to talk that the bill would be reduced to $10 million, that he would rather see the proposal fail than have it become only "a gesture"; similarly, Marvin Jones claimed he would prefer no tenancy bill at all to a "skeletonized" measure.[88] Other supporters agreed; Texas Senator Tom Connally, who had his own billion-dollar tenancy bill, wanted a bill big enough to prove "that we are setting our hands to this plow in a substantial way and are not merely scraping around like we are doing with this Resettlement Administration."[89]

Opposition in Congress came from a variety of directions. Critics like North Carolina's Harold Cooley believed the whole point of the program was just to sneak justification for the RA into law. South Carolina Congressman Hampton Fulmer wondered whether the program would eventually lead to the government's buying threshing machines and warehouses for every client, while Scott Lucas of Illinois asked whether the proposed bill meant "the beginning of Government ownership and control of land in America." Tennessee's John Mitchell saw some aspects of the program as downright un-American, especially the provisions for supervision of clients and government approval of tract purchases. "I do not think that the Government agents should be dictating how he is to do everything," he proclaimed about a hypothetical tenant purchase borrower trying to buy a new farm. "That is how this country was developed and made great. Our granddaddies built it up with individual initiative."[90]

A number of opponents argued that the proposal detracted from what should be the real goal of the USDA: raising crop prices. Fulmer asked whether "the better way to approach this problem, if we are ever going to solve it, would be ... through working out a system of better prices" and went on to assert that the low-income and high-income farmer had identical needs, "a fair price for their commodities."[91] Similarly, another group of critics (and even some supporters, including to a degree Marvin Jones) believed that the program should be tweaked to simply provide easier credit to those at the top of the tenancy ladder.[92]

Perhaps the most significant voice in the debate was one that was not there: the American Farm Bureau Federation, normally busy in every facet of passing an agricultural bill, was almost entirely absent. The AFBF's Washington representative, Chester Gray, sent a letter opposing the withholding of titles to farms that clients could pay off fully, and AFBF president Ed O'Neal provided a statement. But that was it. People like O'Neal were primarily concerned with making sure that the government spent agricultural dollars first and foremost on keeping commodity prices up. A tenancy bill seemed like a distraction at best. But for the time being, the AFBF stayed relatively quiet.[93]

After hearings and much wrangling, Jones introduced the "Farm Security Act of 1937" in April; it died before the House Rules Committee. Yet another version, with much-reduced appropriations, finally passed the House (with relative ease, 308 to 26) at the end of June. Bankhead's bill passed the Senate effectively unopposed, and then the compromise Bankhead-Jones Farm Tenant Act (BJFTA), with some slight changes, easily passed both houses of Congress in July 1937.[94]

Title I of the BJFTA authorized making loans for the purchase and improvement of land. Only farm tenants, sharecroppers, agricultural laborers, and others who made their living from farming were eligible. The final bill kept the greatly reduced compromise appropriations: the tenant purchase program received funding for three years of $10 million, $25 million, and $50 million per year, with only 5 percent a year allowed by statute to go toward administration expenses. This practically required a program emphasizing credit as opposed to a more expansive, rehabilitation-oriented effort. Titles II and III authorized rehabilitation loans and the land utilization program, respectively.[95]

There was a sense of disappointment mingled with the excitement over the passage of the BJFTA. Supporters had overcome political opposition by watering the bill down and drastically reducing its funding (and thus its significance). It certainly did not meet Bankhead's original notion of a billion-dollar corporation carrying out a comprehensive farm security program. Nor did it follow all of the President's Committee on Farm Tenancy's recommendations. The program promised to be regional in character: because the act required funds to be distributed based on the farm population and tenancy rates of each state, about two-thirds of the money would go to the South and about a quarter to the Midwest. More than regional, it would be local, because Jones had insisted on county committees. Instead of a national reform agency, the new program would favor a thin slice of relatively well-off tenants in only a few areas.[96]

Even as the BJFTA was being passed, it was becoming clear that farm ownership, in and of itself, was not sufficient to protect poor farmers. The BJFTA failed to address even a fraction of the problems facing tenant farmers. North Dakota congressman William Lemke put this well during the final House debate. Expressing his surprise at the "fuss," he said, "If ever a mountain labored and produced a mouse, this is it." Estimating that the bill provided only about $5.31 for every tenant in America, Lemke called it "a lot of lip service," whose best feature might be encouraging tenants in the future to put more pressure for action on the government. And yet, "I am going to vote for this bill, because it is a toehold," he said, "in spite of the fact that it is camouflage and make-believe legislation."[97]

For all its shortcomings and all their disappointment, however, supporters recognized the BJFTA as a big step, or at least a first one. Particularly interested were the employees of the Resettlement Administration, which would be handling most of the new act's responsibilities. The RA, now in the USDA, was about to see a small change in name and a big change (supporters hoped) in institutional stability and responsibility. Even if the agency's goals had been reduced a bit and if its vision had become less radical, it finally appeared to be in a safe position to begin its long-term work toward improving the conditions of the rural poor.

CHAPTER FOUR

The Farm Security Administration before World War II

Starling Jackson, a black farmer living in Greensboro, Alabama, wrote to the FSA in 1938 to compare his situation in 1933 and 1934, when "we was lying flat down at that time with out food or clothes," to the present day. The government accepted his "petition and now we are standing with iron props on four sides." Since then, things had greatly improved. "But yet we have food for our selves food for our stock and clothes to wear. And will have money to run our farm with this year," he wrote. Jackson could not know what his circumstances might have been without help, but, he concluded, "I do no that I would not of ben in my present condition had it not been for the [Farm Security] Administration."[1]

Franklin Roosevelt had big ideas for southerners like Starling Jackson, as his tenancy program and talk about the region as the nation's foremost economic problem hinted, but the political tide ran against him. Labor unrest across the nation, an economic recession, and FDR's failed court-packing scheme led to public criticism and a political backlash, dramatically slowing the pace of the New Deal. The push for further reform legislation, much less a massive restructuring of the national economy and the relationship between the people and the federal government, fizzled soon after the 1936 election.[2]

The Bankhead-Jones Farm Tenant Act had been among the final legislative accomplishments of the New Deal, and the new Farm Security Administration was to be among its last administrative creations. Contrary to the apparent faltering of the New Deal in general, the FSA seemed to be only improving its work through the late 1930s. Starling Jackson was a rural rehabilitation borrower, and the relative success of that program (and the hoped-for success of the Tenant Purchase Division) showed in the administration's bureaucratic evolution. The Resettlement Administration officially went out of existence on September 1, 1937. Secretary of Agriculture Henry Wallace pointed out that in practice, the RA treated "resettlement activities as only a minor part of its functions." As Wallace had noted, "it would have been better if this work had been given a name more accurately describing it—Farm Security Administration, or the Tenant Security Administration, or something like that."³ Despite claims by some anti-New Deal critics (like the *Washington Post*'s assertion that the RA "became the first New Deal agency to be voluntary killed by the Administration"), the FSA was basically a continuation of the RA, emphasizing what had become the main work of that agency: rural rehabilitation through supervised credit.⁴

A few functions transferred from the FSA to other agencies, and it picked up some new duties. The FSA's most important new responsibility was the tenant purchase program, created under the authority of the Bankhead-Jones Farm Tenant Act of 1937. The Pope-Jones Water Facilities Act of 1937 also expanded the FSA's power to make loans to farmers, ranchers, and water associations in seventeen western states. Additionally, Wallace transferred the Land Utilization Division, inherited from the RA, to the Bureau of Agricultural Economics. This move, although largely made for administrative and political reasons, had ideological significance as well. Other solutions to rural poverty—resettlement, land use reform—had given way almost entirely to an emphasis on supervised credit.⁵

The FSA's hierarchical system had survived mostly intact from the RA. From the Washington office, FSA administrators divided the country into twelve regions. In each of these, FSA regional directors (assisted by two assistant regional directors, one for resettlement and one for rural rehabilitation) oversaw state, district, and county offices. It was in the district and county offices where the actual supervision and administration of clients and projects took place.⁶

The authority of the FSA, like its predecessors, was confusing, ambiguous, and sometimes contradictory. While the Bankhead-Jones Act appeared to be a legislative mandate for the FSA, in reality it only authorized a portion of the program. Titles I and IV directed the creation of a farm ownership program and provided funds. The act authorized and created standards and regulations for rehabilitation loans and completing resettlement projects, but the president had to allot funds. Through the various emergency appropriation acts, Congress provided some measure of control over farm security programs, but for most of the existence of the RA and FSA, specific requirements were broad and flexible.⁷

Political setbacks and other problems encouraged a decline in emphasis on land use reform and, to a lesser extent, resettlement. But just as much, reformers believed that supervised credit was not only more politically realistic but also that it worked better. The efforts of the FSA's three operating divisions (Rural Rehabilitation, Tenant Purchase, and Resettlement) reflect the evolution of ideas at work in federal farm security activities after four years of existence. The rural rehabilitation program was a relatively large-scale project aimed at providing a small amount of supervision and credit to as many families as possible. The tenant ownership program, on the other hand, had a smaller number of carefully selected clients who received more money and much more supervision. Credit and supervision were the main tools of both programs, but the tenant purchase program used more of both for fewer clients. Resettlement similarly focused on providing more intense assistance to a smaller number of clients, but in contrast to the individualistic approach of the tenant purchase and rural rehabilitation programs, it also took into account the social and community needs of rural families.[8]

The 1940 and 1941 agricultural years provide a good benchmark for the FSA's achievements. Most of its programs were several years old by that point, and the agency was fortunate to act more or less unmolested by its political foes. Moreover, FSA administrators believed they had finally figured out their programs; as one administrator said in 1942, "We could go out now and set up a thousand subsistence homesteads, and 999 would be successful."[9] While this was certainly an overstatement, FSA leaders had a growing confidence in their abilities and in the quality of their field agents. In part, this was because it was clear that so much of the work in setting up the field organization had been done by the workers themselves. Social workers or county agents or whoever had been dropped into wholly new situations, particularly on the resettlement projects, where they became city managers of a sort. When he became the Region IX administrator, Laurence Hewes, getting to know the staff while visiting resettlement projects, realized that "these youngsters had been forced to work out their own methods without help or guidance. The result was that they were almost irreplaceable in their jobs."[10] FSA leaders responded around 1940 with a general trend toward decentralization. The agency began to move away from implementing new rules for their field staff (and in fact began to reduce regulations) and toward making sure that their agents could act autonomously by being properly trained and aware of the goals of the agency. The newly empowered and trained agents got more responsibilities, as approval for loans, officially the task of regional supervisors, moved down to district and in some case county supervisors, who had been essentially making the decisions for most of the period.[11]

TABLE 1: FSA PROGRAMS BY EXPENDITURE, FISCAL YEAR 1941

	Rural rehabilitation	*Tenant Purchase*	*Resettlement projects*	*Labor camps*	*Defense housing*	*Administrative duties*
Expenditures, Fiscal Year 1941	$125 million	$48 million	$5.6 million	$4.4 million	$5 million	$38 million

Source: *Report of the Administrator of the Farm Security Administration, 1941,* 54.

The FSA's peak year was probably fiscal year 1941; that year the agency spent about $241 million (Table 1). For rural rehabilitation borrowers during the 1940 crop year, annual net income increased by about $75.3 million, a 35 percent improvement, and net worth increased by about $75.6 million, a 20 percent improvement. As of June 30, 1941, about 1.5 million families were receiving loans or grants (and the associated farm and home plans), and the FSA had collected just over 80 percent of mature rural rehabilitation loans.[12] A total of 178,737 families had been through the debt adjustment process, reducing loans by almost $120 million and leading to $5.6 million in back taxes repaid, by the end of 1941.[13] The other rehabilitation programs also showed results: in 1941 the medical and dental program covered about 117,000 families in over a third of the nation's counties.[14] And the FSA was doing all this with a limited budget and overworked staff: in early 1941 the FSA had an average caseload of 205 families per farm supervisor (administrators estimated that 125 was the maximum that could be adequately handled by a single agent).[15]

Rural rehabilitation made up the majority of the FSA's work, but not the entirety. The tenant purchase program had nearly twenty-one thousand borrowers and about $120 million cumulatively in loans approved by June 30, 1941.[16] The program kept up an impressive rate of repayment. During 1942, borrowers paid back more than 50 percent more than was due. Seventy-eight borrowers had already repaid their loans in full, and another ninety-one repaid their loans when the government took over their farms for various military purposes. More than two thousand of the borrowers made repayments above $1,000 for the year.[17] The FSA also operated 151 homestead developments as of June 1, 1941, and numerous migratory labor camps, water conservation programs, and similar projects; this totaled (including prior costs incurred by other agencies) an investment of not quite $160 million.[18]

These activities provide a picture of how the FSA, the last major holdout of New Deal rural reform and anti-rural poverty efforts in the USDA, operated in normal, peacetime circumstances. It was a national organization using supervised credit to alleviate rural poverty on a mass scale. Smaller programs like tenant purchase, resettlement, and the various rehabilitation programs (like the debt adjustment committees) functioned as supplements to or refinements of the main approach. The ideology and administration of New Deal antipoverty efforts had evolved, by 1937, into the FSA.[19]

Along with the move to the USDA came a new director, Will Alexander, Tugwell's first deputy and replacement in the RA. Alexander, a folksy southerner who preferred using personal charm and face-to-face meetings to get things done, initially appeared to be poorly suited to running a large bureaucracy. As he later put it, "I should never have been an administrator in Washington. I hate housekeeping and bureaucracy and having to run and see somebody every fifteen minutes and procedure."[20] He was, however, a surprisingly effective change from Tugwell, whose lack of political touch had cost the RA support. And Alexander, among the foremost white southern liberals in the country, was ideally positioned to recruit the kind of reform-minded liberals the FSA hoped to employ. As much as anything else, Alexander appeared to be (unlike Tugwell) someone who respected custom even as he tried to change it. Most important, Alexander was better at building political bridges. When the appropriations bill for the agency came up before Congress, for example, Alexander realized that Tugwell had done almost nothing to cultivate the support of Democratic leadership; he met with Senator Joseph Robinson of Arkansas, who was flattered to be asked for help.[21]

The FSA matured and prospered under Alexander until he left the agency in June 1940. Alexander grew tired of dealing with what he called "sadistic" congressmen who enjoyed abusing their power during the interminable budget presentations and other congressional activity. Although Alexander proved a better politician than his predecessor or his successor, he did not enjoy it much: he never learned how to deal with people like Pennsylvania congressman

C. B. BALDWIN, ADMINISTRATOR, FARM SECURITY ADMINISTRATION. LIBRARY OF CONGRESS.

Francis E. Walter, who had insisted that FSA employees contribute to his political machine. Moreover, with the program in steady hands (particularly, as he saw it, with Paul Maris running the tenant purchase program), Alexander believed that his help was less important.[22]

His replacement was C. B. "Beanie" Baldwin, who had operated an electric supply shop next door to Paul Appleby's print shop; when Wallace brought Appleby into the USDA, Appleby in turn brought his friend Baldwin in 1933. Baldwin, who had been tutored by and was close to both Appleby and Rexford Tugwell, was a crucial part of Will Alexander's early success in navigating Washington's turbulent politics. Baldwin oversaw the financial and personnel side of the FSA under Alexander and was one of the most capable of the reformists still in the USDA.[23] Alexander called him an "expert in organization," saying, "Beanie could see administrative processes—the mechanics of getting something done. He could see it very quickly."[24] Baldwin was also the target of frequent criticisms and widely considered a radical in the tradition of Rexford Tugwell; as one congressman wrote in 1941, "Mr. Tugwell may be gone but it seems to me that his spirit still remains."[25] The American Farm Bureau Federation (AFBF) in particular opposed Baldwin, and attacks on the FSA overlapped with attacks on Baldwin. And to a degree Baldwin *was* more militant about reform than Alexander had been; he was also less willing to accept criticism, which did not help the FSA's relationship with its critics or with the rest of the Department of Agriculture.[26]

The USDA in the 1930s was split between those who wanted to maintain the current balance of power within the agency (and thus its emphasis on commodity prices and the like) and, on the other hand, those who wanted the agency to try more ambitious social and political reforms. The conservatives generally came from the Farm Bureau, the Extension Service, and the land-grant colleges. The reform-minded liberals were generally younger and from a more varied background. Many were lawyers and at least at first followers of Tugwell. Conservatives saw the liberals as unrealistic and ignorant about agriculture; the liberals saw the conservatives as tied to traditional agricultural interests and unwilling to consider the plight of smaller farmers. In the middle, midwestern agrarians maintained their belief in the family farm.[27]

Muddying these ideological divisions were political ones. The USDA was a vast collection of squabbling, overlapping fiefdoms, bureaus, and agencies even before the New Deal. In 1933 it was probably the largest research and regulatory organization in the world, with a budget of more than $255 million and 26,544 employees (not counting temporary employees or those working in state or county extension services and experiment stations), and the New Deal added several new functions and more than tripled the number of direct employees.[28]

By 1937 four agencies were large enough to command serious power within the USDA: the Extension Service, the Agricultural Adjustment Administration (AAA), the Forest Service, and the Soil Conservation Service (SCS); together they accounted for almost a third of the regular annual budget. The FSA immediately joined that group, with a 1938 budget larger than any of them.[29] By January 31, 1940, the FSA had a total of 15,444 employees; in the USDA, only the Soil Conservation Service had a comparable number (15,226). The powerful AAA, in contrast, had a total of 3,150 employees.[30]

Coordinating the policies and direction of all these various semi-independent agencies was virtually impossible. Politics and ideology made the administrative problem worse. Each agency was mindful of any encroachment upon its perceived bureaucratic territory. The Extension Service considered itself the sole legitimate contact between the USDA and farmers. Even as the Extension Service borrowed and used the terminology of other farm programs, including "farm security" as a description of its own efforts, it resented and resisted the other agencies.[31] The FSA, SCS, and AAA were, from this perspective, unnecessary and intrusive newcomers. As FSA assistant administrator Robert Hudgens put it, extension agents and the land-grant colleges saw "the Farm Security people as a bunch of upstarts coming into a field they didn't know anything about."[32] The land-grant colleges and state Extension Services across the country, but particularly in the South, felt threatened by the fact that the FSA could bypass them entirely. The regional offices of the RA/FSA (and other New Deal agricultural agencies, like the SCS) in particular threatened the pattern of state-level cooperation and organization on which the Extension Services depended. The Extension Service turned for help to its closest allies in the American Farm Bureau Federation. The FSA threatened the position of the southern state farm bureaus, and the FSA's efforts to organize poor tenants and laborers threatened the labor supply of the big producers so influential in the AFBF.[33]

The American Farm Bureau Federation developed out of the county agent system in the first decades of the twentieth century. State extension directors acted at first as if they controlled the new and suddenly large movement, encouraging its growth and providing sample constitutions and training, but the local bureaus began to organize beyond the Extension Service's control. By the 1920s the AFBF developed into a kind of semi-governmental agency, connected to and overlapping with the Extension Service, but not identical to it. Its influence was enormous, especially in the House of Representatives. Congress and the AFBF operated more or less as partners in determining agricultural policy.[34]

The national Farm Bureau leadership opposed the growing influence of the FSA. The AFBF had long considered itself to be not so much the best representative of American farmers but the only meaningful and legitimate one. By the 1920s Farm Bureau leaders like president J. R. Howard had already begun describing rival organizations in the past tense. The relatively weak National Grange, stricken with internal divisions and with a large but regional membership base, was little challenge to the AFBF's position as the preeminent national farmers' organization and national leader of the farm bloc.[35] But the Farmers' Educational and Cooperative Union (usually shortened to just the Farmers' Union) could be. Founded in Texas in 1902, the National Farmers' Union by 1940 represented more than one hundred thousand farm families in forty states, with about three hundred thousand farmers taking part in its cooperative associations.[36] The Farmers' Union operated as a self-consciously old-style Populist organization, and its constant spats with other farm groups threatened the AFBF as the accepted voice for the farmers' perspective. The FSA's closeness to the Farmers' Union, a major competitor of the AFBF, therefore harmed the FSA's relations with the AFBF and its allies in the Extension Service.[37]

Beyond such political divisions, the AFBF's antagonism toward the FSA was partly the result of differing outlooks. The AFBF was organized explicitly with the mindset of encouraging farmers to think of themselves as a political lobby and their operations as businesses. Improving the economic and social lives of small and marginal farmers (much less keeping their barely, if at all, profitable operations going) was far outside the Farm Bureau's concern. The AFBF saw farming as a business, and if the inefficient or incapable were forced out, that was simply the nature of the market. The FSA, on the other hand, came from the perspective (going back to its FERA and Subsistence Homesteads roots and, before that, the Country Life Movement and back-to-the-land ideas) that farming had a value in and of itself, and that it was better both for farmers and the community to help ensure that small farmers stayed on the land.[38]

In addition to struggles against older agencies and outsiders, the New Deal agricultural programs also feuded with one another. The FSA was tied up in a number of these fights, most significantly with the farm subsidy and output production program that began in the Agricultural Adjustment Administration. With the support of much of America's agricultural elite, the USDA used farm subsidies to reduce production, limiting supply and thereby raising farm prices. But FSA clients already had low production and a considerable amount of that went toward home use. As a result, price changes at the national level had little impact on their incomes. And the emphasis on increasing overall agricultural income was meaningless to low-income farmers if that increase went only to the largest producers.[39] Small producers and tenants recognized the tension between the agencies, and many saw the FSA as a protection against the harms caused by reducing production. In the summer of 1939, for example, thirty-one Alabama farmers petitioned Roosevelt and Wallace to stop evictions of tenants

A COUNTY SUPERVISOR EXAMINES THE MOUTH AND TEETH OF A MARE BELONGING TO MELODY TILLERY, A RURAL REHABILITATION CLIENT. THE MARE'S MULE COLT IS ALSO SHOWN. LIBRARY OF CONGRESS.

that took place over the winter and to use the FSA to provide "decent shelter" and "relief during the winter months."[40]

On the other hand, some farmers saw the programs as all part of one indistinguishable, hostile entity. Pike County (Alabama) Rural Rehabilitation supervisor John E. Hydrick ran into trouble with a disgruntled client named Z. B. McLendon, who was upset that his brother had received a larger benefits check (the result of an illness) than he had and because his father had some trouble with AAA. McLendon then badgered or misled ten other local tenants to sign a petition asking the FSA to force Hydrick "to show our needs some consideration," a vague request that ended up with the district supervisor dismissing the complaint. McLendon turned his displeasure with two different federal agencies toward the FSA, which to him (and apparently several others) were all closely related. McLendon's confusion does not appear to have been unusual, as many rural Americans had trouble distinguishing the various agricultural initiatives.[41]

The FSA found itself in conflict with other New Deal agricultural programs. For instance, the Farm Credit Administration disapproved of the FSA's easy terms of credit to apparently undeserving applicants and the practical threat of another agricultural agency providing any kind of loan program at all.[42] But more generally, each USDA agency faced the yearly threat of decreased funding, especially those more closely identified with the New Deal. (This was also true for most New Deal agencies other than agriculture; the National Youth Administration and the Civilian Conservation Corps, two of Roosevelt's favorites, both faced enormous budget cuts.) Other than the army and navy, in fact, everyone had to deal with the possibility of annual budget cuts in the years before World War II. But administrators in the USDA believed that they had to fight over a specific, limited pie for the agricultural budget; funding became a zero-sum game when increasing the budget for one agency necessarily meant reducing it for another. Every dollar, then, was crucial.[43]

FSA leaders believed that they had certain advantages in the fight. For one, they could count on high-level support. Secretary Henry Wallace, Undersecretary M. L. Wilson, and Paul Appleby (the special assistant to the secretary) had all proven their commitment to the anti-rural poverty program. Wallace had been one of the key figures during the fight for the Bankhead-Jones Farm Tenant Act. Wilson had spent time as head of Subsistence Homesteads and had obvious concern about rural poverty. Appleby, who in Will Alexander's opinion was just as influential in the operation of the USDA as Wallace, was a close friend of C. B. Baldwin and a strong supporter of the FSA.[44]

This position was precarious, however, since it depended on the support of a few key individuals. When the USDA's leadership changed, so did the FSA's strength. The most significant such change came in 1940, when Wallace became FDR's vice president. One bureaucrat fondly remembered the USDA under Wallace and Appleby as having an excellent "quality of membership, outlook, and identification with the managerial purposes of a top departmental career service."[45] And Wallace, in the words of his biographers, turned the USDA into "a powerful engine for progressive action."[46] However much liberals may have doubted his commitment, especially after the AAA purge, Wallace undoubtedly sympathized with their goals, increasingly so as his tenure as secretary wore on.

Wallace's replacement, Claude Wickard, was both less sympathetic to the FSA and less capable of protecting it. Most important, with a coming shift to wartime production, Wickard was essentially passive, compared to the active and influential Wallace. Under the steady hand of Wallace and Appleby, USDA factionalism had been kept under control. Under Wickard, the USDA became a much more hostile and competitive environment.[47] This forced the FSA to search for new allies, but within the USDA, only the Bureau of Agricultural Economics and the Soil Conservation Service could be considered natural supporters of the FSA. Neither of these carried much political weight.

As for finding allies elsewhere, the FSA had few good options. Roosevelt was less involved in questions of rural poverty, particularly after Rexford Tugwell left and world war threatened. The Tennessee Valley Authority was politically influential, engaged in the kind of community building and rural improvement the FSA hoped to do, and had a similar base of support. Yet FSA leaders made little effort to cultivate a relationship, and aside from some resettlement efforts as a result of reservoir creation, there was little institutional connection between the two agencies.[48] Much of Congress, particularly the House of Representatives, was hostile or apathetic; many simply did not know much about the FSA. As one congressman wrote to Appleby in early 1941, "There has been much confusion in my mind in regard to the Farm Security Administration."[49] These "undecided" politicians, as it were, could easily turn against the FSA. And the support of former defenders weakened; John H. Bankhead, for example, who had spearheaded the tenant purchase program, became increasingly concerned about cotton prices and parity in the early 1940s and, in particular, during World War II.[50]

To address its weakening position within the federal government, the FSA could look to outside allies. One approach, which had been used within the USDA by other agencies, was to live up to the fears of critics and develop its clientele as a constituency. This could be a powerful tool and make agencies essentially independent of the USDA's upper administration. As one anonymous agency official said after Wickard became secretary, "We don't need the Department. We are perfectly able and willing to take care of ourselves."[51] And the FSA, like the other enormous New Deal farm agencies (the SCS and the AAA), appeared to be well suited to create such a base. As Wickard's biographer points out, "Nearly every county in the United States housed one or more FSA, SCS, or AAA offices. It took little imagination to see that whoever controlled these county organizations might control not only the farm programs, but also a preponderance of the rural vote."[52]

The FSA's political base appeared potentially large. By June 30, 1941, the FSA had aided perhaps 1.6 million farm families through one or more of its programs, and there was room for growth as county supervisors collectively estimated that over six hundred thousand more families needed rehabilitation.[53] But these political relationships proved weak in practice. FSA clients were often only temporary and felt little connection to the agency; even if they did, the rural poor were socially isolated and politically impotent. Plus, other USDA agencies and many FSA employees, particularly at the county level, strongly disapproved of any effort to create a political constituency along the lines of the AFBF.[54]

This is not to say that the FSA made no effort to influence public opinion. Besides the activities of the Information Division (which, though smaller than it had been in the RA under Tugwell, remained active), administrators wanted to make sure that the public appreciated their efforts. In Region V, for example, both Regional Information Advisor William H. Dent and Regional Director

Robert W. Hudgens sent memorandums to employees urging them to make their work more visible. Dent recommended sending a short press release to local papers every time an FSA tenant purchase loan was approved or used to purchase a farm. Hudgens was even more explicit, noting that Congress and influential local officials needed to know about the FSA's accomplishments since they made the laws and provided the funding.[55] However, the sort of organized politicking and lobbying that organizations like the AFBF engaged in was not a part of the FSA's approach. Ultimately, FSA clients were not a solid foundation for a political base, and FSA administrators appear to have made little effort to use them as such.

The FSA also failed to make sufficient inroads with the staffs of individual land-grant colleges and state-level farm bureaus, despite the fact that, in the words of one USDA administrator in 1937, they were "friendly to the main purposes of the Resettlement Administration."[56] The RA/FSA had something of a love-hate relationship with the state Extension Services and Farm Bureaus. Local relations between Farm Bureau and FSA employees often went extremely well, and AFBF members supported aspects of the FSA's program. While the land-grant colleges as a whole were hostile, the FSA and its predecessors did share some personnel and information with college staffs; Helen Kennedy, after she went back to work with the Alabama Extension Service from her work as a farm security home supervisor during a furlough in February 1936, reported that "resettlement work has been both worth while and pleasant." Robert Hudgens found "a very close collaboration" in the field between extension agents and FSA employees.[57] The RA/FSA also cooperated with the Extension Services in various undertakings. Officials met with extension agents frequently to exchange information and developments in farm and home management.[58] This grassroots cooperation could not overcome the larger political enmity, however. Part of the agricultural establishment, land-grant colleges were, along with the state and USDA Extension Services, hostile to farm security efforts, at least among the leadership.

In terms of state and local politics, sometimes the best the RA and FSA could do was to simply maintain neutrality. In Region VI, for example, made up of Arkansas, Louisiana, and Mississippi, Regional Directors T. Roy Reid and A. D. Stewart had to deal with influential senators like Pat Harrison, Theodore Bilbo, and Joseph Taylor Robinson, and with difficult state officials like Louisiana governor Richard W. Leche. This generally only worked because RA/FSA officials managed to stay out of local politics. In most cases, RA/FSA officials could manage only nonaggression pacts, not alliances, with influential politicians. In Louisiana, where Huey Long held sway, both the Long and the anti-Long camps sought the RA as a patronage weapon. When the RA worked on a strictly nonpartisan basis, each was unhappy. Long tried to block Robert Hudgens's appointment as state director, and the anti-Long crowd eventually refused to have anything to do with the RA. As it became clear that the RA would not become a weapon of their enemies, both factions were willing to let the agency work mostly unhindered. This was a ceasefire at best, not a political alliance from which the agency could draw strength.[59]

Prior to Pearl Harbor, these problems of political alliance and bureaucratic turf wars were something to be concerned about but nothing to cause a panic. Generally speaking, the FSA found itself able to handle most criticisms and attacks. These took two general forms: first, an ideological criticism that the FSA was a radical, subversive organization threatening to undermine American or regional norms; and, second, more practical complaints that the FSA was wasteful, ineffective, or counterproductive. In the short term, when the FSA still had a good deal of institutional support, these resulted in protest and scattered political criticism, but there was no coordinated attack on the FSA program itself.[60]

The FSA, like the RA before it, had a reputation as a progressive, even subversive, program. Critics particularly linked FSA plans for the creation of government-backed industry to a socialist or communist tendency in agency leadership. In 1938 Michigan Republican Paul Shafer called FSA plans for five homestead hosiery mills in Lansdale, Pennsylvania, (to be owned jointly by the homestead associations and a Pennsylvania mill company) and a farm machinery manufacturing plant in Arthurdale a "socialistic scheme for government-operated industry," the purpose of which was to "Sovietize American industry in the same manner in that Secretary Wallace has been Sovietizing American agriculture."[61] Similarly, Congressman Guy L. Moser of Pennsylvania said it was worth eliminating the FSA's entire appropriation, including loans to tenant farmers and the land utilization program, if that were "the only way to strike at the bureaucracy"; he further claimed that "the end justifies the means, because there are no means that they won't resort to." Such efforts remained a sore spot even after the comptroller general ruled against the construction of such mills.[62]

Such transparent attacks provoked countercriticism and even mockery from FSA supporters. The *Monroe (Louisiana) Morning Herald* noted that while Baldwin, a boyish, blue-eyed forty-year-old who combed his hair and wore blue suits, "looks like a bond salesman . . . you can't trust these Communists and there's no telling what tricks they're up to." The editorial went on to claim sarcastically that Baldwin's communism was evident in his past as a Virginia Democrat who voted three times for Harry Byrd. More seriously, the paper pointed to the obvious fact that the FSA was under attack from those opposed to its program and "Baldwin just happens to be the man who is directing the program and so he has to catch all the brickbats thrown in that general direction."[63]

The FSA also came under intense criticism for its assaults on the southern status quo. There was a sense that the FSA or its allies treated the South unfairly, which came out most clearly during congressional hearings in the early 1940s over the continued funding and role of the FSA. For example, after listening to Robert Hudgens explain the problems faced by impoverished farmers in the South—specifically that some tenants were so poor that their children could not go to school for lack of sufficient clothing—Georgia's Stephen Pace had had enough. "I am pretty well fed up with hearing Georgia and Alabama downed," he said. "The people in those states are as good and as kind and as honest as people anywhere else." If anyone in Laurens County knew about it, he said, they surely would have given the tenants "three times as many clothes as those children needed to supply them and send them to school."[64] Georgia representative Malcolm C. Tarver went further when he claimed in 1943 that the FSA was using the war effort as a pretense to effect an idea that "originated long before the war" to move tens of thousands of farm families out of the South to other areas of the country—"reshuffling of the agricultural population of this country and placing them where certain theorists want them"—stating that C. B. Baldwin apparently agreed with such an idea.[65]

The most potentially explosive criticisms of the FSA, despite the agency's care in avoiding any violation of southern protocols, involved race. Many of the FSA's leaders were southerners, and they knew they had to be careful. At no point did the federal farm security program make a serious effort to directly challenge segregation. In the South, assistant farm or home supervisors were sometimes African American, but while the FSA had a reputation for being willing to push boundaries, the racial makeup of its personnel did not differ much from the rest of the USDA. Most agents were white, and the few black agents almost exclusively served black clients.[66] In many cases, African American farmers were ignored or even pushed off their land in favor of white clients. This was particularly a problem in the Mississippi Delta, where all-white projects could not all be built on virgin land. Tenants were evicted from areas that would become RA/FSA projects either because they were the wrong race for a segregated community or because they did not qualify as an FSA client (because of poverty, generally, which tended to disqualify relatively more black tenants). In some cases, such as around Transylvania Plantation in Louisiana, the agency broke up long-established black communities and replaced them with white FSA clients.[67]

Some southerners who supported the FSA did so on the basis of their own racist, paternalistic beliefs. In creating the tenant purchase program, Bankhead wrote to a constituent that "our primary interest is in the white tenants" and that local control of the application process would ensure that whites could "handle that end of it for their own protection."[68] One South Carolina newspaper editor wrote to Robert Hudgens, asking for a resettlement project for local black farmers forced out of their homes by the Santee-Cooper development. "Probably you are familiar with the Negroes of the Carolina coast," James C. Derieux wrote. "They are a simple, friendly, illiterate people. If paid a few hundred dollars per family, and turned loose, the dealers in second hand automobiles will get the money, and the Negroes will have the cars for a short time. Then what?" Derieux recommended more intensive government aid through a resettlement project in a nearby neighborhood.[69]

Despite the care administrators took to avoid violating southern racial mores, white southerners still criticized the FSA. As historian Donald Holley points out, the FSA's work was "an implied rejection of Jim Crow." There was no length to which the agency could have gone to satisfy its foes in the southern political establishment, because its very existence was a threat to the racial status quo.[70] Any program that even brushed against the color line raised immediate concerns. Senator James Byrnes, for example, complained to Will Alexander about a proposal to create state advisory committees that would include at least one black member. Byrnes asserted that such a committee was unnecessary (even though he admitted, "I do not know exactly what duty will be performed by the Committee"), but if it was essential, he insisted that Alexander reconsider including African Americans. After an ominous request to know the name of any "leader of the negro race in this State" that made "any demand" on the subject, Byrnes turned to sarcasm. He asked whether Catholics, Jews, Irish, Scots, and Baptists were fully represented. He asked "whether this is limited to a negro of full blood, of how much negro blood a man must have in order to qualify." Finally, Byrnes warned that any action that would "raise the race question" would "prove a serious handicap to the success of your work."[71]

For all their care to avoid even a hint of racial subversion, FSA administrators (many of whom had experience with civil rights organizations) still took steps toward racial equality, though usually small ones. Alexander, for example, wanted black women working on FSA projects to make the same amount of money as white women. Baldwin recommended that state committees include black committee members to select the black clients.[72] And on some policies, such as providing loans that could be used for the payment of the poll tax, the FSA refused to compromise.[73] That the FSA could be described by one southerner as the "only government agency that has ever seriously tried to avoid discrimination against Negroes in promoting land ownership" points to the limitations of what the New Deal accomplished for African Americans.[74]

The criticism of the RA/FSA as a subversive organization undermining American norms was important, but more so as the foundation for a more common argument that skirted any ideological issues: wasteful government spending. Virginia senator Harry Byrd was one of the earliest critics taking this line. In a letter to various newspapers in 1937, he claimed that "when the true facts are known of the sinful and absurd waste in the Resettlement Administration homesteads projects . . . the conditions existing will approach a national scandal."[75] In a 1937 radio address, Byrd criticized the use of federal grants and the related "red tape and duplicated effort with which the funds are administered." He condemned alleged waste, claiming the RA spent $1,400 a day for telephone and telegraph messages and $11,300 a day in travel expenses. Byrd's strongest criticism took aim at the homesteads program, "conceived in the brain of Dr. Tugwell" and which took "the first prize in sinful waste." Byrd attacked several projects by name, including the Newport News homesteads ("for the occupancy of colored people"). The RA and its supporters claimed the economic emergency as their rationale, but to Byrd this only hid their belief "that they have divine authority to disregard every sensible mandate of economy and efficiency" to instead spend other peoples' money as they saw fit. The only reasonable solution, he concluded, was to cut such spending and balance the federal budget while the taxpayers could still afford to do so.[76]

A related criticism was that the FSA encouraged "shiftlessness" in farmers—in effect, imparting the wasteful mindset of the program to its clients. Henry C. Taylor, for example, warned that a program helping tenants become owners would undercut "thrift" and make them dependent on relief.[77] This criticism was common enough that the FSA trained its employees to turn it around and argue that such a view was un-American. In a sample speech sent to FSA personnel for use at meetings with committees and county councils, public gatherings, and other events, the speaker was to explain that the FSA "believes in the American people." Those who attacked it for giving loans to "shiftless" people failed to recognize the capacity of American farmers to improve their lot in life. To assume otherwise would mean "there would be no reason to hold out opportunity or hope to our people—and no reason, for example to oppose Hitler and Stalin and all the other enemies of democracy."[78]

Adding to the arguments about waste, critics found the agency too centralized and bureaucratized. With a few important exceptions (such as local committees responsible for debt adjustment or selecting tenant purchase clients), the FSA was administered almost entirely by federal employees, not local farmers' committees or local associations. The picture that its opponents developed, in the words of Minnesota congressman August H. Andresen, was of a top-down, overbureaucratized agency "where they have a lot of personnel drawing big salaries and riding around the country telling people how to live and what to do."[79]

These sorts of criticisms did not pose existential threats to the FSA prior to World War II, when political and ideological changes in the early 1940s put the FSA in a much more difficult situation. The New Deal and American liberalism as a whole underwent significant changes in the first half of the 1940s. The early New Deal had been reform oriented, aimed at changing and improving the very structure of capitalism. Modern capitalism, liberals believed, had essentially failed in some significant areas; government, especially the federal government, would be responsible for building the infrastructure (both real and metaphorical) necessary for society to continue and beyond the capacity or willingness of modern industrial capitalism to build.

In the late 1930s, Roosevelt and his New Dealers had neither the political power nor, increasingly, the political will to institute and maintain such reforms. At the same time, the period of recovery from the 1937 recession and, in particular, the changes wrought by World War II led to a new kind of liberalism. The government would still play a significant part in the economy, but it would do so in order to expand and stabilize capitalism, not reform it. Liberalism went from being reform oriented to being consumer oriented, comfortable with the shape of modern capitalism, committed to expanding mass consumption, and more concerned with the protection of civil liberties and individual rights than with restructuring capitalism or directly acting in the economy.[80]

In this sense, the FSA was, even at the moment of its creation in 1937, almost a relic of an earlier time. As one rural sociologist wrote, "public rural rehabilitation programs achieved a relative permanence as rural poverty came to be viewed as a widespread, deep-rooted, and long-standing problem to which the Federal Government should address itself."[81] But this idea of rural poverty was not to last, and the political climate and ideas surrounding the creation of the Division of Subsistence Homesteads and the Rural Rehabilitation Division in 1933 were already beginning to wane when the FSA went to work in 1937.

PART TWO

Fighting Rural Poverty
Farm Security in Practice

It is necessary to understand the on-the-ground functioning of the Farm Security Administration in order to understand the effectiveness of federal rural reform efforts. Administrators and planners made policy, and the president, Congress, and various members of the United States Department of Agriculture influenced what the FSA could do. However, the same was true for the FSA's employees and clients. Rural rehabilitation in particular proved to be a local affair, as, increasingly, county and district supervisors and local advisory committees made important decisions about supervision and making loans. For all of the programs, the success or failure of any given project depended as much on the men and women who lived and worked there as it did on the particular policies put forward by faraway administrators.

Moreover, while the FSA played an influential role in an ideological and political struggle taking place at the highest levels of the federal government in the 1930s and early 1940s, that was not what was important about the agency for the people directly involved on a day-to-day basis. For most FSA employees, the agency's purpose was embodied in their local work with clients. Similarly, for most clients who received credit, assistance, or homes from the FSA, the agency was not a massive bureaucracy maneuvering against other massive bureaucracies and political organizations. It was the local agent who showed a family how to use a pressure cooker, the county committees who helped decide whether a tenant purchase client was well suited for a particular farm, or the managers of the resettlement community in which they lived.

The FSA's effectiveness as an antipoverty agency depended upon the success of its operating divisions (Rural Rehabilitation, Tenant Purchase, and Resettlement). For this reason, it is valuable to look more closely at the work of those programs. By far the most important was rural rehabilitation. It consumed most of the FSA's budget, and its system of supervised credit became the most important single tool for fighting rural poverty. It also heavily influenced how the resettlement and (particularly) the tenant purchase programs were carried out. By the late 1930s, FSA administrators had settled on farm and home management plans, with their accompanying loans, as the best way to promote rural rehabilitation and thus to solve rural poverty. The FSA should be understood as a local agency, in terms of its county agents and local offices, and as a (supervised) credit agency, in terms of how it extended loans and grants to clients and how supervision operated.

A similar perspective is required for the resettlement program. It had great significance as an ideological ideal or as political football to be used against the FSA by its enemies. But it also had considerable influence on the lives of those who were resettled, both the ones who became successful in their new communities and the ones who failed or moved away. The homesteads around Birmingham, Alabama, eventually became settled and generally successful communities, but only after years of setbacks, false starts, and reduced expectations. Given the limited resources of the program's supporters and the hostility of its opponents, the homesteads could never match the utopian ideals of their early boosters, or even the reduced expectations of later FSA administrators. The Birmingham homesteads, like the resettlement projects around the country, were part of a political failure that still enriched the lives of its residents and communities.

With the outbreak of World War II, Farm Security Administration leadership faced considerable changes in the type of work they did and the political atmosphere in which they did it. The FSA successfully navigated the first change, becoming an effective part of the American war effort, but it failed to survive a concerted political attack in an increasingly hostile environment. During World War II, anti-New Dealers and established agricultural interests, notably the American Farm Bureau Federation and its congressional allies, turned full force against the FSA. The attack was more focused than previous such efforts had been, and the FSA was almost without allies. The Roosevelt Administration typified the change—FDR simply had no time for agencies he had previously championed, such as the FSA, as he concentrated on the war effort. In 1942 and 1943, the FSA's opponents successfully reduced its appropriations to such an extent that the agency was powerless to enact any kind of reform. The FSA dragged on until 1946, when it was replaced by the Farmers Home Administration, effectively a rural credit agency intended mostly for veterans.

CHAPTER FIVE

Regaining a Lost Security
The Rural Rehabilitation Program

"Please send me something dear sir for I aint got nothing now to eat," Rannie Anderson wrote to his local rural rehabilitation agent, a Mr. Coker, in August 1942. "I am looking to you for my help until I can help myself." Anderson, a rural rehabilitation client since at least spring of 1941, had run into various troubles, including the death of a cow in January 1942. Anderson apparently worried that his previous failures might cost him his place in the program. Whatever Coker's response, it seemed to do the trick. Two months later, Anderson wrote again, "to thank you for your services you give me yesterday," closing his letter by saying, "I thank you again and again am all right now." By 1943 Anderson had apparently paid his loan off entirely.[1] Anderson's case sums up many aspects of the rural rehabilitation program, especially the narrow margins of success or failure for the marginal clients, the potential need for continued supervision, and the importance of an individual connection between a client and an agent.

The Rural Rehabilitation Division was by far the most important of the Farm Security Administration's main programs, accounting for about 60 percent of the $1.96 billion that the Resettlement Administration and FSA collectively spent on low-income rural and migratory families.[2] It offered assistance and funds to a larger number of low-income rural families, and in a wider variety of ways, than did the resettlement or tenant purchase programs. Rural rehabilitation also had the broadest geographic reach of any FSA program. The program originated in the Federal Emergency Relief Administration as a set of related projects, the most important of which were supervised loans and emergency grants. Rural rehabilitation also included a voluntary debt adjustment program, tenure-improvement activities, the promotion of cooperative programs, and a medical care/sanitation-improvement program.

Being so broad, rural rehabilitation had a wide and sometimes contradictory set of goals. One goal, and a large part of the reason that the rural rehabilitation program took the shape it did, was to save money while improving the rural economy. Proponents promoted the loan program as more cost effective than direct relief and work programs. And if the commercial agricultural economy was to recover, it had to incorporate small farmers. If the economy did not return to prosperity, or if small farmers did not take part in that prosperity, then they would have to go on relief and continue to cost the government money. Rehabilitation also prevented farmers from becoming migrant agricultural laborers, a national problem largely being handled at public expense.[3]

A related objective was to keep small farmers on the land. This was in part born of pessimism: if there was nowhere for impoverished farmers to get work in urban areas, then they were best kept on the land (a similar sentiment inspired the Division of Subsistence Homesteads and later the tenant purchase program). On the other hand, a number of commentators and planners, including many people involved in the rural rehabilitation program, valued rural life inherently, arguing that it was better than city living and perhaps necessary for national prosperity. M. L. Wilson, who put such views into action while administrator of the Division of Subsistence Homesteads in the Department of the Interior, wrote that he would "keep a much larger number of people on our farms than many agricultural economists would advise. There are values in rural life, even in a rural life without rich material returns, which in my estimation are superior to maximum commercial productiveness without these homely values."[4]

By the 1930s most agrarian language had lost its moralistic streak in favor of emphasizing the economic importance of farm purchasing power, but belief in the inherent virtue and value of farm life remained popular. Groups as diverse as the Southern Tenant Farmers Union and the Nashville Agrarians agreed that the root of American's problems lay in the capitalist, industrialized economy, and that the solution depended on some sort of land reform.[5] In the South in particular, this notion of the inherent value of rural life, and of the necessity of restoring farmers to their rightful economic place if the larger economy were to recover, held sway. Economic conditions, which were worse in the South, were part of the reason. Even in 1937, after four years of federal aid, the average southern income was $314, compared to national average of $604.[6] Southern farmers faced bleak futures: a 1938 report found that almost 60 percent of southern white nonowners and 63 percent of southern black nonowners rated their chances of becoming farm owners in the next five years as "poor."[7]

FERA, RA, and FSA administrators believed that rural rehabilitation would be the key to fixing these problems, or at least keeping them from getting worse, but what exactly rural rehabilitation meant remained a little fuzzy. One set of

FSA instructions defined it simply as "regaining a lost security."[8] Proponents of rural rehabilitation generally had in mind efforts that built on one another, a pyramid to ensure that a family would not again fall into poverty. The foundation was the relief of immediate suffering and misery, the most basic necessity. Permanent self-support and a healthy living standard would allow a family to avoid falling into poverty again. Stable land resources, improved tenure conditions, and a dependable monthly income made it possible for a family to have that healthy self-sufficiency to avoid again being overwhelmed by poverty. Finally, participation in the community and democracy both created a reason for the family to work hard and provided additional resources (such as education and community aid) to help them. During World War II, many definitions of rural rehabilitation included full participation in the war effort.[9] This was both a practical goal, to ensure the continued existence of the rural rehabilitation program, and an ideological one, as the war effort was conceived as a defense of democracy and the American way of life.

That ultimately might be the best description of how proponents (at least, those in the FSA and its predecessors) understood rural rehabilitation: ensuring that the rural poor could participate in the American way of life. But determining the best ways to get clients to that point and to keep them there, much less figuring out what "the American way of life" meant, was a moving target. The Rural Rehabilitation Division thus sought to improve rural life by using a variety of approaches in an effort to restore the sense of rural security that its leaders believed had been lost.

The Rural Rehabilitation Division came to the RA mostly from FERA, but it expanded rapidly in the new agency. From the beginning it was the largest division in the RA, with not quite thirty-one hundred employees (about three thousand of whom were field agents).[10] It had five separate but closely related operations: a loan program built on farm and home management, an emergency grant program, feed and seed loans, a farm debt adjustment program, and a cooperative loan program. It was at first part of a Rural Resettlement Division, handling both rural rehabilitation and resettlement, but in December 1935 the two functions were separated into separate divisions.[11]

The RA provided loans and grants for rural rehabilitation and relief to low-income farmers, farm tenants, and sharecroppers who could not secure credit elsewhere, either through private credit or other government agencies. With loans and grants came a supervised home and farm plan; thus these were known as "farm plan" clients. Grants were used for emergency cases and to help a family get into economic shape to become borrowers. The loan and grant program was by far the largest aspect of the RA program: within the first year of its existence, about 780,000 farm families received aid. Repayment was seasonal, to be made after harvest or sale.[12]

The RA loan program underwent substantial changes from the way FERA had operated it. Instead of repayment in kind or by public work, the RA only accepted cash. Most important, the RA instituted supervised farm and home plans for all clients. This forced county supervisors to make a careful study of eligible applicants. However, being more careful meant that many low-income farmers with insufficient or poor land were no longer eligible for loans. Loan sizes increased and the emphasis moved away from subsistence farming and toward profitable agricultural operations.[13] The RA also improved on the way in which clients understood their loans. One RA administrator found that, at least in some counties, early rural rehabilitation clients (and the public at large) considered their loans to be essentially gifts that did not have to be repaid. After only five months of the RA's operation, however, most clients and local farmers recognized that the RA's loans required repayment.[14]

Agency leaders and outside observers like Henry Wallace saw the supervised loans as the most significant part of the RA's mandate. He commended the RA for its role in helping farmers in emergency situations, for example in drought conditions. But, he wrote, from "a long-time point of view the 300,000 [farmers who had received supervised loans up to that point] are more important," not only for the 5-percent-interest-rate loans (which replaced the enormous interest rates poor farmers had previously faced) but for the supervision and guidance, which promised a new direction for the government's relationship with poor farmers and a new hope for the children of impoverished tenants.[15]

Although conditions in the South were generally the worst in the country, on a per-farm basis the RA spent sometimes twice as much on northern and midwestern farms as it did southern ones. This was particularly true of grants, which were usually made for emergency purposes. This had a number of causes. The harsh winters of some states, especially in the northern plains, demanded that northern farmers receive more fuel and clothes than did southern ones. Southerners also could grow more of their own food—a family in Florida could eat from the garden almost the entire year, while in the farthest northern farms, farmers could not count on anything for more than half the year.[16] The South also had a considerably lower standard of living: problems there did not rise to the level of emergency, because things had been that bad for years or decades.

While relatively popular, the rural rehabilitation program had its troubles, particularly in dealing with poorly planned projects inherited from its predecessors. The program in Coffee County, Alabama, had, in the words of Assistant Administrator R. W. Hudgens, "no sound economic basis."[17] The people there, as journalist Ernie Pyle noted when he visited the community, definitely needed help: they were in deep poverty and suffering from easily prevented diseases like pellagra and hookworm. But that was the problem: even after a great deal of work, the six hundred county FSA clients still mostly lacked amenities like indoor toilets, and the problem still to be handled was even bigger. Pyle estimated it might take ten generations to fix such a problem, which he claimed was "a combination of the landlord and the supply merchant and poor land and low prices and sickness and ignorance—in other words, it was the whole system."[18] The RA's attempts to repair that system were complicated by the necessity of addressing its predecessor's slapdash early efforts.

Credit was the most important facet of the rural rehabilitation program (and the most important part of Farm Security Administration activities overall: by 1943, more than 950,000 loans had been made, although not all of these were for rehabilitation purposes). The standard loans were small (averaging about $240 in 1937, $600 in 1940) and aimed more at general welfare than at meeting the usual standards for risky loans. Higher-risk families who would not have been able to receive tenant purchase loans, much less traditional commercial loans, were eligible for rural rehabilitation loans. If a family was unable to get financing from any other source but could get local recommendations as honest and hardworking and either owned or rented a farm capable of providing a living, then that family was eligible for a rehabilitation loan. Despite having higher-risk clients, the program ended up collecting more than 80 percent of the total amount loaned; by 1945 about 374,000 borrowers had paid up in full.[19]

Rural rehabilitation loans came with strings attached: borrowers had to agree to receive farm and home supervision from trained FSA agents. This program of "loans based on farm and home planning," as the FSA described it, was the heart of the program of rehabilitation. Emphasizing loans and farm management was in part intended to ensure that rural rehabilitation was not treated as relief. FERA planners imagined rural rehabilitation as an alternative to relief, a cheaper and more effective solution. RA and FSA administrators retained this idea of economic improvement but also imagined an even grander mandate to use scientific farm and home management to "help reestablish people on the land with a chance to succeed—to help people help themselves."[20]

In addition to loans, FERA and then the RA/FSA made emergency grants to farmers in response to natural disasters or simple need. The FSA in 1940–41 expanded the program to include grants made for the purpose of rehabilitation. In part this was a response to the tendency of county agents only to approve the relatively most prosperous of low-income farmers. FSA administrators hoped that this new grant program would help lower the eligibility standards for new clients, standards they believed local agents had raised too high. For example, sometimes it simply was not possible for a family to produce enough income to both pay their expenses and repay a loan. A yearly, perhaps annually decreasing, grant could remedy that by taking care of some or all of the household expenses while the family made repayments. Of the $1.15 billion dollars spent by the RA and FSA rural rehabilitation program between 1935 and 1946, 87 percent went toward loans of some kind, while only 13 percent went to grants.[21]

The FSA's standard loan program was not quite as southern-oriented as the original FERA program had been, but southerners still made up a large number of borrowers. Forty-seven percent of standard loans went to the four southern FSA regions, which had 56 percent of the total number of borrowers. The loan amounts for each family were relatively less than those given to non-southern recipients because the South was poorer and its standard of living lower than in much of the rest of the country.[22]

The standard rural rehabilitation program of the RA/FSA was less racially imbalanced than were some other New Deal rural programs (including, within the FSA, the tenant purchase program). Between 1936 and 1939, 13 percent of all borrowers were African American, compared to a 12 percent black population among the farm population as a whole. In the four southern FSA regions, 22 percent of clients were black, while African Americans were 23 percent of the population of farm operators. These statistics overstate the fairness of the program, however, as black operators were a larger percentage of poor farmers than were whites—that is, they needed the program relatively more than did white farmers. Besides overt racism (which played an important part in how and whether loans were made), a variety of factors accounted for the disparity. Black tenants and sharecroppers were even poorer than their white counterparts, so they often fell below the minimum standards of eligibility. African Americans also tended to be less likely to push for resources, based on past experience with similar federal programs. The landlord-tenant relationship was also stronger between black renters or croppers and white landlords; blacks were thus less likely to apply because of pressure from their landlords who opposed any signs of independence or because the furnish system remained in place.[23]

Black farmers who did get into the program often had trouble with racist or disinterested agents. The FSA, for its part, was willing to investigate criticisms of employees, but this rarely had appreciable results. One 1938 case in Calhoun County, South Carolina, for example, began with an anonymous letter from twelve black clients who feared retribution from their local agent or his allies. They accused the agent, James Moss, of refusing to help with bad crops and boll weevil problems. Worse, he made sure that when their cotton was sold, the checks were written to him and to the landlords, not to the sharecroppers. "The rich using the money them self," the writers accused, "and make the poor pay back."[24] Although such allegations had been rather common under the Agricultural Adjustment Administration (widespread enough, in fact, to have been part of the justification for starting the resettlement program), FSA officials had little sympathy. Such complaints had "very little foundation," in the words of the assistant regional director, who still thought it "advisable ... to make a preliminary investigation in this County and determine if there are any discriminations being made against the Negro clients."[25] Unsurprisingly, with this kind of attitude, nothing came of the investigation.

Along with loans and grants to individuals, the FSA also provided loans for cooperative purposes. Starting in 1937, the FSA made loans to client families on a group or collective basis to purchase stock, machinery, farm services, and other things too expensive or unnecessary for a single family to own. In addition, until 1942 the FSA made loans to cooperatives operating creameries, canning plants, and similar rural businesses. Over ten thousand such cooperatives received loans.[26]

Cooperative farming projects reflected FSA officials' concern that small farmers were increasingly at a disadvantage against large, modern industrial farms. Administrators believed that the family-operated farm was a better social ideal than the large-scale farm owned by a business and worked by hired laborers while recognizing the inherent economic advantages of large-scale farming. They intended for small farmers to get the benefits of modern technology and organization through cooperative activity, while maintaining the virtues of family farmers. What was known as a "group-service" enterprise is a good example of such programs. Individual, low-income farmers could pool their resources to purchase expensive equipment or services they could not otherwise afford and that would not be economical if owned individually. Joint ownership and use of heavy equipment was nothing new to American farmers; FSA administrators hoped to promote this activity in a more efficient manner. Generally, either a single farmer owned an item that was used by neighbors for a small rental fee (with agreements made beforehand regarding amount of use, fees, and similar issues), or a number of small farmers owned and shared a particular item or facility. While these were considered cooperatives, they were more like partnerships.[27]

The FSA did promote formal cooperatives, organizations that took responsibility for buying, selling, and processing agricultural production, and for providing farm services on behalf of their owner-members. In some cases, the

FSA provided loans for local farmers to become members of already-existing cooperative associations, or provided loans for farmers already planning to create cooperatives but who lacked the capital. In these cases, direct FSA activity was limited. The FSA's role was largely restricted to strongly encouraging the creation of cooperatives where the situation demanded it and to planning and creating the framework. Beyond that, administrators believed that it was up to the membership to make the program succeed. In a way similar to the Extension Services' group-study approach, FSA leaders also promoted small neighborhood groups; local agents encouraged groups of six to eight families to meet regularly in one another's homes to talk and plan cooperative activities. The program was also intended to alleviate some of the isolation and loneliness that came with farm life and to encourage stronger community bonds.[28]

Because so many southern farms still operated as plantations, improvements for individual tenants were not always enough to promote rural rehabilitation. Particularly in the Mississippi Delta and other cotton-heavy regions, a number of different activities all operated as part of a single economic unit, including the production of cotton and other products, a central cotton gin or commissary, pasture, land for feed crop, and more. It was not feasible to accept a single tenant from such an operation as a rehabilitation client. The FSA instead organized land-leasing associations—cooperatives—owned by the tenants and sharecroppers living there, which leased part or all of a plantation from the owner and subleased individual farms to existing tenants. The association might operate a livestock system, cotton gin, mills, and farming equipment and machinery service. The cooperative effectively acted as the plantation overseer. The FSA handled negotiations for the lease and provided a loan for operating capital, a charter for the cooperative, and an overall farm plan. An FSA official would often act as manager or bookkeeper for the cooperative association.[29]

Most FSA activity involving cooperatives, then, was indirect and small scale. The entities known as cooperative corporation farms were different from most other group-farming projects in that they were generally large farms incorporated as cooperative agencies, with one vote per member and distribution of profits based on the amount of labor contributed by each member. As of March 1942, the FSA had only established twenty-seven cooperative corporation farms with only 757 full-time members.[30] Critics of the FSA decried the cooperative programs as a socialization of American agriculture, akin to Soviet farming practices, but they resulted in very little communal farming. For example, the FSA turned several large plantations into cooperative farms. The plantations shared pasture land, but individual families owned their milk cows and sows. Each family had its own garden. The livestock, farm equipment, seed, and farm supplies were purchased cooperatively but owned by individual families. They farmed the cooperatively held land in individual units. Assistant Administrator Robert Hudgens described the setup as keeping the plantation system under new ownership, with families "working both as plantation operators and plantation tenants, and taking full advantage of all the good that the plantation economy has to offer."[31]

Some FSA officials, like Region VI director T. Roy Reid, believed that "the possibilities of such [cooperative] loans have not been fully realized and further opportunities for their profitable use are many."[32] However, any expansion of the program was difficult. Cooperatives were one of most frequently criticized aspects of farm security efforts, both from without and within the agency. The move of the RA to the USDA and its evolution into the FSA encouraged shifts away from cooperative loans. Increasing regulations limited the way that cooperatives' loans could be created, and by the summer of 1937, administrators told hostile letter writers that farming cooperatives were a small and shrinking part of the federal government's farm security program.[33]

Similar to the resettlement projects, the cooperatives continued to be criticized even as their size and relative importance in the FSA decreased. Congress was strongly opposed to the cooperative farming program in general, and widespread political assault led to the prohibition of all new cooperative activity by the FSA after June 30, 1943, except for the medical, dental, and hospital groups associations. Before then, a large number of families took part in the various cooperative programs despite the hostility. By the end of 1943, over 148,000 families had taken part in the local and state purchasing and marketing programs; 103,000 used cooperative processing facilities like grain elevators, cotton gins, or canneries; and 58,000 families had purchased cooperative-based insurance. Thousands more used migratory health camps, land-leasing programs, cooperative water facilities, and cooperative veterinary care.[34]

A less controversial cooperative enterprise was medical and dental care.[35] Rural living and poverty combined to produce terrible health conditions. For years the Department of Agriculture operated a variety of different programs addressing rural health—some indirectly, like those dealing with animal diseases (which frequently spread to humans) and the elimination of disease-bearing pests. The USDA performed a great deal of work concerned with nutrition, while the Extension Service provided improved medical and sanitation education. However, these programs showed the potential for reform in rural sanitation and health care more than tangible results.[36]

The RA organized a Public Health Section in its Washington office to provide better medical care and a wider choice of physicians for client families. In the FSA, the program expanded to include hospital and dental care; administrators considered good physical health to be important to rural rehabilitation and farming success. For farmers receiving rehabilitation and tenant purchase loans, the general approach was to provide funds for families to pool their resources to create medical associations that could enter into agreements with local doctors and hospitals. These medical associations could maintain better rates for care and a stronger assurance of service than could the families working individually. Moreover, it was the only realistic way to budget for medical care: individual rates and needs varied far too much for reliable planning.[37]

The FSA county or district created a medical insurance fund; members paid annual fees based on the ability of the client to pay. Membership in the medical care plans was voluntary. The doctor sent bills to the trustee, who divided the medical fee fund into twelve parts and paid the doctor bill monthly, in full if possible but prorated if not, out of the medical insurance fund.[38] Physician membership was similarly voluntary and organized: local doctors and dentists often signed informal agreements, usually after FSA employees had come to an agreement with the local or state medical and dental associations. These negotiations with medical professionals did not always go smoothly, as professional associations generally discounted the severity of the rural health problem and, more important, greatly feared any moves that might increase the role of the federal government in health care. The FSA's decentralization and emphasis on local concerns proved crucial in gaining local support; administrators developed good relations with local and county medical societies and agencies and pledged to only introduce plans that had the support of the local medical community.[39] They also cut some of the more controversial elements of the program when necessary: in one county, physician unhappiness with the pooled funds forced a change to a more traditional billing system.[40]

In one sense, the RA and FSA medical program was modest. The problem was simply so large that administrators could only provide a minimal amount of medical care locally to low-income farmers and migratory workers and establish a certain basic level of sanitary facilities (clean water, sanitary toilet facilities, and mosquito screens for home). Little was done to increase the number of medical professionals in rural America, and little was done to improve the number of chronic illnesses and physical handicaps among FSA clients.[41] Still, while limited, the program did have results. If nothing else, it provided (for those able to join a medical cooperative) immediate, tangible improvements in medical care for enrolled clients. For example, officials in FSA Region V had made small ($10–$30) grants for client families to build sanitary facilities to almost 25,000 families by the end of 1939. Other regions followed suit. By the middle of 1943, about 112,000 families spread over forty-five states had received a total of $3.4 million in grants. The program was focused on the South, especially Regions V and VI, where the problem was greatest; limited funds meant that the effort had to be concentrated on areas where the situation was the worst.[42]

The physical improvement for clients was significant. Farming was (and is) strenuous work. Poor farmers lost their farms for medical costs as minor problems became major and then life-threatening ones. Other farmers continued to work, but ineffectively because of pellagra, hookworm, malaria, and other poverty-related conditions. Recognizing this, local FSA administrators and agents strongly favored the medical program. W. L. McArthur, the FSA county agent in Coffee County, Alabama, believed that the health program was foundational to rural rehabilitation. The first thing to do, he told journalist Ernie Pyle, was to get people healthy, and other improvements to the quality of rural life would follow. But, he recognized, the long-term problems could not be fixed overnight. McArthur said it might take generations to make sure that the poor lived well, because, as Pyle put it, "They'd been used to feeling rotten too long to change overnight."[43]

The FSA also improved the situation from the physician's end. Rural doctors frequently had to write off the bad debts of their patients, particularly during the Depression. By 1941 clients had paid only 61 percent of doctors' bills, 81 percent of dental bills, and 74 percent of hospital bills, but this was actually quite an improvement from earlier repayment rates.[44] One observer found that for all the medical program's shortcomings, rural physicians "as a whole received more money under the Farm Security Administration plan than they had been able to collect from the same body of patients previously."[45]

Dealing with families already in such poor condition, the FSA was often playing catch-up. In Geneva, Alabama, the FSA worked with the local health department to sponsor family visits by local doctors and nurses. The home management supervisor and public health nurse went to a client family and met with the family physician in response to the death of one of the family's young girls. Malnutrition and heart disease had been the cause, and an eight-year-old boy in the family was on the same track because of hookworm. After treatment at the County Health Department, the boy lived, and the entire family received hookworm treatments. After additional follow-ups, including a demonstration from the county home management supervisor on food preparation and balanced diets, the entire family's health improved.[46]

Perhaps the biggest problem with the FSA's health-plan program (aside from the obvious problems created by serving a poor, rural clientele) was administrative. The medical program was part of the rural rehabilitation program, so the final responsibility for its operation rested with the county supervisors. But local FSA agents already had trouble keeping up with their standard rural rehabilitation cases. Supervisors tended to let the medical program move along with little guidance, while the member families lacked any understanding of how the plan worked or what part they might play in it. Such apathy, neglect, and uncertainty led to the selection of the quickest and easiest solutions, not necessarily the best ones; for example, this often meant employing local generalists, not rare or expensive specialists. Similarly, county doctors were brought into the program to gain local acceptance for the medical plans, limiting the options for and quality of care. Overall, while rural health coverage could generally be expanded by the FSA medical program, it was rarely improved. Before the government project started, for example, Coffee County, Alabama, had one nurse to serve the county's thirty-five thousand people. By 1941 the county had only three nurses and still lacked necessary specialists.[47]

The reduction in FSA funding in 1942 left the medical program limping along. Enrollment numbers tell part of the story: the peak membership was in June 1942, when nationally 613,854 enrolled; that dropped to 528,094 a year later, 363,443 in 1944, 284,100 in 1945, and 236,780 by June 1946. In many cases, such declines in membership made the plans untenable: by 1944, the average number of families enrolled per plan was only ninety-two, not nearly enough to support a program funded by low-income farmers. Compounding the problem of declining membership and funds was a sense among clients that they had little direct responsibility for the program. Relatively few FSA clients took the lead in managing or creating the medical plans, leaving that work to FSA field agents. Only a third of the associations in 1940 had a meeting of the local board of directors, made up of members, at least once a year. When funding began to dry up and FSA agents could no longer provide leadership and momentum, there were no local leaders to provide continuity. Additionally, many rural physicians and dentists joined the armed forces during World War II; those who remained behind served such a large number of patients that they saw no reason to contract care for a particular group, particularly one made up of poorer patients who might not be able to pay. At the same time, increasing farm income meant that fewer rural families needed to use a low-income program. Finally, the mass reduction of the FSA budget meant that fewer families were a part of the FSA program (a drop during the war years from five hundred thousand to about two hundred thousand); a smaller staff meant there were fewer people, with generally less experience, to organize and maintain such programs.[48]

The FSA also got caught up in a debate over nationalized health care. The agency publicly came out in favor of the 1943 Wagner-Murray-Dingell bill, which proposed universal health insurance as a sort of extension of Social Security. The FSA's advocacy drew the ire of organized medicine, already nervous about the threat of government intervention in health care. Rural physicians, county medical associations, the American Medical Association, and numerous other medical organizations turned against the FSA in 1943, critically wounding the agency at the worst possible time.[49]

In trying to improve clients' lives as part of rural rehabilitation, administrators had to consider their debts. The Farm Credit Act of 1933 created the Farm Credit Administration (FCA), which provided loans to farmers from the Federal Land Banks. But these and other sources of funds proved inadequate. Because of numerous debts incurred before 1929, the general agricultural decline of the late 1920s, and the economic crash, something like one-sixth of all American farmers (about a million people) were on the edge of bankruptcy when the FCA began operations in 1933. Farmers were often in debt to numerous lenders (any of whom might push the farmer into bankruptcy) and behind on taxes; as a result, a debtor might stop all repayments to save money for when one creditor might begin foreclosure proceedings. Unable to help pay for even a fraction of this, the FCA started a program to bring together debtors and creditors in order to work out new terms and to avoid foreclosure and bankruptcy.

The program, which came to be known as Voluntary Farm Debt Adjustment, moved from the FCA to the RA and then FSA. The program was a practical necessity for the operation of a government loan program aimed at helping poor farmers. Otherwise, the receipt of government funds would simply be a signal for creditors to renew their push for payment, and the government would essentially be paying the debt holders. By June 1940 the FSA had overseen more than fifty-eight thousand debt adjustment cases, and thousands more followed the example (which was often both quicker and less costly than legal proceedings). Even though part of the rationale for the program was to protect government investments, about a third of the families involved were not clients of the FSA.[50]

One unique aspect of the voluntary debt readjustment program was that it had essentially no legal power; the committees were effective only because of the personal reputations of the members, the goodwill they generated, and the difficult situation in which creditors and debtors found themselves.[51] The committees, usually made up of five members, were chosen for their reputations as fair and unselfish men, crucial in encouraging a voluntary settlement from both sides. Often it was fairly straightforward to encourage creditors that the current state of affairs—the farm being profitably run by its current occupants—was the best. A prepared agreement between debtor and creditor(s) helped avoid court action, which would cost both sides money. While they did not preclude foreclosure altogether, the county committees urged a friendly settlement whenever possible.

Aiding the county committees was the awareness that even the best farmers took on debts. Knowledge that a farmer could continue to profitably run a farm, along with the support of the committee, was often all it took to encourage creditors to work out a settlement. In other cases, interest rates were dropped, perhaps from 5 percent to 2 percent, until prices recovered. Often simply bringing the creditor and debtor together solved all the problems. In one case, a creditor was unaware that his debtor was taking out a Federal Land Bank loan to partially pay off his debt; the creditor agreed to cancel the mortgage entirely for the value of the loan. On the other hand, committees were willing to see people lose their farms when necessary: the farm of one heavy-drinking debtor, the committee believed, would be better off in other hands, and it recommended that the creditor begin foreclosure proceedings. When the committees became better known, it often became unnecessary even to meet; recommendations from a single member might encourage debtors and creditors to work out a new settlement.[52]

The committees established a good record of debt adjustment. In one of the FSA's poorest regions, Region V, an FSA report found for one eighteen-month period that in 6,044 of the 7,218 decided cases, debts were adjusted for a total debt reduction of $1.8 million. The average debtor in settled cases reduced his debt by $297 (about a 19 percent reduction). These adjustments also freed up just over $200,000 in taxes; about $1.75 was paid in delinquent taxes for every dollar spent on the debt adjustment activities.[53] By the end of 1943, the program had overseen over 187,000 cases, with a total debt reduction of over $109 million. The average farmer's indebtedness (which before the program totaled about $2,700 per person) was reduced by about $585 dollars. The largest single year for debt reduction was 1941, with over 35,000 cases; this number declined sharply with the improving economic condition and other changes caused by World War II, along with the decline of the FSA after 1942. With the FSA's budget cuts, policy shifted; county supervisors picked up responsibility for debt adjustment instead of organizing and overseeing the committees. While many debt adjustment cases in the past had involved non-FSA clients, as a practical matter this could not continue. The FSA simply lacked the funds, and county supervisors—some of whom by then had well over two hundred clients—did not have the time to help nonclients.[54]

Closely related to debt adjustment was tenure improvement. FSA officials hoped to replace vague one-year oral agreements with written contracts for a three- to five-year period and/or an automatically renewing lease. Tenure specialists, along with farm management and debt adjustment experts, trained county and district supervisors in the importance of having standard, written, easily enforced contracts. Some landlords willingly went along; if they believed that farm and home management plans would improve their rental incomes by increasing agricultural output, landlords would cooperate because the loans only went to applicants in good tenure situations (a policy that became official in 1938). In particular, landlords who had made little or nothing on a particular plot of land, or those who were not full-time farmers and had other interests or sources of income, were glad to do anything that might improve the value of their property. In other cases, FSA agents often found it necessary to encourage landlords to go along. They held meetings with local landowners to explain the FSA program and how it benefited both parties to have stable tenure arrangements. Sometimes it was necessary to sweeten the pot; a rehabilitation borrower might also receive a sanitation grant to repair or improve the landlord's home and sanitary facilities to go along with a new lease.[55]

As with debt adjustment, tenure improvement was among the practical changes in tenant lifestyle that FSA leaders believed were necessary for the rehabilitation loan program to succeed. Prior to 1937, federal officials had not done much to remedy tenure conditions, although many people recognized the problem. FERA officials had been busy focusing on subsistence farming. RA officials had been too occupied with the resettlement and rehabilitation projects and then, on passing tenancy legislation, to take steps toward fixing tenure problems.[56]

The tenure improvement program had mixed results. Written leases and provisions for a garden and pastures came easily. Long-term leases, automatic renewal, and other provisions proved more difficult to obtain. However, even the relatively limited change of written leases created a dramatically different relationship between landlord and tenant in the South, where oral leases had been the rule for decades; for the first time, negotiation of rates and obligations became the rule, not the exception.[57]

The trend toward making loans to better-off clients encouraged some to try a new approach. At Will Alexander's urging, the FSA began to develop new plans for experiments with the poorest clients not even eligible for rural rehabilitation loans, the only FSA programs (indeed, practically the only large federal program of any kind) designed to reach such farmers. Late in 1938, the FSA established special staffs in eleven counties across the country to assist a group of families not able to qualify for the regular FSA program. Intensive supervision and experimentation were encouraged for the specially selected personnel. The farm and home specialists were to visit every needy farm family in the area to work out a supervised program for that family or refer them to an appropriate agency that could take care of their case. Only those with physical or mental disabilities limiting rehabilitation were not to be considered. Eventually, about five hundred families were chosen.[58]

The experimental effort for these families reflects the general approach of rural rehabilitation, especially in the South. For example, the biggest emphasis was on increasing food production. In Laurens County, Georgia, only twelve families had cows in 1939, when the program started; by 1940 all fifty did. The program had limitations: cultivating vegetable gardens and especially canning and preserving food for family production came along much more slowly than administrators hoped, in large part because these farmers had never done so before and saw no reason to begin doing so. Still, starting from so little, the increases could appear astounding: in Georgia, Laurens County clients increased their dried-bean storage by 900 percent, and Oglethorpe County clients stored five times as much homemade syrup.[59]

Similarly, FSA agents emphasized proper home and farm management. Supervisors found that their new clients did not understand even "simple farm tasks that are ordinarily taken for granted by farm people who are better educated and equipped."[60] Again, starting from such a low level, it was easy to make big leaps. Farmers in Mercer County, West Virginia, reportedly never used crop rotation, failed to provide enough protein to their animal stock, and used poor, home-grown seeds. Farmers were initially hesitant to change these and other practices, but the evidence spoke for itself. This experimental group had the opportunity for much more intensive supervision than the overworked staff of the standard rural rehabilitation program. Each farm management supervisor was assigned a caseload of fifty families, compared to between one hundred and three hundred or more for the standard loan program, and in a much larger geographical area.[61]

The FSA also promoted mutual cooperation. This included cooperative purchasing, marketing, group ownership of machinery and equipment, and similar programs. This could overlap with educational and farm management efforts. Cooperative seed purchases, for example, provided an opportunity for a lesson in the use of hybrid and certified seeds, as opposed to unselected homegrown seeds. Group livestock purchases necessitated a consideration of proper breeding. And, of course, it had its economic benefits: farmers in Reynolds County, Missouri, saved themselves an estimated $230.98 each in 1940 through cooperative buying. Jointly owned washing machines, pressure cookers, and sausage mills saved both time and money.[62]

The most telling aspects of the five-hundred-families program, however, were the social and educational activities developed for these clients. FSA administrators believed that the physical side of the equation was not enough; in listing the problems faced by the five hundred families, for example, Conrad Taeuber and Rachel Rowe included "Social Difficulties—Lack of satisfactory social and educational services" and "Emotional Difficulties—Engendered feeling of inferiority and distrust in humanity." Like the rest of the rural rehabilitation program, FSA leaders believed they were doing their part to bring these families back into full membership in American society. FSA supervisors were particularly proud, then, to change attitudes. Things like "improved facial expressions, strengthened faith and a friendlier and more cooperative spirit toward neighbors and supervisors" counted just as much as how many more cans of tomatoes a family stored in a given year. Supervisors even imagined themselves marriage counselors. "Constant bickering and strained relationships between husbands and wives, caused to a great extent by the effort involved to obtain a bare existence," Taeuber and Rowe summarized from supervisor reports, "have been alleviated, by frank discussions of problems and differences and by improved living conditions."[63]

With the outbreak of World War II, the experiment shifted toward war production (as did, to an extent, rural rehabilitation as a whole), eventually becoming more closely incorporated in the larger rural rehabilitation program. FSA descriptions emphasized the value of these poorest farm families to the war effort; Rachel Rowe Swiger and Olaf Larson wrote, "Wasted manpower and wasted productive capacity are being eliminated."[64] More work needed to be done, another description explained, as "many farm families do not now have [the] skills, equipment, capital, and productive land" necessary for production. Rural rehabilitation was the key to making even poor families a part of wartime production. More than that, this experiment was linked to the ideals that the war was allegedly being fought for: "Freedom from want is based on production," but neglect due to poverty and ignorance had made it impossible for much of America's now worn-out soil to produce the necessary agricultural output (and thus impossible for it to allow freedom from want).[65] The rural rehabilitation program made it possible for even the poorest families, then, to be a part of the war effort and the fight for American freedom, just as it had earlier opened the door for farmers to both improve their practical farm management and enjoy fuller participation in American society.

From its origins in FERA in 1933 to the end of World War II, rural rehabilitation was increasingly the center of the anti-rural poverty effort in the federal government. Several tools of rural rehabilitation, particularly the farm and home management plans, influenced how the rest of the FSA operated. The FSA's tenant purchase program operated as a sort of more intensified version of rural rehabilitation: large government loans for the purchase, improvement, and operation of a family-size farm combined with more intensive farm and home management. The cost of such a narrowed focus was that the tenant purchase program could only aid a relatively small number of farmers, particularly as compared to the general rural rehabilitation program. But administrators hoped that success could rehabilitate tenants entirely, freeing up resources for the rest of the rural poor while providing a politically popular underpinning for the FSA as a whole.

CHAPTER SIX

Creating Family Farms
The Tenant Purchase Program

The symbolic value of owning one's own farm was evident in the case of Texas farmer James C. Greene. In 1940, according to a newspaper account, Greene "celebrated Texas Independence Day by paying $1,000 on a new farm which, with assistance of the Farm Security Administration, became his own just a month ago." This signaled a new freedom in all sorts of ways. His payment not only freed Greene from tenancy and his mortgage, "but it also marks his independence from cotton, for the $1,000 payment was profit on sale of 80,000 rosebushes" (although he continued to raise both cotton and cattle). Greene's success made him unusual; he was among the tiny handful of tenants provided with FSA loans, and now he was making more than the vast majority of tenants from whose ranks he had risen. But such an opportunity meant hope for any tenant farmer that he, too, might become a farm owner, just as the tenant purchase program as a whole represented an opportunity for tenant farmers to find a new, beneficial relationship with the federal government.[1]

When borrowers described the value of the tenant purchase program, they explained it in terms of the transformation caused by land ownership. During a 1941 interview, John H. McCullough responded to a question about the difference between owning and renting: "Why, there's all the difference in the world. When a man's got his own land he feels like working it and building it up, because he knows he's going to get the benefit of it. And a man's his own boss when he's working his own place."[2] Whatever political and economic changes happened as the Great Depression wore on, the belief in the inherent worth in farming one's own land—the symbolic as well as practical value—remained strong.

The tenant purchase program operated under the idea that farm ownership was a step up from farm tenancy. The living standards for clients were intended to rise with their entry into the program. For this reason, the standards for acceptance were higher, the loans were larger, and the supervision more intense than in other credit programs. Being, then, an intensified version of the FSA supervised-credit program, the tenant purchase program provides perhaps the best picture of how New Dealers, by the late 1930s, imagined post-Depression rural America. The FSA self-consciously concerned itself with the long-term future of rural America, looking ahead not just to the end of the Depression, or even the lives of its clients, but to generations to come. The FSA shows how the New Deal patched together old-fashioned ideas about the small family farm and rural communities and a newer idea of a vigorous, beneficial national government, a combination most apparent in the tenant purchase program.

The Bankhead-Jones Farm Tenant Act provided, for the first time, some direct legislative endorsement of the activities of the Farm Security Administration; the FSA's predecessors had gotten funding and authority through a variety of executive authorizations, emergency and agricultural appropriations, and Reconstruction Finance Corporation loans. The BJFTA provided specific authority to make loans to farm tenants, laborers, sharecroppers, and other destitute rural people in order to purchase, construct, or repair farms and farm buildings. The Tenant Purchase Division was created in 1937 (renamed the Farm Ownership Division in July 1942) to manage and operate this program.[3]

Paul Maris had worked on rural rehabilitation projects in the Federal Emergency Relief Administration and the Resettlement Administration before heading up the Tenant Purchase Division. He believed strongly in the agricultural ladder aspect of the tenant purchase program, the idea that the goal should be to accelerate the move of the most capable tenants into the ranks of farm owners. Additionally, congressional intent in the drafting of the Bankhead-Jones Farm Tenant Act and potential congressional hostility necessitated, to him, a cautious course. To make loans to unprepared or incapable applicants, loans that would never be repaid, would only waste money and risk political fallout.[4] "The law under which we operate contemplates making loans for the purchase of farm on which industrious families can make a living and repay their debts," Maris said in a speech at an FSA conference in 1940. "There is no other intention than that the debts shall be repaid. No other measuring stick is likely to have so much to do with continued congressional support as the collection record."[5]

Maris's primary objective was to have successful clients, defined as those who repaid their loans and became farm owners. He was motivated by a desire to see the program succeed and a heavy dose of political pragmatism. Maris, like the rest of the FSA, realized that Congress considered the tenant purchase program to be the most justifiable part of the FSA; for many critics, it was the agency's only legitimate function. To liberalize lending and lower return rates would risk support for the part of the FSA that had the strongest public-relations advantage. And in that regard, Maris was mostly successful—he could point to a 98.4 percent return rate on loans that had fallen due by June 30, 1943. By 1949, when not quite $82 million was mature, $162 million in principal payments had been repaid—almost twice what was due, plus another $40 million in interest payments.[6]

Maris wanted the program to get off to a careful start, but also a quick one: had it failed to use the entire $10 million allocated for the first year, the administrators or the program itself might look like failures. There was much to be done: forms to create, staff to hire, methodology to be worked out. But the staff had examples to work from. The resettlement and rural rehabilitation programs had experience with farm and home planning, farm budgeting, rural housing, and small-farm record keeping. The FSA learned from its mistakes: for example, the public-relations debacles surrounding the resettlement program and certain cooperatives contributed to administrators' insistence on working only with individual families.[7]

Concerned about the program's success rate and political viability, program staff took great care in selecting tenants, delaying the actual loans despite considerable demand. The agency was busy selecting state and county committees, appointing field representative, training committee members and representatives, developing procedure, advertising the program's availability, and looking over applications. The FSA made its first tenant purchase loan on February 12, 1938, to Wiley J. Langley of Jasper, Alabama (not coincidentally, John H. Bankhead's hometown), almost seven months after the legislation passed.[8]

In its first year of operation, the Tenant Purchase Division provided 1,879 loans, although it received slightly more than 38,000 applications. Even this number underestimates the demand for the program: the first applications came from only the 325 counties eligible for tenant purchase loans, and the call for applications came relatively late in the season, when many tenants were already settled.[9] Compared to the demand for them, the program made few loans. Between 1937 and 1940, the program approved 12,234 loans in 1,633 counties (including, as amused Alabama newspapers enjoyed pointing out, a loan to a Pickens County tenant farmer named Bankhead Jones).[10] These approved clients came from over 280,000 applicants, or about 23 applicants per loan selected. In 1938 the FSA only made an average of 7 loans in each of the 333 counties then designated for loans.[11]

The care of Tenant Purchase Division administrators and their political concerns were not entirely responsible for the slowness in making loans or the small number of clients. Congress played a large part, both indirectly (in, for example, influencing how Maris thought the program should proceed to maintain congressional approval) and directly. The relatively small appropriations limited the number of tenants to whom the FSA could make loans. Other restrictions hindered how the FSA could use those funds. Most notable among these was the Tarver amendment, which sometimes made it impossible to use the entirety of apportioned funds in each state. In 1940 Georgia representative Malcolm Tarver introduced these limitations through the agricultural appropriations bill; tenant purchase farms could not have a greater value than the average farming unit above thirty acres in its locality. Tarver was a proponent of farm ownership, but he was suspicious that poor farmers could not handle large farms and that benefits might not go to the most deserving applicants. "We are dealing here with the very poorest class of agricultural people," he said. "I do not see the point in buying one of these men a farm worth $7,500, when you could buy four farms, for four tenants, at one-quarter of the price, which would buy a fairly good farm in my district for a tenant."[12] Tarver was also suspicious of FSA leadership, particularly C. B. Baldwin, whom he criticized as operating a stealth relocation program. The limit on the size of farms made it difficult to break up large plantations and sell them piecemeal to farmers from another county or state, a procedure Tarver saw as part of a plot by Baldwin to move farmers into new areas and break up the influence of the planter class.[13]

A TENANT PURCHASE CLIENT ON HIS NEW HAY RAKE, MISSISSIPPI DELTA. LIBRARY OF CONGRESS.

Despite this narrow range of acceptable applicants, the tenant purchase program proved popular among farmers. Borrowers mostly approved of their individual farms and the fact that the program moved people from the ranks of renters to owners. Even farmers who were already owners, or who applied but could not get loans, favored the program. In promoting individual farm ownership, it met the approval of virtually all classes of farmers, particularly in the South.[14] The *Farm Journal and Farmer's Wife* found in early 1942 that 75 percent of respondents approved of the job the FSA was doing. Even those who disapproved usually agreed with its goals but found fault with how the program was carried out.[15] Will Alexander estimated that the newspaper response in the first year was about 85 percent positive and that most of the criticism focused on the program's small size.[16] It also met the approval of politicians: by 1939 several congressmen and senators had informally met with USDA administrators to discuss how the program might be expanded.[17] The biggest complaint from Congress may have been the limitations its members imposed: politicians wanted loans in their own states and districts but were unwilling to appropriate the funds necessary to make more loans.

Part of the program's popularity and success came from the fact that it was put into operation much more carefully than rural rehabilitation and could build on lessons learned during rehabilitation. In fact, approved tenant purchase families often had experience in the rural rehabilitation program, so agents were dealing (as it were) with known quantities. But the most important reason for its popularity was that the tenant purchase program was the closest in line with what most Americans thought the government should be promoting: home and farm ownership. As Senator Richard Russell of Georgia noted, the "publicity given the program" was "all out of proportion to the money" spent.[18] It was sort of an anti-resettlement program, in terms of perception: a relatively modest government program using relatively modest innovations to help a small slice of the rural poor, only in this case receiving general approval from both the public and the politicians.

The Tenant Purchase Division had the most stringent requirements for applicants among FSA programs. Regulations in 1938 listed twenty-five distinct steps in making tenant purchase loans.[19] An applicant's existing capital and assets received little attention, though most borrowers were share renters or cash renters—the highest level, in other words, below owners on the agricultural ladder. Applicants who received loans were low income, but they were by no means the poorest farmers. The Tenant Purchase Division focused on the most effective and industrious tenant farmers, judged less in terms of wealth than by their farming skill, willingness and ability to cooperate, history of good character, and reputation in their community.[20]

The number of loans in each state was determined using a formula devised by the Bureau of Agricultural Economics. The FSA created a state committee of farmers, businessmen, and agricultural leaders to select counties in each state, using census figures to determine which had the highest percentage of tenancy.[21] As soon as a county was so designated, local FSA officials created as much publicity as possible, using both newspapers and public notices at trading centers and other public places. County supervisors did not list what land was for sale locally, as the goal was to give applicants the widest possible latitude in selecting their own farms. The county supervisor passed applicants on to the county committees, usually with rankings or recommendations based on eligibility and desirability.[22]

Applicants to the tenant purchase program had to meet a number of requirements. They had to be American citizens who were farm tenants, sharecroppers, laborers, or recent owners who had been involved in farming as the principal means of providing family income. Applicants had to be willing to cooperate with FSA supervisors to develop a home and farm plan and keep up with their records. They also had to have good standing in their community or, in the words of FSA instructions to its employees, to "have a reputation for paying [their] debts and meeting obligations," as well as having shown stability, resourcefulness, and managerial and farming ability. Clients and their families had to be physically able to farm and free of any incurable debilitating illnesses. Finally, an applicant had to be unable to secure a loan through other private or government sources. Preference was given to married applicants or those with dependents and to those clients who already owned the livestock and farm implements. Families with young children got the tiebreaker, all other things being equal. A farmer could be too successful; families who could get credit through other means were ineligible. Rural rehabilitation clients were to receive consideration equally (though in practice, those coming from the rural rehabilitation program often had an advantage because local supervisors and committees already knew them).[23]

Essentially any tenant who could not obtain a loan from a bank, but who possessed the capacity and desire to operate a family-sized farm, was eligible to receive a tenant purchase loan. This did not quite limit the options as much as it may sound; for a bank loan in most of the country at the time, farmers were expected to have 25 percent of the farm purchase price as well as stock, equipment, and operating capital.[24] The purpose, more than in matching any given set of numerical standards, was to find tenants who were qualified to own and operate their own farms but who could not afford to do so.

Whether they were buying the farms they already worked or picking a new one, borrowers went through several steps to obtain loans. Tenant purchase clients and farms provisionally approved by the local committee went back to county supervisors for further study. Farm and home supervisors created a narrative form for the family's history, a list of qualifications of the family, and a description of the applicant's capacity for farm and home management. Provisionally approved applicants received physicals. After a farm had been appraised and a family locally approved, supervisors put together a loan docket. Each tenant purchase docket required twelve different documents before consideration by the regional office. The chief of the Tenant Purchase Section in each regional office received these dockets for inspection and approval. The majority of such dockets were approved or required only minimal changes by the time they reached the regional level; if state or regional personnel contributed, they usually did so early in the process and often at the request of the local agent, not in its final inspection.[25] The client then began purchasing, repairing, and building as necessary.

County committees, including at least two local farmers on each three-person committee, played a large role in determining who would receive a tenant purchase loan. The committees, in the words of the letter appointing each member, certified "applicants for tenant purchase loans" and "farms which approved applicants desire to purchase with the proceeds of the loans," with compensation of three dollars a day for up to five days a month.[26] County committees were a mixed blessing for Tenant Purchase Division administrators. One of the purposes of these committees was to ensure support from local authorities, who could use them to approve or reject candidates they found unsuitable for whatever reason (including race).[27] When deciding on applicants, committee members often violated law and policy in making their selections. They approved families with outside industrial employment or farmers who already possessed considerable capital and land. County committee members sometimes took the view that their approvals were final; they even threatened to resign if the regional office did not follow their recommendations.[28] Generally, however, relations between the committees and the local, state, or national FSA offices were fairly good, and the willingness of county committee members to provide their time and reputation was crucial to the tenant purchase program's operation.

The political cover provided by using locals also made things easier for FSA administrators. Will Alexander recalled one case in which Senator Ellison "Cotton Ed" Smith criticized the FSA for selling a farm to a black farmer in his home state of South Carolina, bringing the complaint directly to Alexander (at that time FSA administrator). Alexander then contacted the committee, which was able to give a detailed explanation for its decision: that farmer had worked that land as a tenant for more than ten years, and he was considered a fine candidate for a loan. Alexander simply forwarded the report to Smith, who could not criticize the decision made by local white farmers from his own state.[29] Using the committees as political cover did not always work: Malcolm Tarver refused to believe that it was the county committees who determined the average value of loans made in his state, claiming that "they have been acting under directions from [Alexander's] office," before Tarver set a limit on the value of farm-ownership loans.[30] In most cases, however, the local decision making was one of the most widely supported parts of the program.

Despite rare cases in which they worked for black farmers, as above, county committees tended to allow local racial bigotry to flourish. As NAACP special counsel Charles Houston wrote to Will Alexander about his "gravest misgivings" over the "despotic power" given to committees, "Can you tell me whether any county committees with similar power over selecting beneficiaries of Federal beneficences have made it a practice to give Negroes fair consideration?"[31] The emphasis on participatory democracy, which the farm security program had inherited from its Division of Subsistence Homesteads and Rural Rehabilitation Division predecessors, allowed those participants to discriminate (something similar happened in the Tennessee Valley Authority, with its commitment to "grassroots" democracy).[32]

Within the boundaries set by legislation, the FSA had some leeway in who it selected. Practically every borrower was the head of a family, usually a large one. FSA leaders believed, in the words of one social scientist, that a "man without a wife is not likely to be considered a good prospect for rehabilitation because having a homemaker is vital from the standpoint of the success of the farm enterprise."[33] Among the most important factors in choosing families was health: farming was hard work, so few major illnesses or disabilities afflicted adult tenant purchase borrowers. Client families rarely had any health problems more severe than the occasional husband blind in one eye or wife with goiter. Children with illnesses were less of a strike against an applicant, as county supervisors often thought of those borrowers as working for a better future for their children, and therefore were more likely to be hard workers. Generally borrowers were young, rarely older than their mid-forties for husbands and usually in their late twenties and thirties, with wives slightly younger. A large, healthy family could make up for old age: one sixty-seven-year-old farmer was recommended for acceptance because his doctor said after the physical examination that the family was perhaps the healthiest he had ever examined, and (as the local supervisor noted) even if the borrower became unable to work, his four sons could take over the operation of the farm.[34]

One crucial qualification for applicants was to be well-thought of in one's community. County committees (which were, after all, composed of local farmers) liked this, as did home and farm supervisors. Along with good health and a family, local reputation as a good farmer and member of the community was a near absolute for approval. One particularly positive family narrative described black farmers as being respected by both black and white farmers and being "leaders among their race" because of their "integrity, industry, and cooperativeness."[35] Another supervisor strongly recommended an applicant because "Mr. Parham's neighbors say of him that if anybody will pay for a farm, June Parham will!"[36] Reputation was just as important for wives. One home supervisor wrote in recommending a family, "Mrs. Harris seems to be the main cog in the wheel of the Rose Hill Community."[37] Even the reputation of an applicant's family could help: the families of both Joe and Mildred Farr had "made a record of ambition and good management" that reflected well on the Farrs' character and thus their chances to become successful tenant purchase clients.[38] Not being terribly involved in community affairs could be overlooked: a county supervisor noted that the eventually approved applicant Otto Moss was not sufficiently involved in community affairs but still had a good reputation among those who knew him.[39] A bad reputation, however, could not be overcome.

Having some formal education was less important in the view of county supervisors and committees. Supervisors in every FSA program, including in the Tenant Purchase Division, worried that too much education might draw a farmer away from farm life. On the other hand, not enough education left him unable to handle farm problems in a practical, capable way.[40] In the tenant purchase program specifically, numerous clients with a fourth-grade or below education were accepted, and often supervisors described clients as having the disadvantage of little education but still recommended approval. More important from the FSA supervisor's perspective was a willingness to do the work necessary to make a farm successful and the clients' innate intelligence. A lack of education could be overlooked if that natural intelligence was evident: one supervisor described applicant Willie Ferguson (who was strongly recommended) as having only a seventh-grade education on paper but in conversation sounding as though he had a high school diploma.[41]

Ferguson's narrative points to what supervisors considered to be far more important than formal education: a willingness to listen to and put into practice the advice of home and farm supervisors. Looking at Ferguson, for example, the home supervisors believed he and his family would succeed because they would "welcome supervision and appreciate everything that is being done to help to own a home and farm and live better than they have ever lived before."[42]

Supervisors frequently used the word "progressive" to describe qualified, desirable applicants—those who had proven that they would take advantage of professional agricultural expertise. Rather than just having a formal education or instruction in farming, good applicants were "hustlers" who liked working hard on a farm and would be willing to take the advice of their farm and home supervisors. Capable supervision and hard work could overcome much, FSA agents thought, if a client was willing to take it to heart: the Page family, described as "a hard working ignorant family of negroes," was recommended by a county supervisor who believed that the family could succeed so long as they had supervision and encouragement.[43]

Closely related to a family's willingness to take supervision, at least in the minds of supervisors making recommendations, was a family's desire to own a home and their gratefulness for the opportunity to do so. One supervisor wrote that an applicant family seemed "to be an ideal family for the tenant purchase program. They are both young and healthy and genuinely interested in owning a home."[44] Mr. and Mrs. Billie Morring were approved in part because "both understand the supervisory program and welcome any help they get from this source."[45] Similarly, another applicant couple, the Hardens, worked hard and maintained a standard of living above that of the average tenant family; they were just never able financially to make the jump to ownership before. The supervisor wrote that "I do not believe I have ever seen a couple more appreciative of a chance to own their home," indicating that the Hardens would be willing to cooperate with their local farm and home management agents.[46]

In fact, appearing cooperative and eager to work with the county home and farm supervisors could help the chances of a farmer who might otherwise not be considered for a tenant purchase loan; supervisors hoping to remake tenants into property-owning family farmers tended to favor those applicants who seemed particularly receptive to their efforts. One borrower, Josh Dowdell, was behind with over two thousand dollars in rural rehabilitation loans, but he was still accepted into the tenant purchase program. His county supervisor noted that Dowdell was cooperative and "always willing to accept any new ideas." Despite Dowdell's history of poor farming techniques, the supervisors' report was optimistic that any client who worked hard could succeed with FSA help: "This family will need plenty of good close supervision."[47] In this particular case that approach almost proved disastrous, as it was impossible for the overworked county supervisor to give the necessary supervision, if even that would have been enough to make Dowdell a successful farmer. Five years after his application, the FSA was on the verge of foreclosing on Dowdell, who was getting too old to work the farm alone, and his two children were getting old enough to move.

The regional director wrote to Dowdell directly, putting him on an informal probation, and he narrowly avoided foreclosure.⁴⁸

Along with a willingness to work hard, there were other intangible characteristics that could carry, or help carry, a family into the program, especially for the mother. The desire and ability to create the right kind of home, particularly for one's children, counted highly. One home supervisor, in recommending an applicant, described the wife's being "reared in a home of good moral atmosphere," noting that her father was a minister for twenty-seven years. Similarly, a lack of education was covered by a desire to have had one: this applicant couple had not finished high school, falling a year short, but the supervisor reported that both "regret very much" not finishing and that they made sure their two young sons made good grades in school.⁴⁹

Although they generally recognized a deficiency in schooling as an unavoidable problem for many poor farmers and thus did not hold a lack of formal education against applicants, county home and farm supervisors brought other notions of class and proper behavior for a farmer to how they described and

NAT WILLIAMSON, THE FIRST AFRICAN AMERICAN TO RECEIVE A LOAN UNDER THE TENANT PURCHASE PROGRAM, WITH E. H. ANDERSON, FARM SECURITY ADMINISTRATION OFFICIAL, GUILFORD COUNTY, NORTH CAROLINA. LIBRARY OF CONGRESS.

selected clients. For example, one applicant was noted as spending a higher than usual amount of money on clothing because he always wore boots and breeches—in contrast, apparently, to the usually (and appropriately) barefoot poor southerner.[50] Similarly, middle-class ideas about personal appearance and behavior were evident in how supervisors thought of their clients. One applicant's wife was described as "probably too stout" and "plump," reflecting the home supervisor's conceptions of what a healthy farm wife should look like.[51] Another family was recommended in part because neither parent used tobacco.[52]

This invocation of class ideals was especially common when judging African American applicants. According to FSA instructions, there was officially "no discrimination on nationality, race, creed or political affiliation" in making loans, but black tenants had a worse experience in the program than did white ones, starting with the application process.[53] Black applicants endured indignities common to African Americans in the rest of the South, such as being referred by their first names by county supervisors while whites were generally "Mr. and Mrs. Applicant." The mostly white local supervisors thought of and described their clients in casually racial terms. Eugene Orr, according to his rural rehabilitation supervisor, obviously demonstrated "some of the slave traits as he invariably removes his cap when approaching and is very gracious for being chosen on the Tenant Purchase Program." The home management supervisor described Orr's wife, Josephine, almost as if she were a child, saying that she was "quite fat" but still presentable with her "very neat red and white print dress" and "large sparkling eyes."[54]

African American farmers were also less likely to receive loans. This was not necessarily because of overt racial discrimination, although that was a factor; African Americans' generally impoverished condition meant that fewer met the standards for a loan. A combination of landlord hostility and apathy made it more difficult to make loans even to those black tenants who were eligible. The problem was particularly acute in the early years of the program; in the first year of operation, for example, Region V (in which 44 percent of tenants were African American) provided 124 of its 505 total loans to black tenants (25 percent). Region VI, where 56 percent of tenants were black, gave out 385 loans, only 67 of which went to black tenants (17 percent).[55] The situation did improve slightly over the years. In 1938, in the eight southern states with the largest black populations, 20 percent of borrowers were African Americans; by 1945 it was 23 percent. But considering that African Americans made up an average of 51 percent of tenants in those eight states, the improvement was only marginal, and the FSA never came close to providing loans equally to black and white tenants.[56]

Even when the best black borrowers managed to obtain loans, they often ended up with lower-quality farms. Wallace Burton and his wife, Lou Ellen, described as an "outstanding Negro woman," whose "yard, orchard, chickens, garden, and general surroundings impressed [the local supervisor] as being far above the average for negroes as we know them in Alabama," purchased a farm which had "been very poorly handled and looks ragged since it has been leased for years to negroes with no supervision." Supervisors, looking as usual at the situation optimistically, described the farm as "a real demonstration of the exact conditions as we find them now," which "offers all of us an opportunity to see what we can do about it," but such confidence could not hide the fact that the Burtons had purchased a subpar farm.[57]

Examples of applicants most strongly recommended by home and farm specialists illuminate what a ground-level FSA employee considered to be a good tenant farmer and a good candidate for farm ownership. Dean Hardy of Cobb County, Georgia, was a white thirty-five-year-old tenant farmer, married with one son. The farm he had selected was, in the view of the farm supervisor, perhaps the best in the county: highly cultivated, with little erosion and no need for further terracing. And the new owner was a good match. As a tenant, Hardy was careful about maintaining soil quality, and he made it clear that he planned to continue this practice on the new farm. The good farm plan, mixing cash and subsistence crops, was entirely his plan. Mrs. Hardy gardened extensively and canned an average of five hundred quarts of food a year, having won prizes in the past despite lacking good equipment. She planned to buy a pressure cooker, curtains, and other furnishings for their new home, and she had been saving money for the chance. In addition to these practical concerns, Mrs. Hardy also told her home supervisor about plans for improving the appearance of the house inside and out, should the family become borrowers. Both Hardys were the children of farmers and had lived on farms their entire lives. They were not very well-educated—he with a seventh-grade education, she with an eighth. But, their home supervisor wrote, "They are both very progressive and through associations and reading they have progressed beyond this level." Both were active in community and church organizations, mostly healthy, and had never been more than a few dollars in debt.[58]

Forty-year-old Alabama tenant farmer William Moody and his family similarly received only praise from their local supervisors. He was born and raised in the community in which he wanted to buy a farm, already owned his own stock and tools, and had no debt. He had been a successful tenant farmer, the county farm supervisor wrote, because he was "industrious, intelligent, honest, progressive and ambitious." Moody adopted the live-at-home, diversification goals that FSA supervisors urged: he planned to get out of cotton in order to raise more cattle, hogs, and feed. His three sons and two daughters were expected to help produce large crops on the quality farm he had chosen. Mrs. Moody, an energetic, "bright eyed woman," was highly esteemed by locals, according to the home supervisor, who commended her clean home and how she kept her bright, popular children in school (which reflected the better life that the Moodys wanted for their children). In sum, the Moodys were "a progressive, wide-awake family, ambitious to own and operate a farm."[59]

With help from local supervisors, families like the Moodys and the Hardys decided on which farm they would purchase. The FSA, through the county committees, employed an unusual appraisal method. Rather than using the purchase price of the farm, Tenant Purchase Division agents emphasized a farm's annual earning capacity. This was in part because, in so many cases, a large part of the loan went toward improving land and farm buildings. Simply considering the purchase price of a plot would not include these factors and thus put the new owner at a capital disadvantage. Furthermore, FSA administrators expected the farm to support a family and produce enough income to repay their loans; this was more important than the selling value of the farm.[60] The program also operated under the assumption that it was creating long-term farmers; since regulations forbade the sale of the farm without special permission for a number of years, the selling value of a farm was not considered very important in comparison to its potential productivity.

In providing clients with funds to improve their farms and homes—in many cases the first time that clients had such an opportunity—the tenant purchase program revealed an enormous shortage of decent housing in rural America. In its first three years, about 97 percent of borrowers in the program used their loans to make considerable repairs to their homes or to purchase new ones.[61] Engineers in the field supervised this construction. The division provided sample plans, but local committees could modify them as necessary; the Tenant Purchase Division's only requirement was that the construction be solid.[62] Often a borrower helped with labor or repairs himself, but in most cases professional contractors did the majority of the skilled labor. In this way the FSA prompted a small wave of improvement in the physical structures of the rural South as farm ownership loans went toward some kind of structural improvement in the farm, be it a painted barn, repaired roof, or a new building altogether.

Administrators considered improvements to clients' homes and farms to be important for more than just physical or economic reasons. Better material conditions were obviously important, but more than that, administrators and local agents looked at what such improvements meant for clients as members of their communities and as American citizens. One of the benefits of one client family's improved home, for example, was in what it meant for their social life, particularly for younger members. This was the case for the Martin family. "The children are now at the age to appreciate a home in which to entertain their friends," supervisors reported, and their living conditions needed improvement for them to have friends over without embarrassment. The FSA would make it possible for the Martins to have the kind of social life necessary for any happy citizen: "Mrs. Martin is anxious for some living room furniture, but does not feel able to purchase any at present. We discussed making a couch out of bed springs. This may be done quite easily."[63] Similarly, having an attractive, pleasant home was almost as important as having a productive farm. One home management supervisor even planned to have experts from the local agricultural college help a client landscape her yard.[64]

Another reflection of administrators' expectations is evident in the efforts to subdivide large holdings, especially in the South. FSA leaders agreed with what, they argued, was the consistent finding of any study on American agriculture, that "small farms provide the basis for a richer community life and a greater sum of those values for which America stands, than do industrialized farms of the usual type." Their entire program showed that administrators and planners believed there was no reason to think that a properly run family farm could not compete economically with large-scale agriculture. Further, in parts of the plantation South it was necessary to break up large holdings if any considerable number of tenants and sharecroppers were to become owners. So, to meet both the practical goal of turning tenants into landholders and to eventually reach the long-term goal of shaping agriculture toward a future built on family-type farms, breaking up already established big farms was necessary. As Maris, head of the Tenant Purchase Division, wrote, "Nothing within the realm of practical achievement that is likely to happen would do more to advance the economic, social, and political welfare of a great region of our country than would such a transition."[65]

There were plenty of candidates for subdivision. In counties qualifying as the "plantation South," which included most of the Deep South plus counties in the border and peripheral states, there were about 140,000 multiple-unit farms with over 550,000 subunits (more than three-quarters of which were occupied by sharecroppers) in 1945. A gradual move from the plantation system to a system of efficient family-sized farms was the goal. However, the scarcity of funds (and limitations placed on the FSA by the Tarver Amendment) meant that the tenant purchase program could only make a dent in the plantation system. The difficulty of getting sufficient funds in a single county proved to be a hurdle in purchasing large tracts for subdivision, but by 1948 the FSA and its successor, the Farmers Home Administration, had subdivided 1,231 large farms among 6,285 borrowers, an average of 5.1 farms to each subdivision. Not surprisingly, Mississippi (at 8.5 borrowers per subdivision) and Louisiana (6.4) had the largest plantations for subdivision. Mississippi also had by far the largest number of subdivisions purchased (227), with Georgia (192) and Alabama (151) following.[66]

The tenant purchase program as conceived in 1935 had a much larger fund than the one that was eventually created in 1937. It is possible that while working on a larger scale, it would have been less effective on an individual basis, because the program would have lacked the supervisory elements that the FERA, the RA, and then the FSA developed between the initial proposal in 1935 and its eventual creation in 1937. Critics in 1935 pointed out that just loans were not enough. J. M. Maclachlen wrote that mere ownership was insufficient and that the program needed to address income and spending. "If we merely wave an official wand over any number of tenants, and change their class-name," he argued, "we shall not accomplish a powerful magic. If we leave the basic inability to buy education, good housing, good clothing, proper medical attention and satisfactory recreation, we shall have left tenant families as unfortunate as they were before."[67]

The Bankhead-Jones Farm Tenant Act of 1937 required that borrowers carry out "proper farming practices," as described in "covenants." This portion of the law essentially allowed FSA supervisors to import farm and home plans and management directly into the tenant purchase program. Thus FSA instructions noted, "Even more important than the law itself, however, is the fact that borrowers who plan their operations wisely in advance may expect, in general, to

CHARLIE MCGUIRE, A TENANT PURCHASE CLIENT, WITH HIS MULES AND CULTIVATOR, PIKE COUNTY, ALABAMA. LIBRARY OF CONGRESS.

be more successful than those who do not do so."[68] FSA administrators strongly believed in the success of supervision in the rural rehabilitation program, and they brought that to the tenant purchase program. Even some of the people came directly from rural rehabilitation. As one observer described it, "certain selected farmers 'graduate' to the tenant purchase plan."[69] Being in the rural rehabilitation program often helped a tenant's chances of receiving a tenant purchase loan: if a county farm or home management supervisor knew a family as rural rehabilitation borrowers and found them satisfactory, that increased the chances of getting a good recommendation and being accepted.

Tenant purchase borrowers filled out farm and home plans, similar to but more detailed than those filled out by rural rehabilitation borrowers.[70] They covered what, where, and how crops were to be farmed, how much food would be produced and stored, the kind of garden and/or orchard the family would operate along with its production, what kind of household goods would be purchased, improvements to the home and farm buildings, financial planning (including paying off the loan), and so on. With their loans received and their plans created, then, clients went to work repaying their loans.

The tenant purchase program was fairly innovative in its form of repayments. Given the fluctuations in yearly income, tenant purchase borrowers could choose a variable payment plan, which eventually about half of borrowers used. This allowed clients to use net cash income (which required careful record keeping) to determine how much to pay in comparison to the amount that a client would have been paying under the standard fixed rate of 4.326 percent. A client behind in payments agreed to put all net cash income toward debt repayment, while clients ahead in payments could meet with local FSA agents to decide what to do with extra cash, perhaps to expand production or make extra payments. This plan depended on evident hard work and progress by the borrowers; if the regional director determined that a client was not making satisfactory progress, that client could be moved to a fixed-payment plan. Furthermore, tenant purchase borrowers could, when necessary, defer their first two years of repayment altogether if approved by the regional director. This was generally reserved for cases when a client had some short-term problem that limited income, for example in building up livestock or for a farm that needed a great deal of clearing or drainage before it could begin operation.[71]

Many clients paid off all or part of their loans early. Sometimes economic success was the reason: timber sales, for example, made it possible for a client to make an early payment. World War II also had an impact. Some clients went to work in defense industries or joined the military and paid their loans off early or transferred them to another borrower. Frequently, one or more sons joining the military meant enough pay was sent home to pay off a loan early. Having a child in the military could also close out a loan in more tragic ways: I. B. Nations's son was among those killed at Pearl Harbor, and the insurance money from his death allowed the family to pay off the loan early. Even this case required special permission, as the FSA forbade selling a farm to a family in less than five years to ensure that it met the BJFTA goals of creating a family farm.[72]

Not every borrower left the program as a successful new farm owner. Clients quit for a variety of reasons. It was not always possible to stick to the farm ownership ideal, and farms even fell back into the old landlord system: for example, Lee Clark, who had purchased a farm with BJFTA funds, found a buyer in Birmingham who planned to operate his farm as a landlord, renting it out.[73] Clients also left the program because they felt they could not make a good living on a particular farm. One borrower, considered by the farm- and home-management supervisors as "an ideal family for the tenant purchase program," complained that his new farm was "cold-natured" and conveyed the farm back to the government to pay his debt.[74] Another made sure to tell his county supervisor that he was, in the supervisor's words, "not sore in the least at the FSA Supervisor of the program. He simply became dissatisfied with place due to location."[75]

Some borrowers expressed their displeasure with the aid, or lack of it, they felt they were getting from the FSA by closing their loans. This particularly became a problem during World War II, when shortages and building restrictions slowed the construction of homes and barns. In April 1942 the War Production Board limited construction work to a five-hundred-dollar yearly limit on the construction of or repairs to a dwelling, with an aggregate limit of one thousand dollars for all farm construction. This five-hundred-dollar ceiling was soon reduced to two hundred dollars, though not before the FSA made extensive plans for constructing temporary five-hundred-dollar residences that could be upgraded or altered at a later date. The result was that tenant purchase loans had to be limited almost entirely to those farms with existing buildings that could be cheaply improved.[76] As a result, many borrowers came to believe that not only had the FSA failed to help them make progress but even that it was holding them back. For example, Waymon Blankenship wrote in 1944 that he had "not got a house to live in and every time I say anything about it they say they can't fix it so that is why I am giving it up." With two sons in the military and a good crop growing, he had enough to cover his existing rural rehabilitation loan. Blankenship sold the farm to another buyer, paid off the FSA from his 1945 crop sales, and ended his relationship with the agency.[77]

World War II changed the situation for farmers in other ways. William Gowan started out happy with his farm when he got it in 1940, but in 1945, after making what his FSA supervisor called "big money" in the defense industry, a small truck farm no longer met his expectations. Gowan sold the farm to an outside buyer, using the proceeds to pay off his rural rehabilitation and farm ownership loans.[78] Another borrower secured a job as a painter doing defense work. In this and many other cases, when a borrower prematurely left the program for reasons having nothing to do with the farm, another tenant purchase borrower could be found, and the farm could be transferred.[79]

Whether it resulted from Paul Maris's desire for care in selecting candidates, the effects of farm and home supervision, the program's small size, the generally improving economic conditions, or some combination of all of these and other factors, the tenant purchase program experienced a relatively low rate of failure. By June 30, 1949, the federal government had made about 62,000 tenant purchase loans; 9,372 (about 15 percent) of those had been foreclosed, sold, transferred, or repossessed by the government. Of these, 2,183 resulted from accidents, death, illness, or old age, making it impossible to continue farming. The second-highest cause was a switch in occupation by the borrower, causing 1,680 liquidations. Third was poor management by the operator that supervisors felt could not be remedied by supervision, resulting in 1,302 liquidations. Unhappiness by the borrower, either with the location or the community (872 cases) or the farm (626) was another reason. It required special permission to sell one's farm within five years of entering the program, but selling the farm for a profit led to 598 liquidations. A change in the family because of marital problems (385) or children leaving the farm (286), dissatisfaction with supervision (255), and various other causes rounded out the rest.[80]

The farm ownership program was the most successful on the balance sheets and most politically popular of the FSA's programs; it was also the most cautious and conservative. Its record as a New Deal model for rural life is mixed. Borrowers, generally speaking, succeeded, and they frequently attributed their success at least in part to their farm and home plans and supervision. Similarly, FSA supervisors believed the programs were effective and found the supervision and plans beneficial for their clients.[81] Congress charged the FSA with the task of turning tenants into homeowners, a responsibility the FSA eagerly accepted, and it successfully carried out its duties.

But the compromises, both legislative and ideological, necessary for the tenant purchase program's creation were evident in its results. John H. Bankhead's initial conception of a billion-dollar government program buying land to resell to large numbers of tenants, sharecroppers, and farm laborers shrank to a government-backed loan program for a small fraction of successful (and, relative to most of the rural poor, less needy) tenant farmers in a handful of carefully-chosen counties.[82] The tenant purchase program's success suggests how relatively modest federal antipoverty goals had become as the FSA discarded the most ambitious reform endeavors. The agency no longer seriously tried to recreate the community-building efforts of the early New Deal, much less the back-to-the-land utopian dreams of the early subsistence homesteads program.

The tenant purchase program was a successful New Deal program, in other words, but that success only indicates how low New Deal liberalism had set its sights in its fight against rural poverty and for small farmers. But this does not mean that the FSA in general or the tenant purchase program in particular lacked any liberal impulses or large-scale reform goals. Most notably in the elements of the program imported from rural rehabilitation, like supervision and farm management plans, the tenant purchase program demonstrated the continuing vitality of a reform-oriented mindset even within the limited parameters of the farm ownership program. In fact, the tenant purchase program carried over many elements of its more ambitious predecessors. In its actual operations, the tenant purchase program was a sort of intensified version of the Resettlement Administration's scattered farms program. That project had moved deserving rural families onto new farms throughout a community. The tenant purchase program did not emphasize the community elements of the scattered farms, but it did incorporate other elements of the RA's and FSA's resettlement and rehabilitation programs. Most important among these was the practice of supervised credit. To an extent, FSA supervisors saw their entire agency as a way to bring scientific farming practices to its clients. It was farm and home supervisors who brought the tenant purchase program to the FSA's clients, just as they brought it to rehabilitation and resettlement clients, and most of this interaction came through individual or group supervision. In this way, even the relatively cautious farm ownership program adhered to both the methodology and ideology of New Deal anti-rural poverty efforts.

CHAPTER SEVEN

Rehabilitation in Action
Credit and Supervision

The importance that Farm Security Administration planners put on supervision is evident in a case that involved a farmer running into trouble despite finding success in farming. A Texan named Tom Carter received a $525.70 rehabilitation loan in 1936. He fell afoul of the law by selling hay that had an RA lien, which he did to avoid having to take out another loan, and by paying his landlord out of an improper account; apparently the landlord believed (or claimed to believe) that the rehabilitation supervisor expected Carter to make such a payment. This took place despite the fact that Carter had enough hay and cotton left to pay off his loan entirely. The client, the USDA general counsel noted, "enjoys a good reputation, but . . . he is very ignorant and is unable to read or write, with the exception of signing his name." The rural rehabilitation supervisor considered him "a good and conscientious farmer" who simply did not know what he was allowed to do.[1] The client's case ended up in the USDA's solicitor's office essentially because of his supervisor's failure to properly explain what the farmer could and could not do with the loan and mortgaged crops. Fortunately for everyone involved, since none of the parties concerned knew they were doing anything wrong, no legal action was taken.

By the time the Resettlement Administration (soon to become the Farm Security Administration) moved into the United States Department of Agriculture, programs like land use reform, resettlement, and community building had fallen out of favor with many liberal reformers in the agency who remained concerned with rural poverty. This was partly the result of politics, as resettlement and land use reform were never popular with the FSA's loudest critics, but it was also because FSA administrators increasingly emphasized rural rehabilitation, especially supervised credit, as the solution to rural poverty. Rural rehabilitation of individual farm families was an important part of the federal anti-rural poverty program from its beginning; by 1937, it had become the foundation of the FSA's work. Farm and home management plans and their accompanying loans were, in turn, the center of rural rehabilitation. FSA administrators operated as if the entire purpose of the FSA was, at least to a degree, farm and home supervision. As FSA instructions told employees, the "purpose of extending credit is to make possible the carrying out of approved practices developed in the Farm and Home Management Plans."[2] Grants and loans were, in the eyes of administrators, only a means to the real end: retraining farm families in improved farm and home management. The assistance and supervision that the FSA supplied via home and farm management plans was, administrators noted, "by far the most important part of its aid to rural needy farm families."[3]

FSA leaders imagined local agents creating a new kind of rural citizen, or, more accurately, restoring rural citizenship. Looking back to a (perhaps imaginary) time when the small farmer could operate on an equal basis with the largest farmers, when small farmers were valued and participating members of their communities and country, farm security leaders sought to renew those elements of rural citizenship by empowering small farmers. As mechanization and factory farming became more important parts of American agriculture, the FSA (like its predecessors) operated on the idea that the small farmer, with government aid via a program of rural rehabilitation, could recover and continue in that idealized agrarian position.

The actual work of rural rehabilitation was carried out by trained farm and home management supervisors. By 1941 the FSA operated 2,270 county offices, with 4,178 rural rehabilitation supervisors and 2,586 home management supervisors for approximately eight hundred thousand rural rehabilitation client families. Reflecting the program's southern roots and the great deal of need in the South, fourteen southern states accounted for 51 percent of the county offices, 61 percent of the rural rehabilitation supervisors, and 66 percent of the home management supervisors.[4]

County rural rehabilitation officers filled a wide variety of roles. Generally men, they handled the farm management aspect of rural rehabilitation. Trained home economists, many of whom were women, handled the home management aspects of rehabilitation. Extension services had occasionally used home economists, but like the male agricultural demonstration specialists, they generally focused on large-scale planters. For the first time, on a large scale, professionally trained specialists were assisting lower-class farmers with their crops and livestock, as well as their homes, budgets, and families. New Deal programs, including the FSA, tended to rely more on male agents than on female home demonstration agents. But no other major USDA program had ever put so much emphasis on home management or production for home use, nor had one ever used professionally trained women to such an extent as the FSA's rural rehabilitation effort did.[5]

Although the FSA's alleged big-spending ways were the source of criticism, FSA field employees did not actually make more money than people working in similar jobs. In the FSA's prewar years, for example, rural rehabilitation supervisors in every state but California made between $1,620 and $2,000 a year. The incomes of home management supervisors varied a bit more among the different regions of the country, but $1,800 was the highest salary in any state. Assistants made less. In comparison, county extension agents were much better paid. There was much greater variance (the Extension Services being state organizations), but nationally, states on average paid extension agents between $2,011.62 and $3,678.42 in salary. The low salary for an Alabama state extension agent, $1,920, was only $80 below the high salary for his FSA counterpart, and it was $120 more than a home management specialist made. No state extension agent started at a lower income than an assistant rural rehabilitation specialist, and no state had a highest-paid county agent who made less than the highest-paid FSA county agent.[6]

The FSA tried to ensure that local agents had help as necessary. The Rehabilitation Division in Washington, for example, had sections for farm and home management, debt adjustment, tenure improvement, and cooperative services, each staffed by specialists in that area. Most regional offices had a similar collection of experts, and most state offices had a specialist in home economics, debt adjustment, and farm management. Local committees also provided assistance. In every county and state in which the FSA operated, there was at least one corresponding county or state committee. These committees brought local farmers, farm organization leaders, extension service employees, and landlords into a formal relationship, at least in an advisory role, with the FSA. Members of most of committees served three years, with compensation for travel and other expenses.[7]

The state committees (limited to only six days of compensated work) were relatively unimportant in the function of the FSA, but the county committees played a crucial role in the agency's work. Members were chosen carefully: "sympathy with FSA objectives," one set of instructions explained, "is the primary qualification for service on FSA councils and committees." County committee members advised and helped FSA personnel interpret the agency to the community, explaining its methods and objectives to local residents, lending institutions and businesses, and others. In many counties a single committee did the jobs of both rehabilitation and tenant purchase. County committees also played a general role as an outside viewpoint. Overworked county agents, administrators felt, sometimes lacked the time or ability to consider their overall work. County committeemen could provide a different perspective on the local situation. They could, for example, recommend to local supervisors a family that had for whatever reason been overlooked or help supervisors get more accurate information about potential clients.[8]

The FSA's county rural rehabilitation agent system was similar, at least on the surface, to the system used by the Extension Service. Extension agents, however, dealt with a relatively stable and well-off group of farmers; thus, their focus was almost entirely agricultural. Dealing with similar but poorer clients, the FSA wanted to hire people with agricultural expertise but also with the right kind of poverty-oriented mindset. Although rural sociology and agricultural economics were rapidly expanding fields even before the Great Depression gave them new urgency, land-grant colleges and universities were understaffed in terms of quantity and quality of such rural experts in the 1930s and 1940s. Particularly in those areas hardest hit by the Depression—the South and the Great Plains—agricultural colleges and experiment stations simply could not keep up with the demand for trained workers. The deep-seated social and economic problems that required trained experts (especially in the South) meant that those regions lacked the resources to improve their research, teaching, and agricultural extension services. That deans and administrators were largely unsympathetic at least to certain aspects of rural sociology and agricultural economics only exacerbated the problem. Responding to the shortage of trained workers and because of its different view of farming and rural poverty, FSA employees had a somewhat different background than did employees of other agricultural agencies. The vast majority of employees in other USDA agencies came from agricultural or scientific backgrounds, but a large number of the FSA's employees were nonscientists and nonprofessionals. FSA field agents in particular came from the rural social sciences, in some states absorbing the majority of college graduates in those fields in a given year.[9]

The FSA put a great deal of emphasis on personnel training. Part of this was because of simple necessity. Especially in the South, many field agents came from state land-grant colleges and the Extension Services. These institutions did not, for example, emphasize a farm management approach, instead focusing on efficiency and increased production. Thus, just-hired field agents had, in the words of RA farm management specialist Bernard Joy, a "comparative unfamiliarity" with the farm management approach. But agency leaders also believed that, as one guide put it, not even a "technical or administrative employee can possibly perform his or her duties adequately without understanding the main problem with which this agency is coping, and how it is attempting to solve this problem."[10] This is why the first qualification for selecting county supervisors, as a 1938 agency manual put it, was a "broad social viewpoint and a sympathetic attitude toward low-income farmers."¹¹ Additionally, because the vast majority of the FSA's work was done by agents in the field, and because that work absolutely required independent judgment, administrators wanted to ensure that agents understood the conditions of rural poverty and the nature of their jobs entirely.[12]

Increasingly, FSA administrators came to believe that the agricultural and economic expertise of their supervisors was secondary. Especially in dealing with poorer clients, the supervisors who had the most success were those with certain personal characteristics: interpretative ability, humility, patience, and confidence in the ability of clients to rehabilitate. Getting clients rehabilitated took something more than mastery of agricultural techniques, many of which did not apply to the relatively small-scale operations run by FSA clients. In this sense, agents were much closer to social workers than they were to extension agents. Furthermore, FSA administrators believed that their clients were as socially isolated as they were economically disadvantaged and hoped that county supervisors might help eliminate rural isolation and alienation by connecting clients to the community. This required personal warmth and tact as much as agricultural knowledge. Supervisors had to consult with local leaders and representatives of the Extension Services, officials of other government agencies, the 4-H Club, and school officials to build community ties for their clients without harming the relationship with clients who were often (with good reason) hostile to local institutions and leaders. Navigating such situations required more than knowledge of the latest farming techniques.[13]

The training that new FSA employees underwent reflects the perceived need for an understanding of rural poverty, practical farming skills, and a willingness to consider larger social questions. By 1941 new employees were going through a two-week training course. Each day began with a lecture on the background of various FSA programs or projects: the FSA's legislative and administrative history, the causes and characteristics of rural poverty, the objectives of various programs, and similar topics. A second lecture handled more practical elements, like explaining the work and role of county committees. The third training period explained the most important FSA instructions and particular aspects of related programs. But training did not consist only of passively listening to lecturers: it also included group discussions about the nature of the FSA's work, such as what it meant to run a family farm, whether or not self-sufficiency was possible and desirable, or what constituted a satisfactory level of rural living. Trainees then tried it themselves in what the FSA called laboratory exercises. They set up model county offices, spoke with clients, made home visits, ran group meetings, and so on.[14]

The FSA also frequently reevaluated its approach, its national leaders questioning their own program and encouraging employees to have a similarly inquiring attitude. The 1940 conference in Washington, for example, saw group discussions around questions like these: "Have we been able to help all racial groups uniformly relative to their need? What do we mean by 'need'?"; "What do we mean by realistic planning and supervision?"; and "What is the relation between the understanding with the client at the time of his acceptance and the subsequent relations with him?" Results from conferences in Washington tended to filter down into regional or district conferences, and then to local employees.[15]

It appears that FSA leaders successfully instilled their beliefs about rural life and poverty into employees. At least some FSA agents and supervisors genuinely believed that it was only because of a lack of opportunity that clients had failed to succeed economically. More cynical observers referred to the FSA as the "Order of the Bleeding Heart."[16] FSA administrators also generally succeeded, as much as was possible, in making sure that their agents had the necessary technical and agricultural background. One 1938 report found that about two-thirds of FSA regions had maintained high standards in hiring qualified local personnel; the problem in the regions failing to meet standards was that the local educational and training facilities simply made it impossible to hire trained personnel.[17]

For most client families, the FSA was the county office: that was where they first met local FSA agents, learned about FSA programs, applied for loans or grants, and went for meetings and evaluations. For this reason, FSA officials tried to make sure that the county office would create a positive impression. Ideally, it was a place that was welcoming and useful. This started with the office's appearance. Planners intended the reception space, for example, to demonstrate concern for applicants and clients. FSA manuals urged that walls not be bare unless required by local rules or ordinances (as in the case, say, of an FSA office located within a post office). Posters and charts were advisable if they were educational and improved appearance. Administrators also recommended a publication rack with FSA pamphlets and USDA bulletins. Home supervisors used wall space to demonstrate techniques like curtaining or making clothes. To give the office a more personal flavor, training materials suggested that when possible the office should put up pictures of successful local families for encouragement.[18]

Before World War II and budget cuts forced reductions, a typical county office had a county supervisor, home supervisor, at least one assistant rural rehabilitation supervisor, and two clerk-typists. The very smallest county offices had at least a county supervisor, a home management supervisor (who might split time with another county), and a clerk or typist. Even the clerk-typist was expected to understand rural poverty and the FSA programs, since clerks needed to understand office transactions and the relevant government regulations, and

INTERVIEWING APPLICANTS FOR HOMESTEADS AT RESETTLEMENT OFFICE IN BIRMINGHAM, ALABAMA. LIBRARY OF CONGRESS.

FSA officials urged county agents to encourage their office staff to make suggestions about home visits and meetings. Given the large number of applicants and clients, the thinking went, it was important to have as much help as possible.[19]

Local supervisors handled almost every element of the client's relationship with the FSA. County agents received applications for loans, grants, and services, and in turn they investigated applicants, consulting with the advisory committee and others in the community. They helped with debt and tenure adjustment, either directly or by aiding and overseeing local committees. Local supervisors helped families create home and farm management plans and stick to the plans or adjust them as necessary after implementation. They encouraged leadership and community among client families. They recommended, when necessary, action on defaults and foreclosures. As one FSA pamphlet for new employees described it, "As far as the community is concerned, the County Farm and Home Supervisors *are* the Farm Security Administration."[20]

FSA officials considered the first meeting between applicants and agents the most important. While filling out an application, the local agent learned about the family, its health, and its relationship with the community. The applicant, administrators believed, subconsciously made up his mind in a way that might help or hinder rehabilitation. At this moment, the interviewer represented the FSA that the applicant was asking for help, and administrators urged agents to present themselves as employees of a federal government gravely concerned about the wellbeing of that particular family. In the first few meetings, a farmer seeking a rural rehabilitation loan or grant had to go over his situation with a county supervisor: the size of the farm, its fertility, what the farmer already owned, existing debt, and similar issues. The county supervisor was to double-check all of this with a visit to the farm. The farmer's wife went through a similar process with the home management supervisor. County agents then had to determine whether the family could become self-supporting with FSA help. County agents made the initial decision, usually consulting with an advisory committee made up of local farmers. Families who did not qualify for rural rehabilitation loans might still use other FSA services. Farmers with too much debt, for example, could be pointed toward the debt adjustment committees.[21]

In contrast to the tenant purchase program, the Rural Rehabilitation Division had much lighter and more varied restrictions on family selection. If a family was unable to get financing from any other source, could get local recommendations as an honest and hardworking group, and either owned or rented a farm capable of making a living, then that family was eligible for a rehabilitation loan. In the earlier years, rural rehabilitation and resettlement tended to select the poorest families, with plenty of variation by local project; in fact, on some projects lower income was the chief basis for selection. As the projects went on, this emphasis on lower income declined, and selections came to emphasize more traditional notions of farmer experience, family health, and apparent capacity for economic advance.[22] Despite all the effort to approve only the most capable candidates, even successful applicants were often a long way from running their own farms. In the words of one administrator, many rural rehabilitation clients lacked "the qualities of character and intelligence to successfully carry out a farming enterprise on their own."[23]

Most clients took a standard ten-year rehabilitation loan. These loans could help a client renting land to change types of agriculture—for example, shifting from cash-crop farming to a diversified livestock and grazing farm. They could go toward the purchase of purebred and high-grade livestock. Or they could be used for improving agricultural land—for example, in the case of a farmer moving to a new farm and getting farm operations started. The FSA made five-year loans for smaller purchases, like the purchase of work stock or farm machinery; even smaller two-year loans were available for seasonal items, tools, and other minor expenses.[24]

Until 1940 the authority to select rural rehabilitation loans was officially assigned to the regional offices. It was on paper a complex thirteen-step process (with many of those having substeps) to apply for a standard rehabilitation loan. Loan renewal submissions and grant applications were a bit simpler, with only nine steps. A standard rural rehabilitation loan for forty-five hundred dollars had to meet the approval of at least six different officials in the regional office alone. FSA instructions provided flowcharts to simplify this process by creating a visual reference, but for the more complicated applications, these charts took on an astrological or alchemical appearance, with numerous, sometimes mysteriously named orbs floating around one another while connected by a series of minutely labeled, overlapping arrows.[25] After an applicant received approval from the district supervisor, the docket went to the regional loan advisors and finance and control units. After regional approval, a voucher went to the U.S. Treasury for payment.

From the beginning, actual authority to approve loans fell to the district and county supervisors. The FSA received an enormous number of applications, many of which were fairly straightforward, such as the same client requesting a follow-up loan as recommended by his supervisor. Because of the administrative backlog and because agency leaders believed that local agents had a better view of the situation, regional directors informally delegated approval to district and even county supervisors.[26] This was the case for years; one RA official claimed in 1936 that the state loan officer's function had become one of "glorified checking."[27] In 1940 the district supervisor's role in approving certain loans became official and, in fact, their primary concern. Many FSA employees wanted to go even further, to have the district and county supervisors work together to approve rehabilitation loans in the county office. By May 1943 C. B. Baldwin estimated that 85 percent of rural rehabilitation loans were less than two hundred dollars and therefore approved by county supervisors.[28]

Most of the interaction between the client family and FSA county agents involved loans or grants and, more specifically, the supervision that came with them. Pragmatically, clients needed training in managing financial affairs, home production, and superior farming techniques if they were going to be able to make payments.[29] Direct loans and grants reinforced the agency's ability to influence and supervise clients. Observers found that among rural rehabilitation clients who continued to use the supply merchants, for example, the FSA had less influence than it did with families who used government credit.[30]

There were far more applicants for loans than there were available funds, and only a fraction of potentially eligible farmers applied for loans, jobs, and relief. Choosing the "ideal candidate" from this enormous pool proved difficult for FSA administrators. The lack of legislative directive made the situation muddier. Before Congress took a more active and intrusive role in FSA activity in 1942 and 1943, the only clear rules were that rural rehabilitation loans would go toward families who could not obtain adequate credit otherwise. What exactly "adequate credit" meant was unclear. So long as FSA loans did not directly compete with private banks or similar lenders (who generally made loans to relatively well-off farmers quite above the FSA's consideration), eligibility requirements were up to FSA administrators.[31]

In particular, FSA administrators had trouble deciding whether they should help the neediest farm families or those farm families most likely to succeed with rehabilitation. FSA leaders believed that the agency existed to help low-income farm families. Giving top priority to the poorest farmers would mean small loans, often replaced with grants, and intensive supervision. However, many administrators, particularly local agents, wanted to help those low-income farm families who could most easily attain economic success. FSA officials did not see their program as a relief program or as intended for the helpless and the hopeless, whom they had neither the time nor money to help in large numbers anyway. And many supervisors feared that loans might only be another burden on the rural poor. To many within and outside the FSA, it seemed best to address the tens of thousands of farmers in the United States who found themselves operating more or less successful farms but were still mired in old debts and in short-term distress. Focusing on these families would have had its advantages—improving the agency's collection record, for example. But this would mean that the FSA ignored those who might not be good traditional credit risks but who had some chance of making repayments on much-needed loans. It would, in other words, fly directly in the face of the purposes of rural rehabilitation and the FSA.[32]

The solution landed in a shifting range somewhere between helping only the neediest or the best-off. Starting in 1940, FSA leaders were confident enough in the training of their field personnel to delegate increasing responsibility to district and even county supervisors to make loan decisions. In the early years, when many clients came from FERA and directly from relief rolls or just barely off them, policy tended toward the relief direction. By the time the FSA had begun to delegate authority to local control in 1940, operating procedures leaned slightly but not decisively toward relatively better-off farmers. From the beginning, rehabilitation programs provided loans and supervision to a wide number of families with varied incomes, quality of land, and management ability. By the time that the RA came into being as a national organization, because of the wide variety of potential clients, regional variation, and agent choice, federal farm security efforts did not focus on a single kind of applicant until 1942.[33]

Second, with the outbreak of World War II, the FSA began to focus on applicants who could rapidly expand food and fiber production. This meant emphasis on relatively large growers (at least, among the low-income class). During the war, the political climate also forced FSA supervisors, starting in 1943, to limit to five years the amount of time a client would have to repay his loan or be clearly well on the way to doing so. Families that did not show signs of progress might be dropped from the lists of active borrowers. Supervisors thus selected the better-off farmers, who were able to repay their loans within that five-year window.³⁴

Evidence of this shift is found in the changing size of the average standard rural rehabilitation loan (Table 2). Between 1936 and 1944,

TABLE 2: RURAL REHABILITATION AVERAGE LOAN SIZE, FISCAL YEAR ENDING JUNE 30

Year	Original loan amount	Supplemental loan
1936	$240	$67
1937	$487	$166
1938	$592	$216
1939	$629	$246
1940	$612	$218
1941	$658	$194
1942	$631	$232
1943	$704	$285
1944	$1,007	$316

Source: Olaf F. Larson, *Ten Years of Rural Rehabilitation* (1947), 168.

average original-loan sizes increased in all but two years, and those showed only a slight decrease. The improving economic situation in World War II, which meant bigger farms and bigger loans, doubtless played a role in such a large increase in 1944, but political pressure also influenced the shift. Before that, the tendency between 1938 and 1942 (after the relief-oriented approach had ended but before orienting to war production) was to make loans between $592 and $658, with supplemental loans in the range of $194–$246. It was in this period that the rural rehabilitation program operated closest to what its administrators wanted and when the process of supervised credit was most important.

~:~

FSA administrators believed that supervision had measurable results. Will Alexander estimated that supervision of tenant purchase clients improved their farm operations by somewhere between a quarter and a third. C. B. Baldwin asserted that the FSA's successes in rehabilitation and farm ownership were "due to the supervision to a larger extent than to anything else." Baldwin also recommended more. "I don't believe," he said, "that supervision has ever been as close as it should have been." R. W. Hudgens estimated that if the FSA did its job of educating and supervising the rural poor correctly, eventually 90 percent of its budget would go toward administrative costs and overhead and only 10 percent for actual loans.³⁵

To succeed, supervision had to fit the family. FSA leaders insisted that county agents take the ideas and desires of the family involved in making home and farm plans: "No plan should ever be prepared for any family," an FSA handbook explained, "it should always be prepared with the family."³⁶ Or, as regional farm management supervisor Sam McMillan told a meeting of forty-five FSA agents in Jacksboro, Texas, "supervision does not mean bossing. It means leadership or a collective education program."[37] Ideally, plans provided a foundation for the relationship between client and agent. The farm and home management plans also represented, in the words of one administrator, a "set of intentions" and a "a process of thinking," tangible proof that a client family had evaluated the situation, thought about what their goals should be, and determined with the help of the county agents what was needed to reach their new goals. Local agents put a great deal of emphasis on the farm and home plans; however, the assistance of local supervisors proved more important in improving borrowers' farming. Supervisors put more emphasis on the home and farm plans than did borrowers, who preferred individual supervision.[38]

In preparing a home management plan, the home supervisor set forward the needs of the family for food, clothing, consumer and household goods, shelter, education, health, and so on. The home supervisors, ideally, should set standards for how much food a family needed to produce and preserve, teach the family how to conserve, preserve, and prepare food, and guide them toward better facilities for serving food. She also was to teach mothers how to encourage the rest of the family to adopt proper habits. Like farm supervisors, home supervisors used both individual meetings and group techniques. They also addressed the types of household furnishings, clothes, and other possessions a family had, touching on questions such as whether a barely furnished, unattractive house was a rehabilitation problem. Home supervisors taught families their ideals of convenience, sanitation, and beauty. One home economist summed it up: "It is not enough to arouse within a family a desire for a better and more satisfying home life. The home economist must go further. She must teach such a family how to attain a better standard, and this means constant supervision, guidance, patience, understanding."[39]

Home management supervisors thus had a more difficult job than did farm supervisors, at least in terms of defining and judging client success. A farm supervisor, in making decisions about which family to select for a resettlement project or which applicants to provide with loans, had objective, if difficult to measure, standards: he could look at the soil, the local market, and so on to determine whether or not a project or client had a good chance of success. Home supervisors, on the other hand, had only the vague idea of what the ideal family looked like. The FSA provided some criteria regarding health and similar issues, but in many cases supervisors were on their own.[40]

Creating a home and farm management plan required a good bit of work. Supervisors inventoried everything from how much food a family was expected to grow to the money necessary for toiletries and underwear. Farm management specialists had to map the farm, including acreage, soil types, drainage, and other pertinent information. They also had to help the client family create a farm budget, comparing the farm's likely income against expenses. Home management specialists had to plan a home budget, determining how much food would be produced and purchased, in addition to expected costs for fuel, clothing, and similar expenses. The FSA provided a variety of forms (like "Budget for Food and Fuel Furnished by Farm," "Farm Family Living Budget," and "Seasonal Distribution of Fresh and Stored Farm-Produced Foods"), but they could not eliminate the burden of gathering all that information.[41]

As much as possible, FSA officials urged a shift away from producing cash crops and toward production of food for home consumption, particularly for the poorest families. A dollar of food produced at home, they asserted, went further than a dollar bought at the store. Under good conditions, agents believed that a family should be able to produce 75 percent of their annual food needs. FSA officials recognized the cost of producing food at home (it required land, capital, effort, and risk), but they believed that careful planning could meet dietary needs at a low cost. Planners estimated that, if a family produced between half and 90 percent of its food at home (along with the cost of forty to fifty dollars for staples that had to be purchased), they could save from two hundred to four hundred dollars' worth of food annually. FSA administrators took this requirement very seriously, to the point that, for example, they sometimes dropped rural rehabilitation clients who refused to plant subsistence crops as well as cash crops.[42]

The FSA succeeded in improving food production among clients. Families active in the program in 1941 increased the production of fruits and vegetables preserved by 114 percent, or 158 quarts per family, and almost doubled production of meat and poultry for home use.[43] In Greene County, Georgia, for example, supervisors required each FSA family to have a garden and asked them to plant at least one new kind of vegetable each year, with meetings to demonstrate and share recipes. In this way, clients came to develop a taste for white potatoes, carrots, beets, and other vegetables. Canning food was a crucial part of the program. Turnip greens, green peas, and sweet potatoes were particularly popular for canning; farm women apparently thought that these did the best job of giving strength to working men. The FSA also made loans for family-sized pressure cookers, commonly called "precious cookers." Supervisors demonstrated that almost everything could be canned: squash, okra, sweet corn, soups, and even veal and beef. The FSA also provided loans for cows and chickens to improve the amount of milk, eggs, and meat families produced at home.[44]

FSA administrators wanted poor farmers to produce far more for home consumption because most families produced so little. By the 1880s and 1890s, most southern farmers found that growing for self-sufficiency was becoming practically impossible to manage economically.[45] Changes in the early twentieth century—everything from higher taxes and changing legal structures to the need to buy new technology like automobiles, gasoline pumps, and tractors—increased demands that farmers grow cash crops for profit at the expense of diversification. The farm of the average family approved in 1937, for example, in the year before approval had produced (per person) about 27 gallons of milk, 35 pounds of meat, 14 dozen eggs, and 22 quarts of fruits of vegetables canned, compared to FSA goals of 90 gallons, 100 to 125 pounds, 20 to 25 dozen, and 80 quarts, respectively.[46] Even worse, the families that needed to preserve the most food tended to do so the least, especially in the South. In Gorgas, Alabama, in the mid-1930s, one study found that the average family among large landowners canned 770 quarts of food and dried 25 gallons; sharecropper families, in contrast, averaged 218 quarts canned and 7 gallons dried.[47] Starting so low, the increases could appear astounding: clients in Laurens County, Georgia, increased their dried bean storage by 900 percent, and to the north, Oglethorpe County clients stored five times as much homemade syrup.[48]

Supervisors argued that whatever could be produced at home should be. FSA guides pointed out, for example, that infant beds could be constructed from salvage materials at almost no cost and that adolescents should be encouraged to build furniture for both themselves and the family. Even tools, if possible, should be handmade: an FSA report proudly pointed to clients in Laurens County, Georgia, making hoe handles out of hickory saplings and building wagons using out-of-date anvil forges and scavenged materials. Ideally, clients would only plan to buy things that could not be made at home, like "customary expenditures for tobacco, barber, and toilet supplies, such as toilet soap, toothbrushes, combs, and razor blades" in addition to medical costs and expenditures for fuel, lighting, and other household items.[49]

For all the FSA's successes in raising home production among clients, this attitude about home production and consumption went strongly against the prevailing trends of farm life. Small farmers had already begun to join the commercial economy. When farmers could afford to do so, they discarded the live-at-home mindset. Even some small farmers by the 1930s did not take their grain to the mill for grinding; they sold it and bought bags of flour from the local store. Few had orchards, which were difficult to maintain and whose crops (especially frost-vulnerable plants like peaches) were easily destroyed. Instead, they bought truck-shipped fruit from other parts of the country. Few still made their own vinegar. Most farmers kept chickens but rarely hatched their own; most of their flock were sold as fryers in town, not consumed on the farm. The crops most likely to be produced at home were those easiest to grow or impossible to obtain before the widespread introduction of electricity and refrigeration. Other sources of food were rare; hunting, for example, was more a hobby than a reliable

A HOMESTEADS CLIENT WITH SOME BABY CHICKS, BANKHEAD FARMS, ALABAMA. LIBRARY OF CONGRESS.

food source, especially considering that by the 1930s most large game had been hunted out (wild populations of animals like deer rebounded only well into the second half of the twentieth century).[50]

The home-production message was also muddied by the FSA, which was not consistent about the value of the live-at-home mindset because farm and home supervisors carried a certain middle-class perspective that valued store-bought consumer goods. This mindset is apparent in the Resettlement Division's furniture catalogs, which promised to ease the transition for new home-owners by helping them choose their décor using standard sets. For living rooms, clients could apply for easy chairs, coffee tables, and day beds in "Amber Maple" or "Brown Oak" that promised to match "any other model of living-room furniture shown." Clients could choose from a variety of dining chairs, armchairs, dining tables, and sideboards. White oak rockers and wicker side chairs were available to fill out the spacious porches intended for the resettlement projects. Single- and double-sized beds, chests, and dressers were available for the bedroom, helpfully labeled to match the other bedroom furniture available.[51]

So while the FSA emphasized the importance of home production, administrators recognized that it was neither possible nor desirable to adhere to an entirely live-at-home program. County agents were instructed, for example, to plan for education, both formal (books, paper, and other supplies for children in school) and informal (such as the goal that every farm family should subscribe or have access to at least one newspaper and one farm magazine). Supervisors also had to keep in mind recreation and socialization. Church and charitable contributions had to be planned. More generally, FSA leaders saw social isolation as one of the greatest dangers of rural poverty, and an entirely live-at-home program would make it difficult for farmers to be part of their community. Agents, then, were to emphasize home planning and production but to "be very careful to recognize its limitations."[52] As with their general perception of the place of the family farm in the new agricultural era, administrators hoped that clients would be able to get the savings of home production while still enjoying the benefits of interaction with the larger economy.

A large part of supervision was simply explaining basic budgeting and record keeping to lower-income farmers; a client family had to plan their income for the year or years ahead, for perhaps the first time, when making farm and home plans. Supervisors considered record keeping essential. It allowed a family to see their strong and weak points. To encourage good record keeping, the FSA provided farm record books. It recommended that families list everything they owned—their current real estate and machinery, food on hand, and so forth. Then the family could list their debts and so determine their net worth (helpfully defined as "what you owe" and "what you own"). As the year progressed, the family was encouraged to keep a record of all receipts from farm product sales, loans, and a record of all money spent for farm and family operations, as well as records of how much, and what kinds, of crop and livestock had been produced. The record books also included space to summarize the family's yearly progress, with forms for keeping up with open accounts, monthly cash balances, stored food, maps of the farm, and more. It also contained information useful in daily farm record keeping or to encourage the right kinds of behavior by client families—lists of how many quarts of fruit could be canned per bushel, for example, or standard weights per bushel of farm products.[53]

In many cases, the wife or an older child (often an older daughter) kept the records. One visitor to Coffee County, Alabama, found that every account book inspected was kept by the wife or one of the older girls in the family.[54] For the first time, farming became a business venture for many of these families. Some even went so far as to build their own business centers for their record books and other important documents, made from easily obtained materials. Orange crates were common: they were turned into shelves and partitions, then painted. A plywood cover—or, for a poorer family, a dyed salt or flour sack—acted as a door.[55]

Supervisors often had difficulty imparting the value of record keeping to clients. Many low-income farmers had only the vaguest idea of their income and expenditures, and they had little experience keeping track of such things. One client, when asked to show his records, brought out a hanging wire basket full of partially read letters, contracts, and other mailings sent from the federal government (separated from the other mail by the fact that they had been sent without stamps). For barely educated farmers, any change could cause problems: some supervisors found that just altering the placement of items on the page or using a differently colored book required again convincing the family of its value.[56]

Maintaining a good relationship meant a steady level of contact; county agents found that clients who, as one county agent put it, had "not had benefit of as close supervision as they should have had" were much more likely to complain about their treatment by the FSA. To ensure that home and farm plans were being understood and followed, county and home agents took (ideally) at least monthly trips to visit their client families. While the most frequent and important visits were designed to develop and stick to home and farm management plans, other visits investigated for loans, grants, medical needs, and releases, or to plan cooperatives, farm debt adjustment, or medical group services.[57]

On visits, county agents were supposed to both praise clients for the work they had done and encourage them to do better. Agents and clients filled out a farm visit report listing the client's responsibilities, what the client had done well, and what the client needed to do better. One typical report noted that the client had done a good job of preserving food and pointed to such successes to encourage better practices in the future. The client needed, the agent said, to give more attention to the "fall garden" and to "save seed and all the feed you can." These reports emphasized both the farm and family: one client was commended for meeting the agent's recommendations to keep "home and yard clean and attractive."[58]

Given the demands on their time, it was not always possible for agents to meet with families individually. Monthly group meetings had a number of virtues. They saved time, and they helped skeptical farmers accept new things by showing that others were also doing something new. Supervisors would remind clients of particular needs for the month—make sure to fix fences in January, plant gardens in March, and so on—and give demonstrations of skills like preparing and canning meat, making peanut butter at home, or properly balancing a nutritious meal. Group participation, administrators believed, also promoted community feeling, perhaps even encouraging a family to join a cooperative group enterprise, and made it more likely that farmers would share experiences and recommendations among themselves.[59]

∾⋮∾

From the start, the supervised-credit program had its problems. Its emphasis on a scientific approach to farming conflicted with clients' traditional knowledge of farming. The FSA's conception of rural relationships differed from that of its clients; administrators thought mostly in terms of individual families while tenants relied on social, economic, and kin networks that government supervision could not replace. The FSA wanted clients to diversify and plant food crops when they (and their landlords) wanted or needed to plant cotton.[60] Many southern farmers were still committed to the idea of cotton as the key to prosperity. As James Edward Rice wrote in 1931, southerners would "dream of selling their 'white gold' at twenty or twenty-five cents a pound. Then, they can buy plenty of food and clothes."[61]

These sorts of big-picture issues were a problem in terms of planning and creating a system of supervised credit, but a bigger headache for administrators was implementing the program and making sure that FSA employees carried it out. Both supervisors and borrowers, Will Alexander noted in 1940, failed in some degree to understand the meaning and purpose of the farm and home plans. Administrators even toyed with more rudimentary forms of teaching and rewards. C. B. Baldwin thought that the FSA might provide a narrative to each client as an example of what was to be done—the local agent interpreting the plan to the client by describing the number of hogs that needed to be vaccinated or the amount of corn to be planted. All administrators agreed that it was necessary to move the farm and home plans beyond being seen as just another "document" by the clients and instead understood as a better way of farm management and rural production.[62]

Getting a family to fill out plans was a difficult and time-consuming task (one version had seventy-eight separate lines of information to be entered in the summary alone), not to mention the struggles of imparting to clients the ideal that the plan was more than just a budget.[63] Clients often had little education, and agents sometimes found it easier to simply tell them what to do. FSA leaders had trouble getting local administrators to recognize the larger purposes of the plan and then in getting them to undertake the difficult task of teaching those purposes to clients. Increased and improved training could overcome some of those problems, but it could not remedy a shortage of agents. Because of the caseload, it was not even always possible to create a plan each year for rural rehabilitation clients, although these families still received supervision from some form of in-person contact with county agents.[64]

In fact, some supervisors saw the creation of farm and home plans as an obstacle to rehabilitation. If a supervisor believed a loan to be sound and necessary, the need to prove it with a farm and home plan felt like unnecessary red tape. It was no surprise, then, that their clients did not appreciate the plans after being told they were an unnecessary requirement of the national office.[65] The results were rarely ideal. Robert Hudgens visited a farmer who had been dropped as a client. The county supervisor who had made the client's budget had realized it was short fifty dollars for the year and so had included a line of income for selling "sorghum syrup" to make up the difference. The client had been short of money, sold a bale of cotton to buy food even though the crop had a government mortgage, and was dropped for not repaying his loan. In this case, at least, it provided something of a learning experience: Hudgens reported that the supervisor came to understand the relationship between a grant and a loan, and between success and the farm management plans.[66]

For a variety of reasons, many families failed to complete a farm and home plan. This was more often the case for rural rehabilitation borrowers, who got relatively less supervision than did resettlement or tenant purchase clients. For all the FSA leadership's emphasis on supervision, in July 1941 only slightly more than half of rural rehabilitation clients had a complete farm and home plan, although this does not include partial plans, and the number rose through the early 1940s.[67] Even for the tenant purchase program, which had much more intensive supervision, a 1945 study found that 46 percent of midwesterners and a mere 14 percent of southerners had an "adequate knowledge of details" of the farm and home plans; only 57 percent of midwesterners and 56 percent of southerners found the plans helpful, and 23 percent and 12 percent, respectively, said the plan was a bother. Almost a quarter of southerners did not have a plan at all.[68] While this came a couple of years after the FSA had begun a decline because of budget reductions, it still suggests some limitations in the ability of FSA leaders and local agents to get their clients to produce the farm and home plans.

Part of the problem was that plans and clients sometimes just did not match up. FSA administrators constantly urged supervisors to tailor their farm and home plans to fit their clients. One farmer, for instance, disliked farming and instead wanted to make a living by itinerant trading and selling. His supervisor found success by developing a farm plan that emphasized the production of garden truck vegetables and similar crops, along with money to convert the family car into a truck. The farmer and supervisor worked out a plan to sell crops from other local FSA clients in a nearby town with a good market. This was successful, but such creativity and willingness to work with clients was not always in ample supply. When supervisors ignored family wishes, plans failed. Another supervisor developed a plan that completely ignored a client family's dedication to raising chickens (they kept their small chicken flock in the house because they lacked a henhouse). The family tried to use their loans for chicken production anyway and refused to initiate any contact with their supervisor. It took years for a new plan to develop that improved their chicken operation and renewed the client-supervisor relationship. And this example was, in a sense, a success; untold other clients never returned.[69]

The FSA had particular trouble when its ideals ran into what clients saw as reality. The Casa Grande project in Arizona had trouble with unrest and factionalism among the settlers. Myer Cohen, a regional staff member described by political scientist Edward Banfield as "an intelligent young liberal of an urban background," who "was expected to bring the techniques as a social scientist," went to the project to help. At the introductory meeting, Cohen spoke grandly of democracy and how important it was for community building. The speech did not go over well: one man claimed that management made all the decisions, not the settlers; an argument began over who had begun the drinking that had led to ending the Saturday night dances. Finally, Harry Olivier, a settler known for his unhappiness with how the government ran the project, stood up and said that they would be glad to cooperate with the FSA "after we get our bellies full. How can you expect us to talk to you tonight when our bellies aren't full? Do you think that a family can be happy on about $45 a month? We are discouraged on that account."[70]

FSA administrators worried about the local employees and the client-supervisor relationship. Good, creative supervisors could keep a family involved even through tough times. One supervisor, for example, won over a reluctant farmer by lavishly complimenting his beloved hunting dog, eventually becoming the supervisor's most successful case.[71] But families who resented their supervision would not stay with the program for long. If the local supervisor identified too much with the elite or upper class in a locality, that could discourage potential clients who had found themselves on the wrong end of that elite. It was worse for a supervisor to adopt a condescending or overbearing posture, and experience taught that clients could have long memories. The regional chief of home economics in Region V found, for example, that the insistent "dictation" of the earlier Social Division made it difficult for home supervisors to be anything more than "tolerated" by their current clients.[72]

On the other hand, the problem was not always too little supervision; many of the more paternalistic elements of supervision created tensions. Borrowers were expected to stick to their plans in a way that did not always seem fair. This was especially true for tenant purchase borrowers, who could not, for example,

A FARM SECURITY ADMINISTRATION–SPONSORED "KNOW YOUR FARMER" TOUR STOPS AT THE HOME OF A TENANT PURCHASE BORROWER. LOWNDES COUNTY, MISSISSIPPI. LIBRARY OF CONGRESS.

"buy a tractor, a radio, a refrigerator, incur a new debt, or pay off an old debt" unless it was included in the farm and home plan.[73] The requirement for joint signatures on checks for those clients with a shared bank account, one observer found, "was extremely irritating to many settlers."[74] This was a hassle, as making sure to get approval for each purchase was difficult and time consuming, but more than that, it was demeaning. The requirement implied that a client could not be trusted to use funds properly.

Administrators also left much to be desired in their individual relationships with both black and white clients who had been approved. James L. McCamy, assistant to the secretary of the USDA, remembered noticing on the part of local supervisors "an unconscious but obvious lack of regard for the dignity of the farmers who had gotten loans from FSA" when he toured resettlement communities in Mississippi and Arkansas in 1939. Some of this was a racist tendency to dismiss African Americans, but much more was simply thoughtlessness. On one instance McCamy, Regional Director Roy Reid, and Henry Wallace visited the home of a borrower. They were followed by "seven shining black cars . . . each filled with a complement of officials." The very uncomfortable farmer being visited was then peppered with questions on how the program was going. In another case, a Mississippi project manager and six other men tramped mud across a freshly washed porch and into an unprepared woman's home; the local manager later responded to complaints by insisting that "the more you gave these clients the more they fussed and the harder it was to get them to cooperate."[75]

The FSA's warnings about what a supervisor should not do provides a window into the sorts of problems that FSA administrators ran into with local supervisors. The agency discouraged any sort of coercion or threat to force a family to take action. Administrators opposed any insistence on the client family showing gratitude for supervision, which implied that a client had lost some measure of control over their situation. Supervisors were enjoined to be careful about overemphasizing one particular aspect of supervision, such as focusing on just farm production at the expense of other needs. FSA leaders worried about supervisors who could not adjust their procedures to handle relatively worse-off families less capable of making the sorts of lifestyle or economic changes they recommended.[76]

Problems with county agents also resulted from personal greed and malice. As administrators feared, local farmers often recognized that misdeeds by local agents related to the agents' connections with the local elite. One anonymous complaint from Roberts, Georgia, came on behalf of farmers "who have been so unfortunate as to have made farm and crop loans through 'Rehab' Agent Padrick," described as "a thorough rascal and a grafter" who required borrowers to make purchases at high prices from his friends. The writer remained anonymous for fear of becoming "the victim of any revenge from this carpet-bagger Padrick." The letter requested that FBI agents come to investigate and jail the supervisor, as the local office only ignored complaints.[77]

Sometimes, though, failures of supervision and rehabilitation happened for reasons beyond the control of the FSA or the client. Health problems were the most significant example. Farmers had to be able to physically work in order to repay their rehabilitation loans or put supervisors' suggestions into action. One study of Butler County, Alabama, found that, of the 155 clients liquidated during 1937, 79 percent had "a physical defect that required medical care," 52 percent had a disability that required (but usually did not receive) medical care, and "all families visited showed definite signs and symptoms of hookworm." Untreated epilepsy and mental illnesses among young and old further contributed to a failure in some families to earn enough money to support themselves, much less pay back loans.[78]

~:~

The repayment rate of the rural rehabilitation program was fairly good. In April 1938 the FSA had collected not quite half of all loans that had come to maturity— about $48.7 million out of $97.6 million loaned. But this included areas receiving a great deal of emergency loans because of drought, where the repayment average was only about 14 percent. In nonemergency areas, the average repayment was 60 percent. In three states (Indiana, New Hampshire, and Wisconsin) the repayment level of mature loans was over 100 percent.[79]

Collecting on loans depended on a variety of factors. Official policy was that a client would make a good-faith effort to repay loans on schedule and to protect the value of any property held as security. County agents had to do the actual collecting, so the amount of collections depended on the leniency or strictness of individual agents and their immediate superiors. Overall the FSA was rather loose with repayments. Variable repayment installments eased the pressure on clients by making it possible to make larger repayments in later years, when presumably they would be better able to do so. Clients behind in payment for good reason might find their loans renewed and extended or see the agency declare its intention to take no actions on collection for a limited time while the debtor family got back on its feet. Renewals were also frequently made in states in which local law required a repayment schedule for loans shorter than the FSA considered reasonable or in situations in which a borrower received one or more supplemental loans and it would be more convenient to combine all notes into a single debt.[80] Just an apparent willingness to try harder was often enough for local agents. For example, rural rehabilitation borrower Octave Fongemie went bankrupt, threatening the FSA's security. The agency decided not to pursue a foreclosure, however, after Fongemie signed a "Waiver and New Promise to Pay." The value of such a document signed by a bankrupt small farmer might have been rather limited, but for FSA administrators, it indicated the right attitude on the part of the borrower, and so they were willing to give him another chance.[81]

Borrowers who could not or would not receive supervision or financing but who had not yet completely repaid their loans were classified either as collection-only, if it appeared possible to get anything, or dropped if there was no chance of collecting. Defaulting borrowers might find the county supervisor recommending voluntary liquidation. If that proved impossible, legal efforts to recover property or liquidate had to be approved by the regional director and undertaken by federal attorneys. This last resort was rarely used.[82]

One reason that county supervisors tended to shy away from taking legal action was that, frequently, nothing much happened. Local officials complained about delays in legal action. Often, what county officials believed constituted a good reason for moving against a client did not match what upper-level officials thought. Even when state or regional or national officials agreed, there was often further delay before the case was forwarded to the USDA solicitor's office (although by the end of 1943, the FSA was mostly organized enough to eliminate that sort of problem). Once the case reached the solicitor, that office often lacked the ability to take immediate action. When the case got out of the USDA, the district attorney took weeks, sometimes months, before taking action. This was one situation where the FSA's instinct toward local activity did not pan out: county supervisors and committees requested that local attorneys handle the cases in county courts, rather than at the federal level, but this was not possible for a national federal agency like the FSA.[83]

Another practical problem was that the FSA was dealing with some of the poorest people in the country; if they failed or went bankrupt, there was usually not much that could be recovered. Generally speaking, the FSA did not treat clients as if they were trying to defraud the government. They had simply been unable to succeed as farmers or borrowers. Usually, that meant the client was left with practically nothing, and the FSA could only recover a part of its original loan, if that. As Monroe Oppenheimer, the USDA's solicitor, wrote about one case, it should be closed because "the possibility of collection was too remote to justify instituting legal action"; that could almost be the slogan for FSA debt recovery efforts.[84]

One client in Texas owed the FSA $660.99, plus interest, as a result of taking out four separate loans. The FSA turned the case over to the USDA attorneys, who then referred it to the district attorney's office. After the client allowed the sale of what property could be recovered, the FSA received $66.65, and the matter effectively ended, as any deficiency judgment against the client would be uncollectable.[85] In the case of a client from Goggins, Georgia, who owed $439.66 plus interest, the FSA (through the attorney general's office) managed to force the sale of a horse, various pieces of equipment, and fertilizer for $35.75 (minus court costs). The U.S. attorney's office saw no point in further action. An assistant U.S. attorney wrote to the attorney general's office on the matter: "this defendant . . . is a negro who owns no property and who is insolvent. A judgment against him for the deficiency would be absolutely worthless."[86] The FSA ended up receiving a check for $0.75.

One borrower from Wedowee, Alabama, summed up the problems of delay on the government's side and poverty on the borrower's. He owed the FSA $147.51 (having taken loans from the Alabama Rural Rehabilitation Corporation in 1935 and the RA in 1936, which became overdue to the FSA). On October 13, 1937, the matter was referred to Thomas D. Samford, a U.S. district attorney in Alabama. Samford recommended (in light of the small sums involved) that the FSA try to obtain the money voluntarily, as any cost of repossessing mortgaged property would almost certainly exceed the value of the property. The only recoverable chattel was a steer worth $66, which was finally sold on October 7, 1939, just short of two years after the case was sent to the district attorney. After expenses relating to the sale, the FSA recovered $41.20.[87]

The desire to ease repayments for rehabilitation led to difficulties with other government departments and sometimes compromised the FSA's lending role. FSA Instruction 758.2 was changed in 1941 to explicitly allow FSA clients to deposit money earned from the sale of crops or other farm operations into their supervised bank accounts regardless of the clients' debt status. FSA supervisors could, in other words, allow clients to save money for the next farming year without consideration of repaying their debts. This was in line with the FSA's attitude toward debt, farming, and rural poverty: a family had to be able to feed itself and operate a farm successfully in order to make rehabilitation possible. But as W. R. Fuchs, the USDA's assistant director of finance, wrote, such an approach was "somewhat difficult to reconcile . . . with our concept of the obligations of a collecting officer of the Government and of the purposes of a chattel mortgage." Fuchs was not, he claimed, against the government remitting debts for "for humane reasons" in certain cases, but he wanted these to be "the exception rather than the rule."[88] Otherwise, the FSA was forfeiting any claim to be a lending agency.

It was easy to identify the failed clients—they were unable to pay back their loans—but recognizing or defining the successful ones proved a bit harder. The FSA never specifically answered the question of what exactly being rehabilitated meant. For tenant purchase clients, it generally was the point at which they could be safely trusted to make payments without needing supervision. For a barely literate or illiterate tenant with almost no experience in money management, even the best supervision and the right amount of loan money would not, in the short term, produce an independent farmer.

How long the process should take was unclear. The proposed amount of time necessary to rehabilitate a client varied depending on who was talking. Perhaps being optimistic, at least outwardly so, Will Alexander said in 1940 that experience had shown that a family could be rehabilitated in five years.[89] An RA employee in 1935 estimated that new farm plans and annual loans might be necessary for more than ten years for some rural rehabilitation clients.[90] Robert W. Hudgens went even further, writing that there were "very few families in our Region [V] who will be carried from 100 percent supervision down to no supervision, but rather all of them should be carried down from 100 to 75 percent, to 50, to 25 percent and then lower. I doubt if there are any families in our program who should, at any time within the next two generations, be left without at least enough advice to cover the making of a farm plan."[91]

For Hudgens, talking in terms of years and dates was beside the point. He argued that most of the proposed time limits for rehabilitation, particularly relatively short periods like five years, were "not sufficient time in anybody's language to remove the imprints of several generations of rural poverty."[92] More important than lasting a certain number of years in the program, in his view, was reaching a set of standards for a rehabilitated family. FSA personnel generally believed that a family should receive assistance until its standard of living had reached a point where healthy living (which could mean a variety of things) was certain. It was also important to continue supervision in order to ensure that any FSA loans were fully repaid, especially if other sources of credit became available. And FSA personnel wanted their clients to reach the managerial potential (again, a nebulously defined concept) needed to maintain their farms without need for more assistance or relief.[93]

It is difficult to estimate exactly how many families the federal farm security programs assisted, much less how many it rehabilitated. Between the state Rural Rehabilitation Corporations, the RA, and the FSA, the federal government provided about 912,500 standard rural rehabilitation loans. However, the same family may have received loans from multiple sources that were still recorded as "original" loans. Some families might have repaid the entirety of one group of original and supplemental loans but later been forced to take out more. Families moving between different regions may have been recorded multiple times. Further, a family might have received both rural rehabilitation loans and tenant purchase loans. According to one estimate, something like 875,000 families received loans from the RA and FSA based on the standard home and farm plan supervised model. Forty to fifty thousand of these were probably long-term borrowers in the resettlement or tenant purchase program.[94]

Rehabilitation and supervised credit did show results, especially in the South. FSA efforts went toward solving that region's lack of capital, which held so many small operators back. One tenant purchase engineer pointed out the case of a farmer in Hale County, Alabama, who took a loan of about $9,000, part of which went toward a new hay barn. The client could then plant and sell surplus hay, making a possible yearly profit on hay alone of over $900, more than double his annual repayment of $390.[95] Even with such big improvements, the South was still relatively much poorer than the rest of the country. For example, total cash receipts in FSA Region V almost doubled between 1941 and 1943, yet that region was still the poorest. Average annual income to operators in Region V in 1943 was only $604, making it one of only three regions (along with Region VI at $611 and Region IV at $889, also southern regions) with average incomes below $1,000 a year.[96]

Farm security efforts also improved the social life of many rural southerners. One study of seven resettlement projects found that families on the projects were more likely (compared to where they previously lived) to exchange work and borrow items from other families, which improved their lives not only materially but also socially.[97] And by getting tenants, particularly mobile and ecologically destructive southern tenants, away from soil-depleting crops and by providing more consistent tenure agreements or the possibility to own land, the FSA encouraged a sense of permanence and connection to a particular community.[98]

Equally important to many observers were the improvements in morale and outlook the FSA created. Rural sociologists Arthur Raper and Ira De Reid noted that the "democratic philosophy of the FSA" had challenged the old plantation style. "By demonstrating that the South's landless people will respond to sympathetic assistance and scientific guidance, and that the money used in such programs is a public investment rather than a public expenditure," they argued, "the FSA is doing much to free the South of its greatest handicap—lack of faith in its own people."[99]

CHAPTER EIGHT

The Resettlement Communities around Birmingham, Alabama

In 1941, while questioning Will Alexander, then administrator of the Farm Security Administration, about its progress in getting rid of resettlement projects, Georgia senator Richard H. Russell called the resettlement program a "black eye" for the agency. Years later, one resident of the Palmerdale community near Birmingham, Alabama, Jean Walker, said (in an interview about how she felt about the New Deal project) that "it was good, very good, and I think that everybody that came out here [Palmerdale] really profited by doing so."[1] These two quotes reflect opposing conclusions about the resettlement program. In one sense, most individual projects and the program as a whole were utter failures. Poorly conceived, most of the early resettlement communities were planned in too great a hurry while at the same time constructed extremely slowly, all at higher costs than expected. Even in the earliest stages, planners, administrators, observers, and residents had different ideas about the goals of the communities and the makeup of their residents. Such problems only exacerbated the controversy surrounding the projects. Consequently, the resettlement program, even though it was only a minor part of the Farm Security Administration's efforts, was a major part of the reason that the FSA came under attack during World War II. For the administrators of the FSA and its predecessors, resettlement was a financial and political disaster. In Birmingham specifically, the projects never lived up to the subsistence farming ideal, and a variety of efforts to improve the lives of residents (like cooperative services) turned out poorly.

At the same time, most residents who stuck through the early difficult years (made difficult in part because of that poor planning and political instability) strongly approved of the program. Homesteaders saw improvements in their housing, diets, and financial situation. Even if they were not successful as subsistence farmers, residents were proud of their communities (pride that still exists in many of them to this day) and glad to get a chance for new, better lives for their families. Overall, it is fair to say that for most homesteaders, particularly in Birmingham, the communities eventually turned out well, after a few years of instability and false starts. But by the time that the resettlement programs had overcome those difficult early years, their apparent failures and expenses had irreparably harmed the reputation and political position of the Farm Security Administration as a whole.

In the 1920s, in contrast with the sagging agriculture in the rest of the state, Birmingham was experiencing unprecedented levels of economic success. Its iron and steel industry earned the city the title "Pittsburgh of the South," and even in 1930 U.S. Steel responded to the growing demand for steel, iron, and coke production by investing $25 million in its subsidiary, Tennessee Coal and Iron Company, which operated in Birmingham. But, as the Depression showed, the city's economy was fundamentally weak. And, like most major cities in the South, Birmingham faced significant urban problems magnified by the city's recent rapid growth.[2]

Birmingham was hit harder by the Depression than any major city in the South. Relying heavily on a single industry, steel, made the city's economy vulnerable to large-scale shocks. Steel production in 1932 dropped to a quarter of its 1929 high. Coal production (closely connected to the steel industry) dropped to about a third of the mid-1920s highs.[3] In Walker County (just northwest of Jefferson County), which depended heavily on coal mining, employment in 1934 was only 60 percent of the pre-Depression level.[4] Birmingham's leadership held the traditional perspective that economic growth would solve such problems. Jefferson County had legal responsibility for caring for the poor, but that mostly consisted of a punitive poor farm and an almshouse that a 1928 grand jury found unfit for human habitation. When the Great Depression hit, city leaders had little inclination to play a greater role in relief; by the time they were willing, around 1932, they lacked the funds to do so.[5]

As a result, large numbers of Birmingham residents fled the city to try their hand at farming. The number of farms in Jefferson County increased from 3,349 in 1930 to 6,491 in 1935 as families streamed out of the city.[6] Nationwide, the number of farmers had risen by half a million, and Birmingham was one of the places where the greatest increases took place. This encouraged government intervention; as one RA official later explained, "Far from embarking on a new and untried venture, the Resettlement Administration is simply reinforcing, organizing and channeling a spontaneous, unplanned movement of the local population which was already under way."[7]

Subsistence homesteads planners believed that workers had fallen on tough times as a direct result of the Depression but also because of larger changes like mechanization. While in previous years industrial work had been constant, in the early 1930s it was intermittent, and things appeared to be getting worse. If such conditions could befall steel, "one of the great basic industries," as the project book for one homestead project explained, it could happen anywhere and last for years. The expected periods of "idleness and leisure" resulting from a lack of industrial employment could be filled, homesteads supporters claimed, by farming. Home ownership would promote stability, education, culture, and a whole host of other improvements in the lives of the homesteaders. And while Birmingham was currently in bad economic shape, planners emphasized that the steel industry would continue to be important and provide at least some income for many workers; the government would not be throwing money away on a hopeless situation.[8]

The cities of Birmingham and Jasper obtained approval for five projects, likely reflecting Senator John Bankhead's influence.[9] The first application, for the Jefferson County homesteads, was approved, and a local corporation was organized in December 1933. Erskine Ramsay, a Birmingham industrialist and philanthropist, was elected its president, and the corporation received a loan of $750,000 dollars. Soil surveys began soon after, and tracts were tentatively selected in late 1933 out of more than seventy considered. Planners hoped to begin developing the project in January and perhaps get occupants in early spring.[10] Under the Division of Subsistence Homesteads, the emphasis was (at least at first) on decentralization. To encourage local control, two corporations were formed, one each for Jefferson and Walker Counties, to direct the projects. Sites were chosen by committees of local citizens selected by M. L. Wilson, head of the Division of Subsistence Homesteads; many committee members became members of the first local corporations. On January 9, 1934, FERA regional director Oscar Dugger (owner of the *Andalusia Star*) announced that Birmingham Homesteads, Inc., was officially in operation. A few days later the corporation announced that family selection would soon begin and that application blanks would be distributed following approval from Washington.[11] The projects were named Palmerdale, Mount Olive, Greenwood, Cahaba, and Bankhead Farms.

Observers believed that Palmerdale Homesteads, located nineteen miles northeast of Birmingham, had the best chance for farming success. The rolling, loamy soil had been used primarily for dairy farming, truck crops, and poultry in recent years and was considered perhaps the best agricultural land of the Birmingham projects.[12] However, it took time to get what had been known as the "Kent-Cannon" tract prepared because Cannon Dairy, the previous owner, was slow to move. This delayed everything. For example, the Homesteads Corporation intended to use existing fencing material for development work and to use the existing residence as a storage house and a field office to save money. But at the end of July 1934, those fences still enclosed the pastures, and Mr. Cannon still lived in the house. Eventually, the 689-acre project came to include 102 units in two parts: Palmer, having 60 units, and Palmerdale, with 42.[13]

Bankhead Farms, located five miles north of Jasper and forty miles from Birmingham, eventually covered 2,096 acres and consisted of one hundred homesteads, along with apartment buildings, a park, and a community building/school. Each homestead was wired for electricity and had a frame barn and an individual well. Applicants came from those employed in local lumber mills, coal mines, and textile mills. The project began in January 1934, when ninety-two thousand dollars was allotted for the project, but construction did not begin until the end of January 1935.[14]

Greenwood Homesteads was composed of eighty-three units about five miles southwest of Bessemer. It was originally the Martin Farm, a plantation-type operation employing about twenty black families who lived on site and received their pay in scrip for use at the Martin Commissary (and who, like other black Birmingham residents, were not eligible to live in the new community). Hit hard by the Depression, the Martin family sold their land to the Birmingham Subsistence Homesteads Corporation in May 1934. Greenwood's 402 acres appeared to be a good deal at the relatively low price of twenty thousand dollars (about fifty dollars an acre), but much of the land was useless for farming or badly eroded; the homestead corporation had to drain, terrace, and improve much of the land. Even then, the first homesteaders had to invest in intensive soil improvement before good yields were possible. Much of the land was never good enough to be used for anything but pasture.[15]

About ten miles away from the center of Birmingham, Mount Olive in Gardendale was perhaps the least promising of the projects around Birmingham. Not only was it relatively far from Birmingham, but the hilly land made it difficult to farm. The soil on its 512 acres was sandy and easily worked, but it was prone to leaching nutrients and required heavy fertilization. Residents complained that their farms missed summer showers, as its valley location made the site a relatively dry pocket north of Birmingham. Despite its apparent shortcomings, once underway Mount Olive did well. Managers began accepting applicants on April 16, 1936, and quickly received far more applicants than there were houses available. By August 1, 1937, every available house had been assigned.[16]

The Cahaba Project, near the city of Trussville about sixteen miles east of Birmingham, sat on 810 acres formerly known as "Slagheap Village," which had old mill houses and slag heaps that had to be removed before construction began. It cost about $109,000, or $135 per acre, a fairly high price considering that the land was entirely unsuited to subsistence-style farming. From the beginning, Cahaba was expected to be a residential suburb for the fairly well off: almost half of its houses, for example, were either brick or brick and frame construction. A majority of the houses had five rooms, and all had baths, porches, and a garage.[17]

Sponsors of the Birmingham projects were optimistic about the possibilities for rapid results: in the summer of 1934, for example, John H. Bankhead wrote the director of the Division of Subsistence Homesteads with plans to start producing tomatoes for the fall.[18] Planners hoped that the land would need little preparation, as most of it had been cultivated at some point in the previous few years. And given the existing trend in subsistence homesteading and community gardening in Birmingham, supporters believed that the government would only be building on what industrial leaders in Jefferson County already encouraged.[19] Most supporters and planners were completely unprepared for the variety of obstacles the projects would face.

In June 1934, the secretary of the interior dissolved the local subsistence homesteads corporations, and administration of the projects became centralized in Washington. Regional Supervisor Oscar M. Dugger became most directly responsible for the Birmingham homesteads. Between July and November 1934, the Subsistence Homesteads Corporation carried out additional surveys of the soil, topography, layout, and others necessities.[20] The projects were federalized between approval of the purchase and finalizing the sale, which created difficulties. With administrative changes and unclear ownership, attorney Philip M. Glick noted, "more than 100 different instruments needed to be secured in order to clear up the title."[21]

The Birmingham Homesteads did not get off to a quick or inspiring start. Palmerdale, Mount Olive, and Greenwood required substantial terracing because of their hilly topography. The Jefferson County Soil Conservation Association helped terrace approximately six hundred acres, requiring the construction of seventy miles of terrace lines and six thousand feet of outlet ditches, plus three thousand square feet of Bermuda sod. As a result of such delays, the Division of Subsistence Homesteads completed only eighty-four houses around Birmingham: twenty-four at Bankhead and sixty at Palmerdale. These houses were built at about two thousand dollars each, initially considered a good deal. It turned out, however, that they were poorly constructed: floors were cold, pipes froze, and improper weathering allowed rain to seep into the side walls and ruin the inside finish of the homes. The Division of Subsistence Homesteads did not survive long enough to see even this modest achievement come to fruition; the RA officially opened the homes at Bankhead on September 1, 1935, and at

FIELDS AT GREENWOOD HOMESTEADS, ALABAMA, TERRACED
TO PREVENT SOIL EROSION. LIBRARY OF CONGRESS.

Palmerdale on November 1, more than two years after Roosevelt authorized the secretary of the interior to begin the projects.[22]

When Roosevelt created the Resettlement Administration in 1935, it took responsibility for most of the federal government's community building. The RA's Resettlement Division took over more than thirty subsistence homestead projects from the Department of the Interior, including the Birmingham-area homesteads efforts, and a roughly equal number from the Federal Emergency Relief Administration. The move to the RA, the second step in centralizing the subsistence homesteads (after federalization of the projects), created new problems. The Birmingham projects, for example, were subject to the Management Division, the Construction Division, and the regional office in Montgomery, meaning that at least three lines of authority existed for the projects within the RA.[23] The RA also had trouble using relief work. A former section chief for Region V, which contained the Birmingham homesteads, described the process at its worst: development work went to FERA's Work Division, which then procured the necessary relief labor. Planners assumed that the allocated money for land and improvements would pay for the cost of labor. But then the money ran out before the improvements were finished. The workers moved on to a new job with new funds while the job was not completed until well into planting season.

Families moved into half-finished houses or farms with no outbuildings, already weeks or months behind preparations for the crop year.[24]

When the RA took over the homesteads around Birmingham, planning was well along, except at Cahaba. However, little construction had been done—the first units begun at Palmerdale and Bankhead were not even completed. The RA's Construction and Management Divisions took over the projects already underway. The Construction Division did the building, generally with better results but at a higher cost than under the Division of Subsistence Homesteads. In addition to houses, the RA built roads, a store, and a combined school and community center for each community. At Greenwood, planners constructed a central water system. Houses ended up averaging about sixty-five hundred dollars each in Greenwood, Mount Olive, and Palmerdale. The Bankhead Farms homes, which were generally slightly larger, were more expensive, about seventy-two hundred dollars each.[25]

The RA also selected most of the families who would live in the projects it was constructing. Community managers worked with family selection specialists, who helped choose client families and then moved on to other projects. It was a decentralized process: administrators and experts on the ground, RA leaders believed, would do a better job of understanding the issues on a particular project than the Washington office could.[26] The Washington office did set standards for training and education of family-selection personnel and for the kind of family that would be selected. Applications for the subsistence homesteads were generally only accepted from families making less than sixteen hundred dollars a year. Each family was considered according to its prospects of economic stability, health, age, number of children, and similar concerns.[27] Applications were very detailed: Cahaba residents joked about having to count out how many pillowcases the family had before getting approved.[28]

RA leaders wanted to approve the most capable of the eligible clients. Since applicants from this group often lacked anything like the traditional notion of "credit history," administrators instead focused on "credit character," the personality and integrity of a family in, say, paying bills when due. These were the kind of families that would do what was necessary to make their payments.[29] The ideal applicant family had that personal integrity and some (but not too much) income to ensure a good start. As one resident remembered when her family was chosen for the initial sixty homes at Palmerdale in 1935, "Our main credentials were character, being a family, being employed because they did want someone out here who could start making payments on the homes back to the government."[30]

When the RA moved to the USDA in early 1937 and became the FSA in the fall of that year, management of the projects had become its most important concern. Much construction remained to be done: when the resettlement program was transferred in the fall of 1937, only thirty-eight projects had completed construction.[31] But FSA administrators were mainly preoccupied with two issues: managing the already completed homesteads and selling them to residents.

The biggest problem in managing the homesteads was residential turnover, as clients sometimes decided they did not like the management, farming, or some other element of life on a subsistence homestead. Administrators solved this problem mostly by ignoring it. The regional FSA office in Montgomery appears to have paid little attention to the Birmingham project; instead the community managers made policy. The result was not necessarily fiscally sound: for example, homeowners who fell behind on rent were not evicted but were given a new schedule of payments. But once the process of self-selection was finished—that is, once the residents who decided not to live in the communities had moved away (amounting to perhaps a bit over half the total number of families in each project over the course of several years)—the communities achieved a measure of stability.[32]

The FSA also oversaw the sale of the homesteads to residents, which it accomplished via homestead boards. It was clear from the beginning that the Birmingham homesteads would not even come close to paying for themselves. Except for Cahaba, which could be operated profitably as a rental project, the Birmingham projects would "have a substantial write-off in their capital cost in order to meet the income group for which they are built," the director of the FSA's Resettlement Division wrote in 1938.[33] Delays increased costs and homesteader unhappiness: the first people who moved to the projects did so believing that they would get the selling price within the year and then have the chance to buy their lots, but it actually took over four and a half years for the first prices and sales contracts to be announced. The process of selling the homesteads to residents began in 1938, when the local homestead board purchased Bankhead Farms and most of the occupants signed purchasing contracts for their homesteads. Palmerdale, Mount Olive, and Greenwood followed in 1940. Most administrative problems were resolved by these boards, which were made up of local homesteaders; they and the occupants were collectively responsible for total rent and purchase payments of all the units.[34]

FSA leadership wanted to influence the homesteads after the purchase. One idea was to allow the homestead associations to purchase the projects with the understanding that they would only have to make payments and accrue interest when the associations had received sufficient rental payments to do so. Officially, the FSA would make such allowances to protect the homestead associations from a shortage of money due to vacancy, unemployment, or illness. This would draw out the repayment schedule and make the homestead associations more dependent on the FSA. The USDA's solicitor's office nixed this idea as legally impermissible, given that it effectively relieved the association of the requirement to make payments.[35]

Another policy, quite unpopular with residents and critics, allowed additional monthly payments but withheld full title to the units until after twenty years of occupancy. The policy had its purposes: it preserved the strength of the homestead associations and secured many of the benefits of zoning and regular subdivision restrictions. Administrators feared that full ownership without adoption of special regulations might result in new construction like unsightly barns or hog shelters, commercial poultry houses, or cheap quarters for black workers (several of whom already lived in barns and structures without sanitary facilities in Palmerdale when the sales contracts were announced). The boards also wanted to avoid speculation, which would be likely and be heavily criticized because the sale prices were not based solely on construction costs. This proved to be a controversial policy, condemned both by residents and members of Congress. Homesteaders, naturally, wanted full title to the land they occupied and had entirely paid for. Congressional critics saw a transparent ploy by the FSA to keep homesteaders dependent on the agency (which was true, to an extent) and to maintain government control of the farms as part of a scheme to further the socialization of American agriculture (which was not true).[36]

As the FSA slowly conveyed the homesteads to new owners, the United States' entry into World War II reduced the need for subsistence farming and the homestead communities, as well as the FSA's ability to oversee them. By the summer of 1944, most Birmingham homesteaders were working full-time, even overtime. Some young men joined the military. Many women also found jobs; Palmerdale residents, for example, went to work as nurses, truck drivers, and industrial workers, among other occupations.[37] Local farmers who had acted as hired labor for the homesteaders found themselves with more work than they could handle. There was less surplus farm labor available in and around Birmingham than there had been in years, if not decades. In addition to a shortage of labor, war industry created a shortage of equipment. Cooperative farm service equipment went to the military, war industries, or other agricultural pursuits where the need was greater. As a result, only about 20 percent of homesteaders had all of their land in intensive use in 1944; about 17 percent had their land in no agricultural use; and another 15 percent had only a garden.[38] War also undermined the projects as communities, as residents were drafted or left for defense-industry jobs.[39]

The homesteads also lost administrative support after a presidential order moved them from the FSA to the Federal Public Housing Authority (FPHA) on October 1, 1942. The Birmingham office stayed the same, with the community manager renamed the area adviser. But the FPHA was not a farm-oriented organization; it was concerned with housing and the sale of the projects. Its policy was to keep prices down and get the projects ready for self-management. The FSA continued to supervise the cooperative associations on the transferred projects until June 30, 1943, and farm and home management assistance continued for some clients. But generally, the projects began to move away from the federal government's influence: after the Birmingham area office closed on August 20, 1944, the homesteads had no direct connection with Washington. The FPHA instead functioned primarily to audit the homesteads, insuring that they were more or less paid up and that the houses were well maintained.[40]

The organizational transition did not go smoothly, and the FPHA does not appear to have been a conscientious auditor. By the time its successor, the Public Housing Administration (PHA), went about figuring out exactly what it had responsibility for after World War II, many parts of the project had effectively come to an end. Francis Brill of the PHA's Finance and Accounts Division tried to determine the exact conditions of the industrial and subsistence cooperatives. He found the records barely usable, "in a deplorable condition. Some of the records located there are not even in cabinets. It appears that they were placed at random on the tops of boxes and cabinets ... it would take one person weeks to completely inventory the records, check the subsidiaries, and generally certify whether or not the books are posted to date." Brill found that some records from the FSA, by then the Farmers Home Administration, had never been sent to the PHA (including the records of the Bankhead, Cahaba, and Palmerdale cooperatives). The records that were in the PHA's possession (like those from Mount Olive) had simply lain in storage. Only those that had been kept by residents or local employees, like those at Greenwood, were mostly up to date.[41]

Resettlement was plagued by a variety of problems, the most obvious of which was political. One congressional report described the FSA, "beginning with the administration of Rexford G. Tugwell and continuing throughout the administration of C. B. Baldwin" (conveniently skipping the more popular Will Alexander), as "financing communistic resettlement projects, where the families could never own homes or be paid for all that they made or for all the time that they worked."[42] Resettlement was the most controversial FSA program, despite the fact that resettlement was not the largest part of the FSA's efforts and that resettlement families were fewer than 2 percent of those served by the agency.[43] Resettlement's failures even pushed away some FSA supporters: Bankhead, for example, became a less vocal supporter of the FSA for a number of reasons during World War II, but his disappointment with the subsistence homesteads around Birmingham could not have strengthened his commitment.

Resettlement was a political disaster despite the care taken by FSA administrators. The cautious approach on the issue of race is a good example. Many employees, perhaps most notably Will Alexander, worked hard to ensure that FSA benefits, as much as possible, profited blacks and whites equally. But the FSA never directly confronted Jim Crow. In some cases, including during the construction of the Cahaba community, black families were forced to move off of land and out of homes that were often destroyed so that homes for white residents could be constructed.[44] The resettled communities were segregated, either officially or in practice. At Palmerdale, for example, the procedure book announced officially that selection of applicants did not discriminate based on "race, nationality, or creed"; however, planners did say that "consideration will be given to the homogeneity within the group."[45] The result was an all-white community. There was some discussion that one of the sites around Birmingham would be used for a homestead project for African Americans, but such a plan never got off the ground. Administrators feared, as Oscar Dugger put it, that such an effort would "invite antagonism" and threaten the other projects.[46] The board of directors of the Birmingham Homesteads instead adopted a fairly meaningless resolution for the Division of Subsistence Homesteads to authorize a black homestead project around Birmingham "at the earliest possible convenience."[47] Similarly, around the country a number of proposed black communities were canceled because of the protests of local white residents.[48]

None of the Birmingham-area projects met their original objectives. One issue involved the size of the homesteads; deciding on acreage allotments for each community depended on the type of homesteader on each project. Administrators faced questions of how many acres low-income industrial workers could really operate, what a part-time worker could devote to farming, and how to accommodate full-time but low-income workers. Decisions had to be made before answers could be found, and the result was a poor allocation of land. Initial expectations, generally, had been that less than half of the occupants in the four projects around Birmingham would be white-collar workers and that these would be low-salary and part-time employees, but this did not turn out to be the case.[49]

Bankhead Farms, for example, was planned for irregularly employed miners, so its occupants were expected to have a rounded subsistence farming program on units of nine to twenty-five acres, the average being twenty acres. Coal miners found the location bad and the rent high, so instead middle-income families moved in. While some had good gardens, only a few used all of their land profitably. Those residents who did have a full, effective farm did so by hiring labor, occasionally using a son or grandson. Bankhead Farms fell between two stools—too large to be effectively used for intensive, part-time cultivation but too small for commercial agriculture with hired labor.[50]

MR. AND MRS. J. A. BRITAIN IN THEIR NEW HOME AT BANKHEAD FARMS, ALABAMA. LIBRARY OF CONGRESS.

Palmerdale, Greenwood, and Mount Olive had similar problems. At first, most occupants were from the underemployed working class who might follow through with subsistence farming out of necessity; about two-thirds of the first sixty families selected for Palmerdale, for example, were industrial or manual laborers.[51] But this apparent need to farm did not translate into the ability to do so successfully. Residents at Greenwood, largely unemployed or underemployed Tennessee Coal and Iron workers, had little knowledge of agriculture, despite the fact that a rural background was supposed to be a requirement for applicants. Allegedly one farmer destroyed his entire corn crop and instead cultivated Johnson grass, a weed that resembles young corn stalks.[52] A Cahaba resident knew so little about farming that he failed to weed his first garden.[53]

The suburban Cahaba project strayed far from the subsistence ideal. Its plots had much smaller acreages for gardens or orchards on a fraction of an acre. Planners expected that owners would have subsistence gardens and orchards, but few did, since the typical resident was a white-collar worker with an average yearly income high enough to practically eliminate any need for gardening. Those who did keep good gardens considered it a leisure activity or hobby, not a foundation for economic wellbeing. Most of the residents by 1940 were middle-class workers who could have lived comfortably in Birmingham, but at Cahaba

TABLE 3: DISTANCE IN MILES TO INDUSTRIAL CENTERS FROM BIRMINGHAM HOMESTEADS

	Cahaba	Palmerdale	Mt. Olive	Greenwood
Birmingham	15	21	14	18
North Birmingham	16	19	10	22
Tarrant	14	15	12	23
Fairfield	18	27	18	11
Hueytown	32	37	31	6
Woodward	29	34	28	8
Brighton	27	32	26	10
Bessemer	32	40	30	5
Ensley	20	25	14	13

Source: Walter M. Kollmorgen, "The Subsistence Homesteads Near Birmingham," in *A Place on Earth*, ed. Russell Lord and Paul H. Johnstone (1942), 68.

the rent was about half of what it would be in the city.[54] Marion Ormond, for example, was an office worker employed by the federal government, working in an accounting department in downtown Birmingham.[55]

The Birmingham-area homesteads were inconvenient for working-class residents. To save money, the homesteads board wanted to buy land cheaply (less than one hundred dollars per acre) and in large tracts (four hundred or more acres), which eliminated many sites close to the city. This also meant, as Table 3 indicates, that many projects ended up quite far away from Birmingham's industrial centers. Mining operations were spread out across the area, and job opportunities outside the industrial centers were scattered. As a result, most people could not carpool to work. Public transportation hardly solved the problem. Palmerdale, for example, had bus service, but irregularly so; one resident remembered that the "Blue Goose," the privately owned bus that first served the community, was unable to navigate the community's bad roads during heavy rain, with the result that the bus driver had to get out and walk to pick up passengers and bring them back to the bus. Similarly, Greenwood faced a delay in getting paved roads, and the dirt roads were often in such bad shape (particularly in the rainy first year) that no car or bus, including the school bus, could get there.[56] Residents who owned automobiles discovered that the new dirt driveways were prone to trapping cars.[57] Three out of five families who had left the project by May 1944 cited financial problems or living costs, employment, and the location or transportation costs of the homesteads as their reasons for leaving.[58]

The homesteaders, then, had good reasons for not living up to the early objectives of the program. They needed a cash income, for example, because of transportation costs and other expenses, so they could never devote sufficient time to subsistence farming. But then, too, homesteaders never really understood the goals or nature of subsistence farming. They never shook their old habits, in particular hiring help and the idea that the field was particularly the place for African Americans. Farmers who could afford to (and many who could not) shifted as much as possible to black labor. Families with monthly incomes over one hundred dollars considered a black servant and aides indispensable. Since most occupants had incomes in the range of one hundred to two hundred dollars per month, they were accustomed to black help in and out of the house.[59]

Sponsors and planners initially believed that locals' farming experience would increase the effectiveness of subsistence farming. One planner wrote in 1934 that Birmingham was different from similarly sized industrial centers because "a very large percentage of its citizenry came from farms... there is a great urge on the part of a great many people in Birmingham to get back to the soil, where they believe they will have greater security."[60] However, the experience of most unemployed industrial workers, as with most of the rest of the South, was with one-crop and cash-crop farming, usually meaning corn, cotton, and perhaps an undiversified garden, and not with subsistence farming. The projects had four agricultural experts and two home-demonstration experts to provide an agricultural education, but that accomplished little because hired laborers did most of the work.[61]

Thus most families did not make good use of their land, even during World War II, when wartime conditions raised the economic value of home production. Those who could most benefit from the subsistence homesteads were men engaged in manual labor. They were used to doing physical work, and so gardening or farming did not strike them as recreational or enjoyable. It was just more work.[62] Further, the rhythms of farming and the schedules of industrial employment did not mesh well together. A farm requires time, energy, and the possession of certain tools and equipment. Rarely did the three always come together just right for men with full or mostly full-time industrial employment.

In 1943 the homesteads in Birmingham were apparently full and fruitful, but even success stories pointed to unsustainability. Palmerdale resident Tobe Missildine worked at the *Birmingham Age-Herald* from 9:00 p.m. until the early morning. After work, he drove home and farmed until 11:00 a.m., then went to sleep. His wife kept the house and raised four children. The Missildine family had erased their debt and stored away canned goods, but it required a great deal of work, as success did for other residents. About half the families at Palmerdale remained there seven years after construction. As one observer described it, "All of these families had to go through hell and high water in those experimental months from 1935 to 1937. But now, in 1943, Palmerdale is one big, happy family, thriving on its own prosperity."[63] Residents could thrive, then, but effectively working two full-time jobs (as Missildine and others did) did not promise to be a blueprint for success that everyone could follow.

The struggles of the cooperative services at the Birmingham-area homesteads embody the problems of the subsistence homesteads and resettlement program as a whole. In Birmingham and around the country, homesteaders liked the idea of cooperative services, but they failed to overcome a combination of poor planning, poor management, and bad luck. In every community around Birmingham but Cahaba, residents organized cooperative farms services, and in every community but Bankhead, they operated cooperative stores. None of them lasted long beyond the end of World War II.

Cooperative farm services matched almost everyone's idea of what living on a homestead community should be like. Except in Cahaba, community barns were built and the necessary farm machinery and animals were purchased. Prices were set at seemingly reasonable levels. However, with hired labor, the cost of field operations was too high for most homesteaders. Scheduling was a problem. Homesteaders often had to hire help seasonally, but the numbers of mules and implements available from the cooperative were limited; everyone wanted them at the same times, such as on the weekends, since many homesteaders worked during the week. The farm services therefore failed to meet expenses. They took in enough to operate during the heavy work periods, but the mules and implements were idle much of the year. As a result, the farm services often had to be sold to individuals on or near the projects with the requirement that the tools and mules be available to homesteaders at a modest charge.[64]

The Bankhead Cooperative Association, formed in June 1937, stuck to farm services, and it seemed to have the best chance of success of the four homestead farming cooperatives. Its residents had the biggest tracts of land, and mining schedules potentially fit comfortably with part-time farming. The association put a twenty-thousand-dollar government loan toward mules, horses, a tractor, and related equipment. But this effort in cooperative business failed. Despite the fact that it paid only a dollar a year to lease the property from the federal government, the cooperative could not operate at a profit. Between June 12, 1937, and October 15, 1938, the association earned about $1,350 in farm services, but spent almost $3,300. In-season, residents found the service unsatisfactory: they demanded prompt service and grew impatient with the delays of a sharing program. Off-season overhead cost more than in-season profits. The association voluntarily liquidated on March 21, 1940, with close to the original loan amount still owed to the federal government.[65]

Administrators also planned cooperative stores and filling stations. There were already some stores located close to the projects; but they generally had a limited stock of goods, and planners believed that homesteaders coming from

Birmingham would be used to a high standard of service and facilities. When the first homesteaders arrived in Palmerdale, Mount Olive, and Greenwood, they were asked about establishing cooperative stores. Administrators encouraged them, and FSA representatives gave meetings to talk about the opportunities of a cooperative project. These talks did not go well. Homesteaders got the impression that they could buy for less from the cooperatives than from other stores and still earn enough to buy the enterprises.[66]

The Palmerdale Cooperative Association received its charter in September 1936. It used a $20,000 federal loan for a store, filling station, and cooperative farm service. The association constructed a barn, purchased work stock and farm implements, and opened for business in March 1938. Like the other cooperatives, the Palmerdale Cooperative Association got the use of its buildings at extremely low cost. The store at Palmerdale cost $30,000 to build, but the contract gave a value of just over $5,600 dollars, and the building and site were made available at an annual charge of $12 for the first three years and $175 for the next two, with the option of buying at the listed book value with a forty-year payment period at 3 percent interest. Similar arrangements were made with the other associations. The cooperatives, then, paid almost no rent and had the opportunity to purchase their buildings at a greatly reduced price.[67]

Despite these advantages, business did not go well. In-store management was a problem. Some managers were not trained or qualified, hired at low salaries on the assumption that anybody could run a store. This created problems like poor stock maintenance, messy and disorganized stores, and improperly kept accounts and inventory. Even the experienced managers who took over failed to get the stores out of the red.[68] Palmerdale had perhaps the worst management. Its store was not run according to good accounting standards: money was withheld from daily receipts and used for petty-cash expenditures, employees were reimbursed and checks cashed in excess of the petty-cash fund, and the Palmerdale Credit Union's funds mixed freely with the association's monies. The bookkeeper often backdated checks so that the expense would be recorded in the same month as the unpaid purchases had been made, making it difficult to accurately determine available cash and unpaid liabilities. At the end of 1939, the association had assets (including cash, inventory, and fixed assets) totaling $12,360.02 and debts of $17,593, with a net deficit of $4,178.58.[69] After the store closed in February 1943, the association sold its inventory to Greenwood but still owed almost $16,000 on its loan.[70]

Similar mismanagement plagued the Greenwood Cooperative Association. It was in an even more advantageous position than Palmerdale: in addition to operating nearly rent free, the store was far away from potential competitors. The association formed in June 1937, took out a loan of about $18,500, and opened its store in March 1938.[71] With so much money, the association made the mistake of buying a large inventory, hiring a big staff, and otherwise accumulating heavy overhead costs. The farm service never showed a profit and had to be sold to a private operator after ending service in October 1940.[72] At the end of 1939, the Greenwood Cooperative Association had a debt of about $2,400, though it did at least show a profit of $20.20 in 1939, after losses in 1938 of over $2,500. Like those of the other associations, Greenwood's store was burglarized several times in the spring and summer of 1939.[73]

The Mount Olive Cooperative Association, organized in June 1937 using a federal loan of about $18,000, had even worse luck.[74] The store opened in March 1938 and burned down in October. Despite monthly increases in sales, the rebuilt store failed to meet expenses. It was also robbed repeatedly. Burglars broke into the store four times in 1938 and 1939, including a failed attempt to dynamite the safe, stealing $609.66 worth of goods (including $66.66 belonging to the cooperative baseball team), although theft insurance meant that the association (and the baseball team) suffered no losses. By the end of 1939, the association owed $17,041.60, versus total assets of $12,964.[75]

Befitting its more suburban nature, the Cahaba Cooperative Association was organized in 1938 to operate a general store, filling station, and similar enterprises.[76] Despite high hopes, the Cahaba Association had trouble even maintaining local interest in its operations. In the summer of 1937, the Resettlement Division could not find any residents to become members of a proposed Cahaba Mutual Association, so that idea was dropped. In 1939 the association had to amend its voting rules, changing from requiring a majority to allowing one-third of the members as a sufficient quorum to conduct business at its poorly attended meetings.[77] The community manager threatened to close the store in February 1944 to force more residents to shop there.[78]

Overall, only the Cahaba filling stations produced a modest profit. Otherwise, the Cahaba Association did not plan or run operations very well. Projected sales were wildly optimistic, on the expectation that homesteaders would buy almost everything from the store. But only a minority of occupants proved to be regular patrons, and sales to nonoccupants were limited. Unlike the more distant cooperatives, Cahaba residents could easily patronize the A&P in Birmingham, often buying a large amount of groceries on a monthly trip to save money.[79] Numerous experiments were made to save the stores, including rallies and campaigns to popularize them, repeated changes in management, and changes in purchases and sales methods, all of which were unsuccessful. Milk and coal delivery were introduced to make use of a truck purchased for grocery deliveries, but neither was successful and the truck had to be sold, resulting in a 30 percent drop in grocery sales. As a result, even though the association rented a building from the federal government for only eighty dollars a month, it consistently failed to pay; rent was reduced to fifty dollars, which it still consistently failed to pay.[80]

After World War II, the cooperative associations came to an end, at considerable cost to the federal government. By the end of 1947, the Mount Olive Cooperative Association was officially liquidated. However, it was essentially defunct by September 1945, when the Public Housing Authority's Field Accounts Section closed its books (as much as was possible), leaving an unpaid debt to the federal government of almost $16,000. Similarly, the Palmerdale Cooperative Association, after paying off its creditors, wrote a check for the remainder of its bank account ($389.95) to the federal government on September 25, 1945, leaving an unpaid debt of about $19,000.[81]

While the Birmingham-area subsistence homesteads failed to meet most of their original goals, residents approved of the projects almost from the beginning. The *Birmingham News* noted in June 1936 that while there was controversy surrounding the Resettlement Administration, "there is one place in Jefferson County where the resettlement agency has no opposition. That is Palmerdale, first of four resettlement projects in Jefferson County to be occupied."[82] For one thing, the program promoted a general improvement in living standards. The biggest change for most families was in diet. Almost all had reliable, year-round access to milk, eggs, and fresh vegetables. Unsurprisingly, the overall health of residents improved dramatically in their new homesteads. The communities also provided instruction in cooking, sewing, quilting, budgeting, and other skills that administrators believed wives could use.[83]

Clients found their financial situation much improved. While later arrivals were economically better off than the first ones or the ideal ones, they were still far from wealthy, and even small economic improvements could greatly improve their condition. Once the program got settled around 1941, almost all applicants found steady jobs and decent pay. Many who were in debt or barely above water developed a net worth of over a thousand dollars. Those who came early and stayed had equity in their homes by the end of World War II, and some had paid off almost their entire debt. While not all or even most residents took advantage of the subsistence farming opportunity, those who did had a fairly good record (as even hostile congressional committees recognized).[84] For example, Otis Ledbetter, a machine shop foreman, succeeded at part-time farming in Mount Olive. He devoted his afternoons to it and managed to produce almost all the meat and vegetables his family consumed, selling surplus to others on the project or to men with whom he worked. His rent was about half of what a

PLOWING A FIELD AT PALMERDALE, ALABAMA, WITH A NEW HOMESTEAD IN THE BACKGROUND. LIBRARY OF CONGRESS.

similar home in Birmingham would be, not including the three and a half acres of farm.[85]

Just as important as the physical improvement was a psychological one. There was a strong sense of pride in the homestead communities: residents liked the way they looked, believed in the quality of the people living there, and were proud of the community they had built. Of course, not everything went ideally. The slow start hindered community feeling: Greenwood, for example, did not have a local school until after the Christmas break at the end of 1937.[86] Other formal and institutional bonds were lacking. Forty-four percent of women from the four industrial communities interviewed around the end of World War II said they belonged to no club of any kind. But interviewers still frequently found that residents mentioned the neighborliness that prevailed in the communities. Of the 171 interviewed women, 167 responded that they liked their neighborhoods.[87] Residents prided themselves on their ability to work together. In 1943 one reporter noted that despite a shortage of pressure cookers, residents of Greenwood had canned thirty thousand quarts of food the previous year and expected to break that record by sharing the cookers. "That's the way these homesteaders do," a 1943 newspaper article explained. "They share their troubles and their successes alike."[88]

Once the projects settled down, after the initial period of uncertainty and delay, resident turnover declined, and the projects filled to capacity. For example, at the end of 1939, all one hundred sites in the Bankhead project were occupied, with tenants owning the tracts via the operating company, and with several civic organizations and a large community school. Residents enjoyed the level of control they had: while the government provided a community manager, the homesteaders' Bankhead Homesteads Company could, for example, reject any applicant for any open homestead. Even the local economy looked to be improving—a hosiery mill opened on November 8, 1939, employing sixty residents on a two-shift basis.[89]

Homesteaders also considered themselves part of a successful social experiment that, if nothing else, played a part in overcoming the Depression. Critics argued that the resettlement communities were unrealistic or un-American, but to homesteaders, they seemed to be just the opposite. Residents claimed that Roosevelt was correct to build the communities and put people to work and that such programs were necessary, given the circumstances of the Great Depression.[90] A *Birmingham Age-Herald* editorial in May 1936 pointed out, "To some the sight and the whole idea may be somewhat suggestive of regimentation and standardization. But doubtless to the many families already living in Palmerdale, it is a great escape into what is at least a great freedom compared to the regimentation that adversity had forced upon them."[91]

Residents also believed that their successes held enduring lessons for the country. After World War II, one recalled that the Cahaba project had dramatically changed the area. "You should have seen that site before the project was begun," he said. "It was largely a wasteland. And you should see it now—and all the good people who have come to the Trussville community and made good neighbors and helped things along. I tell you I'm for that sort of thing. The spirit is right when the government tries to help people along—not spoil them, but so that they can help themselves."[92]

The most successful residents combined the material gains of subsistence farming with the psychological ones of living in a proud community. Nolan McRee, who grew up near Central Park in New York City, moved onto 3.98 acres at Greenwood in October 1937. He planted sixty-five fruit trees and a small vineyard, raised hogs, cows, and chickens, and grew corn with the help of USDA publications. McRee managed all this in his spare time while working as an electrician for the U.S. Pipe Company in Bessemer. His wife canned four hundred jars of fruit, vegetables, and meat the previous summer; she won first prize at the Alabama State Fair for her canned bread and butter pickles and for a jar of chili sauce and second prize for butter beans and for peach pickles. The McRees patronized the local cooperative store, sent their children to the community school, and enjoyed the neighborhood community life: basketball games and dances at the auditorium, taking their children to the playground, and neighborhood gatherings. Mrs. McRee told reporters, "We're not afraid of the future now, either for ourselves or our little boy."[93]

CHAPTER NINE

The Farm Security Administration and World War II

The Farm Security Administration, its leaders believed, had made great progress as world war loomed in 1941 and 1942. The agency had developed a viable plan for addressing rural poverty, using supervised credit to promote farm security. But problems remained. Writing in 1941, sociologists Arthur F. Raper and Ira De A. Reid described sharecropper Seab Johnson, who with a single mule produced about two bales of cotton a year. The absentee landlord for whom Seab worked had gained the property after its previous owner fell into bankruptcy despite planting cotton everywhere he could, an endeavor that included removing the garden and most of trees around the house. The Johnsons had lived in a tenant cabin on a neighboring plantation until they found themselves unable to get credit. They had been lucky to move into what was formerly the home of a great planter, but veiled threats from their landlord left the husband and wife wondering where they could move next. Their son in Pittsburgh had trouble getting work, so the old man surely could not. Similarly, a daughter in Jacksonville was working hard and still falling behind. Their other children were all too poor to support their parents. Old and alone, the Johnsons hoped to stay in their current home in case one of them got hurt or sick, because the other could ring the bell until help came.[1]

The needs of those struggling with rural poverty remained obvious across the country, but political concerns were also growing. Although the FSA's upper leadership was unprepared for an attack threatening the agency's survival, they understood that World War II would dramatically alter the FSA's situation. They took steps to strengthen their position within the United States Department of Agriculture and the federal government, while the FSA's foes worked through a series of congressional hearings and appropriations bills to weaken and eliminate the FSA. Wartime changes gave new impetus to budget cutters and, within the Department of Agriculture, created an atmosphere of survival of the fittest.

In making their case for the agency's importance during World War II, FSA administrators argued that programs such as rural rehabilitation were crucial to the war effort. To win the war, the United States would need healthy farmers who could produce the necessary soldiers, manpower, and food and fibers— farmers who were, in other words, rehabilitated. A vast rural peasantry on the edge of starvation and unrest (made up of people like the Johnsons, those too old or too poor to flee the limited options of a life in rural poverty) could not aid the war effort. With a little help from the FSA, however, the rural poor could be transformed into a powerful asset. Such arguments came naturally to FSA administrators, who had settled on the rehabilitation of individual farm families using targeted loans and expert farm and home management as the key to ending rural poverty. It seemed only natural that the FSA should help the war effort by doing the same thing.

The FSA generally succeeded in its contributions to the war effort, but it failed politically for two reasons. First, its opponents made compelling arguments about how to best harness rural production. Proponents of the agricultural lobbies hitched the traditional belief that only higher prices could improve farm production to the argument that only large operators had the means to expand production in time to help the war effort. From this perspective, the FSA's activity was irrelevant at best and probably harmful, wasting money and reducing the supply of cheap agricultural labor necessary for big, efficient farms. The "arsenal of democracy" atmosphere of World War II only made these sorts of arguments stronger.

Additionally, the FSA met political defeat because in wartime the rules of politics changed. Franklin Roosevelt explained that "Dr. New Deal" had been replaced with "Dr. Win-the-War" when the work of reform had been done and a new patient showed up. But well before the war began, the New Deal had ground to a halt, and surviving agencies like the FSA worked in an increasingly hostile political atmosphere. Opponents of the New Deal managed to slow any new reform efforts; with the Roosevelt Administration more and more occupied with foreign affairs after 1939, anti-New Dealers took the offensive. Some New Deal reforms, like the Social Security Administration and agricultural price supports, had become too entrenched or popular. But others, like the FSA, the Civilian Conservation Corps, and the National Youth Administration, found themselves without political cover while facing longtime political foes operating under the pretext of making necessary wartime sacrifices. Critics of the FSA (and other New Deal agencies) had powerful new rhetorical and political weapons just as the FSA's allies became less willing or able to defend it. Ultimately, World War II provided the opportunity for the FSA's foes to gut it almost entirely. Though it survived until 1946, the agency was effectively defeated by 1943.[2]

The outbreak of war in Europe in September 1939 initially meant little change in the USDA. Given the continued surpluses of so many commodities, Secretary of Agriculture Henry Wallace recommended that American farmers avoid changes to production. With America's recent history of overproduction and the New Deal's emphasis on reduction, conservation, and raising prices, most USDA leaders agreed. This policy stuck until the first of several reorganizations in 1941, when Claude Wickard (Wallace's replacement as secretary of agriculture after Wallace became vice president) created the Office of Agricultural Defense Relations to coordinate food production, built on the Agricultural Adjustment Administration.[3]

Despite Franklin Roosevelt's best efforts, America was slow to get its economy on war footing, but once it began in earnest, the change was rapid. In 1940 America's military leaders pleaded for additional funding; in 1941 that funding flooded into the armed forces. But the sudden influx of government cash into the economy did not turn the United States into a military power immediately, nor did it mean immediate economic recovery. Policymakers hesitated, uncertain about what kind of military strategy the United States would be implementing, who its potential enemies were, and where America would be focusing most of its economic and military effort. Organizational and political problems mounted: business leaders did not want or know how to shift to military production, and it was far from obvious what sort of administrative structure would be best for either the war effort or the Roosevelt Administration's political position. Bureaucratic turnover and organizational reworking was not limited to the USDA.[4]

Reformers worried, though, that the gains of the New Deal and the opportunities for further change would be lost, partly because of the war atmosphere but even more because of the growing emphasis on promoting the war effort at any cost within the executive branch, especially by Franklin Roosevelt. Despite occasional rhetoric, Roosevelt by 1940, and certainly by 1941, showed his willingness to concede liberal reformers and agencies in exchange for conservative support for the war effort. At the same time, the totalitarian nature of the combatants in World War II—Fascist Italy, the Soviet Union, and Nazi Germany—led to self-doubt among liberals about the value of an active state, plus even greater hostility from political and ideological foes who pointed to the totalitarian dangers of a potent federal government.[5]

In such an atmosphere of political and bureaucratic uncertainty, FSA leaders readied for the coming conflict. Preparation for war meant significant changes in FSA policy. For the Resettlement Division, land purchases for resettlement increased to make farms available to rural families displaced by or moving to military bases and defense industries. With the growing threat of war, an overabundance of labor turned into a labor shortage. The division diverted migratory labor camps (which it had operated since 1935) to defense and wartime use. The FSA built more temporary and permanent camps and shelters; by the end of 1942, the FSA had built ninety-five camps capable of holding seventy-five thousand people. It also administered a program that brought almost thirteen thousand Mexican workers into the United States and moved eight thousand domestic farm laborers within the country by May 1943. The FSA cooperated with the U.S. Employment Service to make sure that laborers knew where camps were located and where labor was most needed, and FSA supervisors contacted borrowers to determine whether labor could be spared on larger nearby farms at peak harvest. The Rehabilitation Division also moved to a wartime footing. Eligibility standards for rural rehabilitation loans began to favor those who could most rapidly expand production, and emphasis shifted more toward the production aspects of farm rehabilitation. The division reorganized into six branches, which demanded new personnel and the shifting of existing employees (and took up valuable time).[6]

The outbreak of war raised bigger issues beyond administrative shuffling. FSA leaders had to consider the problems that the war would create for the rural poor; dislocations could undo much of the work of the previous years. And they had to win a political fight by providing a strong rationale for their agency during wartime. The FSA had walked a fine line for years, balancing New Deal reformers' belief that reduction in agricultural production was necessary with American farmers' instinct to increase their production as much as possible. Wartime conditions threatened to make that balance impossible. There were a number of approaches, based on wartime responsibilities the FSA could claim. The agency had some specific jobs: for example, it handled the agricultural side of evacuating Japanese and Japanese-American farmers from the West Coast in early 1942. But the FSA had to find a more general role. Administrators argued that the FSA could strengthen America's rural citizens for the war effort and that the agency had experience moving and rehousing laborers, but the most important argument for the FSA's wartime relevance was the issue of improving small farm production.[7]

FSA leaders were careful to publicize the agency's value to the war effort.[8] In January 1942, for example, an FSA regional office claimed that its female migratory farm workers were doing more "in the way of war production" than any other group of American women.[9] Administrator C. B. Baldwin echoed the earlier debate over the Bankhead-Jones Farm Tenant Act in noting that the FSA dealt with "the most under-privileged, poverty-stricken part of the farm population. This group might provide fertile soil for those agitators and sympathizers with foreign philosophies who are trying to plant seeds of discontent and doubt about our American system." More positively, Baldwin argued that America needed "manpower," but the only way to get that was to have healthy citizens. The FSA helped by improving the diet, medical care, and living standards of the rural poor. The FSA took farmers off relief, "changing them from national liabilities into national assets."[10]

FSA administrators also claimed that the agency had a role to play in relocating and housing the labor necessary for the war effort. The FSA was involved in defense housing even before war broke out: the all-steel prefabricated houses it had developed, for example, became commonplace for building defense housing.[11] The FSA claimed to be the ideal agency to handle the 14,500 farm families already displaced in the fall of 1941 from land purchased for defense purposes: after all, it had built or repaired more than 37,700 houses and was ready to do more. The agency was also already active in the South, where military projects were particularly concentrated. The FSA organized the Hinesville Relocation Corporation for one of the first big moves, involving 800 families near Hinesville, Georgia, who lived on 350,000 acres used by the army.[12]

While the FSA argued that it had an important role to play in housing, labor, and strengthening rural citizens for the war effort, Baldwin and the FSA chose to emphasize the importance of small farm production. This made political sense. USDA administrators increasingly worried about possible food shortages; Wickard especially feared a shortage of farm labor (matched by worries that a solution might lie outside the USDA); other administrators, like Howard Tolley in the AAA, agreed. Most everyone believed that wartime manpower demands threatened the farm labor supply. Baldwin and other administrators argued that the FSA was the key to solving that problem. While large commercial farms found themselves suddenly short of labor, small or marginal farms, with FSA assistance, could rapidly expand production while at the same time improving the lives of the rural poor; otherwise, the poor would leave their farms, worsening the shortage of agricultural production. As Baldwin pointed out in a Senate hearing in April 1942, rural rehabilitation clients increased their food production, considering food consumed at home and produced for sale, by around three hundred dollars per family. On a national scale, that represented the possibility of an enormous increase in food production.[13]

In a series of talks in 1942, Baldwin asserted that to win the war, the United States "must mobilize the energy that is going to waste in 1,200,000 of our farm families." He pointed out how the role of FSA borrowers had changed. Previously, "these people were important," he said, "because they needed help. Today they are important because their help is needed." He went on to emphasize the

FSA's importance in improving the conditions of thousands of small farmers ("a great reservoir of unused manpower").[14] Baldwin claimed the FSA was uniquely ready to improve war production: unlike decentralized war boards or committees along the lines of the Farm Credit Administration, the FSA possessed the capacity for rapid, single-minded activity. Wartime food production, Baldwin wrote, was "no job for debating societies."[15]

FSA employees and supporters down the line made similar arguments. Even before the United States entered the war, one project manager wrote to his clients that America's allies fighting for liberty needed guns, ships, and airplanes, but for ultimate victory "it comes down to the fact that we in America must raise more FOOD, not only for ourselves, but for our friends whose fight is our fight."[16] Even health services took on a martial tone, as one publication phrased it: "temporary illness can sabotage production efficiency and seriously hinder the cause of victory."[17] Eleanor Roosevelt wrote, "Our great hope is to increase the production on small farms. The Farm Security Administration, working with the lowest income farmers, has proved that this can be done through wise advice in management, small loans and assistance in marketing produce."[18]

While this ultimately failed as a political argument, it was successful as part of the war effort. Both as a percentage and in absolute numbers, FSA borrowers, especially dairy farmers, increased production dramatically during World War II.[19] Between 1941 and 1942, 38 percent of the total increase in American milk production resulted from FSA borrowers. Other gains were less impressive but still considerable: of the total increase in American production of dry beans, beef, and peanuts, for example, FSA farmers were responsible for 17 percent, 11 percent, and 10 percent, respectively.[20] Beyond supervision, the FSA helped clients acquire laborsaving machinery such as tractors and combines. American farmers and agricultural policymakers generally failed to make proper use of farm machinery during World War II, with the exception of the FSA—at least, until Congress began to eliminate funding for the cooperatives that made it possible.[21]

While the FSA made its case for a role in the nation's war effort, critics and political foes launched increasingly effective attacks on the agency. The FSA's opponents criticized it for various and sometimes contradictory reasons, but the foes generally fell into three groups. The most vocal opponents were representatives of agricultural interests who had long resented the FSA's position in the USDA and used World War II as cover to eliminate a political rival. A second group was made up of those who saw the FSA as an un-American or unconstitutional exercise of power, including many anti-New Dealers emboldened by the war and the declining political strength of New Deal reformers. A final, disparate group of critics disapproved of how the FSA operated because of its alleged snubbing of congressional dictates, or because they wanted to turn the

FSA into an agency of their own design. Critics in these three groups at times overlapped, but many found common ground only in their opposition to the FSA. These varied attacks worked together to reinforce one another to make something that had seemed rather unlikely in 1940—the elimination of the FSA—an accomplishment in all but name by 1943.

The most important anti-FSA faction was organized agricultural interests, most notably the American Farm Bureau Federation and its congressional allies. Prior to 1940 or so, the AFBF had been generally hostile toward the FSA but had taken few concrete steps against it. AFBF leadership began to worry, however, about the bureau's relationship with the USDA, particularly with Secretary Wallace. As World War II got underway, the war effort provided the opportunity to eliminate a troublesome political rival. Responding to FSA administrators' argument for an expanded wartime role, the AFBF countered that the FSA had the situation backwards. AFBF president Ed O'Neal asserted that no one besides the FSA seemed to expect low-income farmers to produce much, as the USDA called for the biggest wartime increases in states with high agriculture incomes.[22] Before the war, half the farmers had been producing 89 percent of the country's commercial farm products. Instead of increasing production by FSA clients, critics argued, it would be better to increase the production of large farms and encourage small farmers to become agricultural laborers or work in the defense industry. AFBF leaders also stuck to their traditional belief that price increases would do the most to raise output, while the FSA argued that the best solution would be to maintain the supply of fertilizer, equipment, and other assistance to small-time farmers.[23]

The agricultural lobby's spokespeople barely considered what the FSA was doing as real farming at all. Oscar Johnston, president of the National Cotton Council, criticized the fact "that the F.S.A. has purchased many thousands of acres of alluvial land in the valley of the Mississippi River.... When I say the F.S.A. has purchased them, I mean that F.S.A. has provided the funds with which to consummate the purchase, holds a mortgage on them, and directs the operation of them." Further, he asserted, "in this very fertile area," farms working under FSA direction "have produced less than they were producing under individual private ownership."[24] It was probably true that farms under FSA supervision produced less cotton for sale, because the FSA instructed its clients to grow subsistence crops. But big producers did not see it that way: production of food for home consumption was tantamount to no production at all. Tenant ownership via government credit did not count as ownership, while tenants renting via landlord or merchant credit did.

Large producers approved of the FSA when it helped them, of course, particularly in providing labor.[25] But when the FSA threatened their supply of cheap labor, big farmers changed their tune. This had been the case since the RA had begun sponsoring migrant camps, which critics claimed promoted "radical agitation" and raised labor costs.[26] Sometimes the FSA could not win: cotton producers in Alabama, Mississippi, and Georgia complained that the FSA made it easier for their laborers to move to Florida, while citrus growers in Florida protested that the agency was moving too slowly and not providing sufficient numbers of workers. And growers balked at having to meet any standard on how to treat imported laborers. One congressman received complaints about requirements to feed Mexican laborers meat six days a week.[27]

The testimony of L. L. Chandler, representing South Florida vegetable growers during the House farm labor hearings, demonstrates how many large producers perceived the FSA's labor program. The laborers provided by the FSA, Chandler said, came from "the riffraff of the human family." The FSA forced him to sign a contract that "guaranteed minimum wages, maximum hours, fancy housing and sanitation, transportation and feeding" for his workers. As a result, the worker became "petted and pampered" because the "Government feeds and sleeps him and then spends days chasing around to see what kind of sleeping and cooking facilities and toilets are provided." Workers refused to work, Chandler said, especially black ones. He claimed that he saw "time and again the Negroes . . . send in to town for a case of I. W. Harper whisky." Then "the loose women of that outfit follow right on out into the field and pick up their men, and it is just a grand orgy."[28] With this kind of perspective, it is not surprising that many big producers thought so little of the FSA's work.

The earliest clear anti-FSA effort came in 1941 budget hearings before the House Agricultural Appropriations subcommittee. In February AFBF president Ed O'Neal appeared before the committee to insist that parity was the key to agricultural (and thus American) prosperity. Part of the problem, he argued, was the "duplication of effort and overlapping and even conflict that has developed among the various agencies" in the USDA.[29] O'Neal recommended that everything (including FSA supervision) be streamlined under Extension Service control. The FSA's credit activities would move to the Farm Credit Administration. This would divide the credit and educational elements of the FSA for the benefit of both. As O. O. Wolf of the Kansas Farm Bureau Federation put it, "there can be a lot of consolidation made" by eliminating the FSA's supervisory role and moving its loans to other agencies.[30]

There was not much evidence for this claim, and the whole of the AFBF's charges against the FSA—mostly a collection of unverifiable personal reports and unsupportable allegations—amounted to a weak case. Georgia representative Malcolm C. Tarver, hardly an FSA ally, pointed this out: "If you are going to continue the work and merely shift the personnel from one agency to another, then you have to have as many people... I do not think it makes any difference whether John Jones works for the Soil Conservation Service at $3,000 a year or whether he works for the Extension Service at $3,000 a year, so far as the Government is concerned."[31] The AFBF, even FSA critics noted, was transparently demanding more power for itself and its Extension Service allies. But the atmosphere was such that even weak arguments could be dangerous.

A good example of how even the most transparently thin attack could prove harmful came during testimony by the AFBF general counsel Donald Kirkpatrick regarding the Farm Bureau's investigation of the FSA. His description, taken from brief looks at states in the Southeast, the Midwest, and the Northwest, was not terribly damning or convincing. Most telling was the study for the South: William C. Carr of Chicago was tasked with looking at Alabama, Mississippi, Arkansas, and Louisiana. Kirkpatrick admitted that Carr only had a limited time, less than three weeks, to study the situation and prepare a report. He was specifically told to look at FSA cooperatives, solicitation of clients, excessive or duplicated efforts, and other misdeeds. Carr spent eight days in Alabama, looking at one of several "socialized farming projects" in Hale County. Carr claimed that the FSA was soliciting clients for these projects but confused the issue by presenting as evidence an advertisement for the tenant purchase program. Carr used a few allegedly representative farm plans as proof that the FSA was overburdening clients. He spent even less time on the other states, only a day in Louisiana, but drew a similar picture of mismanagement and abuse.[32]

C. B. Baldwin testified after Kirkpatrick and pointed out that it would have been much more useful for such allegations to have been sent to the FSA, rather than sprung at a hearing. Refuting Carr's attack, Baldwin noted, was practically impossible: it was based on a small subsection of the program, on interviews with unnamed and unknown sources that could not in any meaningful sense be objectively assessed. Kirkpatrick himself was obviously biased and had only the most tenuous grasp of what the FSA was doing. But countering such criticism was fruitless, since the actual merit of the FSA was not the point, as demonstrated by the discussion of how the FSA's work related to Rexford Tugwell and his sympathies for the Soviet Union.[33] Attacking the FSA, not the pursuit of fact, was the goal.

The AFBF's criticisms of the FSA did not change considerably over the next few years; rather, they intensified. Tellingly, by 1942 the AFBF argued that Congress should reduce agricultural spending as a whole. When O'Neal appeared before the House Agricultural Appropriations subcommittee in 1942, he recommended the coordination of agricultural programs under the Extension Service and reducing all other expenses, mostly by ending or transferring programs. He favored abolishing the FSA generally and the farm-tenant program specifically (along with several other agencies), citing the need for wartime reductions. Maintaining control of agricultural policy had become more important for AFBF leadership than even raising farm incomes.[34]

In addition to the political machinations of the agricultural lobby, a second group of FSA critics argued that the FSA was an illegal, unconstitutional, or un-American agency. Other political foes made a similar argument for tactical reasons: O'Neal gave Harry Byrd's Congressional Economy Committee a list of nine charges regarding the FSA's "insidious and indefensible" behaviors in Alabama, including soliciting for clients, "collective farming projects, which make you think you are in Russia," politicking, and more; and in another set of hearings, O'Neal criticized the FSA for possessing objectives that were "inconsistent with the spirit and genius of the American way of life."[35] But this was a tactical move for O'Neal; other critics were primarily motivated by a belief that the FSA's activities were illegal and wasteful, not because it overlapped with someone's political turf.

The most prominent such critic was Virginia senator Harry Byrd, a longtime foe of the FSA and, indeed, of any federal intervention in agriculture. He chaired the Joint Committee on Reduction of Nonessential Federal Expenditures (better known as the Byrd Committee). Although Byrd and his committee had a much different conception of the federal government's role in agriculture than did the AFBF, they shared the belief that the FSA was a wasteful duplication of other government programs. Byrd asked Baldwin during the hearings, for example, whether the FSA was not doing exactly what the WPA was supposed to be doing: providing help for families in catastrophic or disaster situations. North Carolina representative Robert Doughton followed up by asking whether state and local responsibility had been ignored and whether local authorities had been let off the hook by the federal willingness to provide relief. Some criticism of the FSA's wastefulness targeted individual administrators: before a Senate subcommittee in 1942, Assistant Administrator Robert W. Hudgens defended a trip he took to New England in part to look at Canada's extension service. His having taken the trip in the summer, critics argued, indicated that Hudgens was just looking for a way to escape the Washington heat at taxpayer expense.[36]

A number of congressmen supported Byrd because the FSA seemed from the outside so expensive and therefore wasteful. Part of the problem, as Georgia senator Richard Russell (an FSA supporter) pointed out, was that many in Congress did not know how it worked. The program looked "like a tremendously expensive program to a man who has never been out in the field to see how it operates," Russell said, and "you would be astonished at the ignorance of some Members of Congress on the actual workings of this program."[37] Confirming Russell's claim that many members of Congress did not understand the mechanics of the program, Virginia's John W. Flannagan criticized the FSA for using yearly appropriations for rehabilitation work, when the Bankhead-Jones Farm Tenant Act provided no money for such activity. He characterized this as ignoring or circumventing the will of Congress. But the Bureau of the Budget told the FSA to go through the appropriations process to get money for rural rehabilitation, and Congress had passed such appropriations, but this did not suit Flannagan. Told that the Appropriations Committee had provided such funds, Flannagan responded, "I did not, and I am protesting now... That is a direct violation of the clear provisions of this act."[38]

While Byrd opposed on principle a great many federal activities, other critics opposed the FSA specifically because they believed it to be un-American and perhaps even communistic. There was a sense that some projects were anti-individualistic, socialistic, and therefore illegitimate. In 1942 Malcolm Tarver asked Baldwin, regarding the Lake Dick Farm cooperative in Arkansas, "In what respect does that enterprise differ from the system of agriculture practiced in Soviet Russia?" He pointedly avoided the word "cooperative" in favor of a more radical-sounding term: "When did you start this collective project?" and "To what do you attribute the brilliant success of your experiment in collectivism in Arkansas over that in North Carolina?" Just the idea of what the FSA was doing seemed ridiculous. Tarver said, "I do think this whole fantastic idea of trying to furnish everyone in this country with low-income decent quarters in which to live at a reasonable price at Government expense ought to be abandoned."[39]

Baldwin, long seen as a radical from the school of Rexford Tugwell, came under particular fire. Tennessee's Kenneth McKellar told the Senate in May 1942, "I think Mr. Baldwin is a communist. I do not think he is really in favor of our American institutions." Other senators came to Baldwin's defense. Claude Pepper asked for evidence, and John H. Bankhead called Baldwin "a gentleman, a Virginia gentleman, and if any criticism could be made of him it is because of his generosity, possibly, his tender heart toward relieving the sufferings of the poor farmers."[40] Such defenses did not stop the accusations; the FSA even got caught up in anticommunist hysteria and the House Un-American Activities Committee in 1944.[41]

Charges that the FSA not only was acting illegally but was also anti-southern followed news that the FSA allegedly paid southern farmers' poll taxes. Robert K. Greene, a Hale County, Alabama, probate judge, told the Byrd Committee that at the end of 1941, he found an FSA employee, Curdin McGill, examining the poll-tax records of Hale County. McGill allegedly told Greene that he was looking for white clients who might need the FSA to help pay their poll taxes. Upon further investigation, Greene found that in at least two other Alabama counties, FSA employees carried out the same procedure. Greene contacted E. S.

Morgan, director of Region V (which included Alabama), who defended the loans as similar to loans for any other tax. The FSA only did so for white clients, but critics feared that the agency might soon extend it to black clients, threatening the political balance of power in many rural counties.[42]

Advancing money for the payment of the poll tax was not technically illegal in Alabama unless it was done with the purpose of influencing that person's vote. To prove this was the case, Greene provided affidavits from people who claimed that the FSA had made efforts to force them to vote, that Morgan pressured local supervisors to make sure their clients had paid the poll tax in case their votes were needed, and that the FSA was taking on as many cases as possible in order to have a larger constituency. Baldwin responded by pointing out that the USDA solicitor, Mastin G. White, had approved making such loans in Alabama. This did not satisfy the committee members (except an obviously irritated Senator Robert La Follette), who tried to maneuver Baldwin into admitting that the FSA was forcing clients to vote a certain way in an effort to build up a political organization.[43] Much of the criticism was openly racial, despite the fact that black clients could not vote whether they paid their taxes or not and thus were not eligible for such loans. Virginia Senator Carter Glass told Baldwin that he doubted the connection between voting and rehabilitation, saying, "Our Appropriations Committee room is small and I hope you bring before us all your rehabilitated Negroes from Alabama so I can have a look at them."[44]

FSA administrators did not back down, arguing that rehabilitation included both paying taxes and ensuring that its clients were "self-respecting members of their community."[45] As Region V director E. S. Morgan put it, FSA leaders believed that "you can never rehabilitate a family that is qualified to vote until that person pays his poll tax and exercises his rights of citizenship."[46] The distant but still sympathetic Roosevelt hesitated before openly supporting the agency. Asserting his lifelong opposition to the poll tax, FDR accused critics of believing that the poor should not vote. H. L. Mitchell of the Southern Tenant Farmer Union and Howard Tolley of the Bureau of Agricultural Economics joined Baldwin before the Byrd Committee, criticizing the AFBF as politically motivated and for failing to help or represent small farmers.[47] Eventually the issue blew over but not before further cementing the FSA's reputation (for some) as an un-American, anti-southern agency.

A third group of FSA opponents was the broadest and least rigidly anti-FSA: congressmen who resented the FSA's refusal, in their view, to adhere to congressional dictates or to sufficiently respect congressional wishes. The most important example was the apparent FSA foot-dragging over the liquidation of the resettlement projects. The delay, Malcolm Tarver told Baldwin, constituted a "policy of defiance" that would "be exceedingly harmful to the Farm Security

Administration."[48] Congress wanted the FSA to liquidate resettlement projects as quickly as possible, but it was going slowly. Worse, as war preparations began, the FSA appeared to be creating more resettlement projects for families forced to move by defense activities. To congressmen opposed to resettlement, it looked like the FSA was sneaking additional projects into a wartime effort.[49] Similarly, Tarver asserted that the FSA was using its migratory labor program to stealthily, and against congressional intent, "bring about the migration by persuasion and by paying money to them also, paying expenses, of farm families of the South who are satisfied with their surroundings ... to other sections of the country."[50]

This criticism of the FSA's tardiness and underhanded behavior in liquidating the resettlement projects indicates a difference in perspective between the FSA and its critics. Congressional foes attacked what they saw as intentional delay, but FSA employees felt that things were going at breakneck speed. James Heizer, acting director of the FSA's Management Division, worried at a conference in 1943 that in the haste of liquidating the communities, FSA employees needed to keep in mind "the important phase of doing an orderly, intelligent, and constructive job of withdrawing Federal supervision from these communities." Congress demanded a speedy transition of ownership from one party to another, but for the FSA, ending ownership did not mean ending responsibility. Administrators expected community managers to stay as the FSA's field representative, continuing supervision and making sure that the opportunities provided by the community setup would not be lost.[51]

The most serious attacks on the FSA began in early 1941, in hearings for the House agricultural appropriations (although there were earlier skirmishes in 1940 when Tarver introduced several budget amendments to restrict FSA operations and eliminate some of its programs).[52] Many members, particularly Representative Clarence Cannon of Missouri, used their time to criticize the FSA. Cannon also went out of his way to force Baldwin to admit that raising prices was the real key to solving the problems of American farmers.[53]

At the corresponding Senate agricultural appropriations hearing for the fiscal year 1942, taking place in March 1941, AFBF president O'Neal continued his case against the FSA when he recommended that the USDA be reorganized both philosophically and administratively. The argument was not very strong, but the support from his congressional allies was, and if nothing else, the AFBF was prepared to add a heavy dose of repetition. The Senate agricultural appropriations hearings for 1941, however, were still friendly to the FSA: when Baldwin appeared, he mostly received questions from two of the FSA's biggest supporters, Senators Russell and Bankhead. Most of the discussion involved the shortage of funds, requests for specific information, and the like.[54] With FDR's support, the Senate restored funds to the FSA that had been cut by the House and removed most of the proposed limitations.

The hearings on the reduction of nonessential federal expenditures, held by a bipartisan committee chaired by Virginia senator Harry Byrd, represented another line of attack. Agricultural interests and their congressional supporters strongly opposed any cuts to parity or soil conservation payments; Byrd and his committee generally agreed. Since this made up 57 percent of the agricultural appropriation for the fiscal year 1942, the rest of the USDA would be forced to make all of the Byrd Committee's contemplated cuts, including the elimination of entire agencies like the FSA and the Rural Electrification Administration.[55] These hearings raised some of the most prominent criticisms of the FSA: that it illegally paid poll taxes, was communist-inspired, and was generally an unnecessary and wasteful program.

The House hearings on agricultural appropriations for 1943, taking place in 1942, went almost as poorly as they could have for the FSA. Very little new information came from these hearings: in the words of economists Lee J. Alston and Joseph P. Ferrie, "supporters of the FSA were less persuaded by these one-sided presentations [from FSA opponents] than they were worn down by them year after year."[56] Again, the AFBF and its allies launched an attack on the FSA and similar programs as, at best, a distraction from the real need for parity pricing and, at worst, a socialist plot for land reform. Some FSA programs, like tenant purchase, had some general support. But the tenor was decidedly hostile, particularly regarding resettlement projects, subsistence homesteads, and cooperative farms. Much of the hearing involved general criticisms that the projects cost too much or were poorly run, or petty criticisms of Baldwin personally. Tarver upbraided him, for example, for failing to bring the right kind of documents every day to the hearings.[57]

Edward O'Neal again appeared before the committee to represent the AFBF, accompanied by several state presidents, with little in the way of new arguments. Some specific criticisms of the FSA already felt like a word-for-word recitation of earlier attacks. "We have had so many complaints of waste, extravagance, and abuses," O'Neal said of the FSA, including quotas for the number of clients, client solicitation, "impractical collective farming projects," and more.[58] In March the House came within twenty-three votes of repealing the farm tenancy program. Reflecting the AFBF line, Illinois representative Everett Dirksen said that he wanted to eliminate the more than five thousand FSA farm and home supervisors, who received some $18.5 million, because there were "already 7,000 on the Extension Service payrolls now throughout the country, capable, ready and eager to advise, and Triple A committees in every county of the nation."[59]

The tide was strongly against the FSA, but other issues helped drag out the USDA budget fight through the summer. The USDA even became technically without funds on July 1, 1942. The House and the Senate split on spending for the FSA, but far more contentious was the issue of parity prices. The House insisted that the government pay full parity prices to dispose of the huge potential surplus, which senators feared would raise prices and lead to inflation. FDR thought both options were insufficient but sided with the Senate's relatively generous FSA funding and eagerness to keep food prices low. Behind the scenes, some believed the whole fight was a smokescreen for AFBF machinations. Finally, after Roosevelt publicly criticized Congress for harming the war effort and threatened a veto if he did not have freedom to take care of farm surpluses, Congress worked out a compromise.[60] The FSA survived, but with a slashed budget. One measure of the FSA's decline was the agency's loss of personnel, dropping below ten thousand employees by 1944 from a peak in June 1942 of over twenty thousand people.[61] And in this the FSA was fairly lucky: other New Deal agencies, like the more popular Civilian Conservation Corps, failed to survive the congressional budget battles of the summer of 1942.[62]

USDA reorganization hurt the FSA further. At the end of 1942, the Farm Security Administration moved with several other agencies into the Food Production Administration (FPA) within the USDA. In a surprise choice for the powerful position of director of food production, Secretary of Agriculture Claude Wickard selected Herbert W. Parisius, an ardent New Dealer with no loyalty to any USDA faction. Parisius's term as head of the FPA was a disaster. He thought in terms of food production, so within the USDA, it was natural for him to turn to the FSA. Parisius even selected C. B. Baldwin as his deputy director. Wickard, however, rejected the nomination; he respected Baldwin as an administrator and as an influential member of the liberal wing of the USDA, but he also recognized that Baldwin was unacceptable to conservative legislators. In the resulting fight, Parisius's plan was shelved. He resigned from the FPA in January 1943, part of a round of firings and reshufflings widely seen as a defeat for the FSA.[63]

That cycle of USDA reorganization went poorly, and Claude Wickard took most of the blame. Wickard had apparently been chosen by Roosevelt solely on the recommendation of Henry Wallace; Wickard recalled that at his first cabinet meeting, it was obvious that Roosevelt did not know who he was. Things only got worse. Wickard proved unable to arbitrate among the competing USDA factions or to enforce his decisions.[64] He faced criticism from both sides: the FSA and its allies worried that he had no sympathy for them, as his rejection of Baldwin showed. Other saw him as a pie-in-the-sky reformer: Congressman J. William Ditter of Pennsylvania, for example, criticized Wickard's "star gazing" and "socialized philosophizing."[65]

Roosevelt lost confidence in Wickard after the Parisius affair. He therefore ordered, on March 26, 1943, that the War Food Administration (WFA) take over the FPA's administrative functions. Like the Food Production Administration, the WFA was in the USDA, but its head, Chester Davis, onetime agricultural adjustment administrator and friend of the AFBF, answered directly to the president. Although Davis too would be gone within months (citing his own lack of authority and disagreement with administration plans to control inflation), the entire affair indicated that the FSA had little support within the USDA, particularly as the much-weakened secretary of agriculture had little ability or desire to defend it.[66]

The FSA's remaining claim to wartime relevance was the labor program. Across the eastern United States, the FSA built migrant labor camps to remedy local shortages of labor, and it had been housing migrants in the West (increasingly including Mexican laborers) for years. Events seemed to prove the FSA's usefulness: In May 1942 officials announced that hundreds of tons of crops (particularly asparagus, the first major crop harvested in the Northeast) were rotting in the fields due to labor shortages. Announcing the construction of two migratory labor camps to address labor shortage in the area, FSA officials indicated that with the draft and the defense industry's growing demand for labor, American agriculture could expect further such shortages. And the situation was actually better in the Northeast, where income for field hands had doubled since the previous year. The message was clear: American agriculture needed the FSA's proven skill to provide labor.[67]

Agricultural leaders greatly feared the possibilities of labor shortages, but many big farmers still opposed the FSA. For example, with his crop having lain in the field for several weeks, one Arizona cotton farmer refused to hire FSA-provided labor at the price of thirty cents an hour. This was not because of the cost—the farmer had been willing to pay that kind of wage—but because of unwillingness to accept, on principle, the concept of an agricultural minimum wage.[68]

Politically motivated spokesmen also attacked the FSA's role in the labor program. The National Cotton Council's Oscar Johnston criticized the FSA as a "spearhead to drive a labor wedge into agriculture." Instead of improving agriculture, he charged, the FSA only gave a hand to the CIO's organizing efforts and those trying "to build a giant bureaucracy which can be used to prosecute a philosophy of state land socialism."[69] W. R. Ogg, the AFBF's director of research, attacked the FSA during farm labor hearings in February 1943 as redundant and unnecessary. He recommended that the entire responsibility for organizing and transporting farm labor be placed in the Extension Agency, with each state developing and running its own labor program when possible. Tipping his hand, Ogg also recommended that the USDA be prohibited from making any kind of regulation regarding minimum wages, housing standards, or maximum work hours, or imposing or enforcing any collective bargaining agreements.[70]

The FSA was in a weak political position for this fight. For one thing, even supporters recognized that there were flaws in how the migrant labor program operated. Corporate farms made enormous profits by treating labor poorly, and the migrant labor program, in taking over the care of these laborers, simply allowed large farms to continue to do so. As Georgia's Senator Russell said, these big farms, "having been willing to let these people live like dogs when they had no work for them, and now that they can use them, they want the Government to provide their living facilities." Russell worried that the program, if expanded, would encourage further "predatory" practices. Baldwin could only respond that the conditions already existed and that he did not believe that the FSA was doing anything to encourage migratory farm movement, but was rather only addressing the problem as it stood. Further, while Baldwin did not want to expand the program, it had become necessary because of wartime demands for "maximum agricultural production."[71]

A second problem was that Franklin Roosevelt had essentially given up on old favorites like the FSA and the National Youth Administration, conceding to Congress official responsibility for making cuts in non-war-related government spending in late 1942. Well into the spring of 1943, the farm program was up in the air; the FSA's $65 million labor plan was cut by Congress to $26 million, while FSA opponents pressured FDR to set up a separate farm labor agency. The FSA finally lost any influence in the farm labor program when FDR signed House Joint Resolution 96 into law on April 29, 1943, moving the farm labor program to the Office of Labor in the WFA (along with many experienced FSA personnel).[72]

During the 1943 appropriations planning, the House again tried to cut FSA funding. The Senate Appropriations Committee restored the FSA's funds, but it hinted at its willingness to compromise. New regulations limited the FSA by prohibiting land purchase, capping loan sizes, and recommending the liquidation of the resettlement projects; the Senate also drastically cut the agency's funding.[73] The FSA only survived at all because of in-fighting among the Rules, Appropriations, and Agriculture Committees. The House Appropriations Committee drastically cut funding for the USDA, including eliminating the FSA, but it went too far for those who feared its power to step on their own toes.[74] North Carolina representative Harold Cooley, though an FSA critic, attacked the move, saying the cut was action that the Agricultural Committee or legislation should have covered. Agricultural committee members resented "the fact that the Committee on Appropriations has apparently constituted itself into a super-committee."[75] The impasse was only temporary, but in the tussle, the FSA managed to live for another year.[76]

The FSA survived, but these drastic reductions in the summer of 1943 essentially finished off the agency as a political force. Funding cuts had begun to show. The resettlement program, always a small part of the agency's budget, was almost entirely gone; by 1943, 99 percent of the FSA's budget went toward its two other field operations, tenant purchase and rural rehabilitation.[77] The FSA rapidly sold off resettlement projects in 1942 and 1943, selling 500 family units a month by the spring of 1943. After the nonfarm projects were transferred to the National Public Housing Authority, almost half of the remaining 10,109 units had been sold by June 1943, when the FSA switched to fee-simple sales to quickly liquidate the projects.[78]

The lack of funds and the new regulations forced other significant changes in FSA policy. Congress reduced and limited the uses of loans for rehabilitation, which deemphasized supervision. A shortage of funds forced the FSA to eliminate almost its entire staff of technical experts in farm and home management, debt reduction, cooperatives, and tenure improvement. Farm and home plans were considerably reduced in complexity and purpose. The influence of local farmers' committees expanded, which encouraged approval of relatively higher-income farmers.[79]

The appropriations fight left the FSA a shadow of its former self, and C. B. Baldwin in particular faced a difficult time. Some of the criticism was professional, regarding his inexperience as a farmer or failures of the FSA attributed to his administration. But he also dealt with what National Farmers Union president James Patton called "a highly personalized smear campaign."[80] Baldwin tried to defend the FSA as best he could, and he turned down a job with the Lend-Lease Administration, he said, until "the atmosphere clears a bit."[81] However, the total lack of support from the USDA, especially Secretary Wickard, encouraged Baldwin to leave. He was doubtless influenced by rumors swirling that he would be forced out anyway; in June, WFA chief Chester Davis was widely believed to be planning Baldwin's removal, and Davis's replacement, Marvin Jones, told Baldwin that the FSA could not survive with him as administrator.[82] As one reporter put it, "Some congressmen were heard to mutter that if Baldwin were ousted they'd gladly vote F-S-A all the funds it asked for."[83] Baldwin resigned in September 1943 to take a job in the reconstruction of postwar Italy, a position that never actually materialized. To most observers, Baldwin's resignation reflected the victory of anti-New Deal politicians like Harry Byrd and the AFBF.[84]

Had the appointment followed the policy of promoting from within the FSA's top leadership, the new administrator probably would have been the capable and experienced Robert W. Hudgens. He had served as both a state and regional director in the RA and had been in the Washington office of the FSA for four years, including as acting administrator after Baldwin's exit. Hudgens, however, was unacceptable to many in the USDA and to several members of Congress, who associated Hudgens (like Baldwin) with the early, Tugwell-influenced New Deal. Marvin Jones in particular had little respect for Hudgens, although he had trouble finding a replacement, being rebuffed by a number of different people before finding a candidate willing to take the job.[85]

Eventually the post fell to Franklin W. Hancock, a moderate southerner and former congressman committed to at least the Bankhead-Jones Farm Tenant Act. Hancock's background reflected an interest in home ownership: he had served for three years on the Federal Home Loan Bank Board as director of the Home Owners Loan Corporation.[86] He had no connection with the early New Deal or farm security. He believed that the FSA had gone too far in promoting expensive social experiments and changed FSA policy along these lines. Rural rehabilitation loans were limited to fifteen hundred dollars, and qualifications for all types of loans were raised. Grants dropped dramatically.[87]

It is easy to overstate the importance of the change in leadership. Hancock (whom Baldwin later called "not a bad guy") and Baldwin had different visions of what the FSA should be doing, but under both, the FSA emphasized rural rehabilitation and farm ownership as the agency's primary activities.[88] Hancock did not oversee a massive shift in the nature of the FSA or most of its goals. FSA publications under both administrators, like the 1942 pamphlet *Farm Security* and the 1945 pamphlet *Services to Rural Families*, shared many similarities. The latter pamphlet emphasized the tenant purchase program more, and the 1942 pamphlet went into more detail about other FSA programs like medical-aid loans and tenure improvement, but both described supervised loans as the chief work of the FSA. In 1945 Hancock listed the FSA's four major activities in order as rural rehabilitation, tenant purchase, water utilization, and the liquidation of resettlement and cooperative projects.[89] Baldwin's list likely would have looked much the same. More important than leadership philosophy was the decline in available resources. The change from Baldwin to Hancock, and the different approach Hancock took, was a symptom of the FSA's decline, not its cause.

Hancock recognized the problem the FSA faced in terms of congressional support, which meant, as he wrote to regional directors, that the "present and the future well being of Farm Security is, therefore, being weighed in the balance." He was turning all of his attention to "meriting the confidence of the people's representatives in the Congress." Toward this end he urged every employee to make whatever changes possible without harming the FSA's primary goals, for example, by substantially reducing their expenses on travel and communication.[90] Hancock made the case to critics in Congress that the FSA was valuable but also that it was sticking within its bounds. "To the best of my knowledge, gentlemen," Hancock said during the House appropriation hearings, "we are conducting no program, carrying on no activity, that is not definitely within the authorities given to us."[91]

Hancock recognized the congressional mood and tried to work with it. In 1944, when Cooley first proposed a bill replacing the FSA with a more credit-oriented agency, Hancock said that "there is no substantial difference in objectives between myself and the members of the subcommittee which drafted this bill" and praised the proposed bill for stopping projects that had been "ill-advised and misguided practices which, however well intentioned, tended to defeat rather than promote the advancement of the family's industry and independence."[92] Hancock also told Cooley's committee that he supported the Bankhead-Jones Farm Tenant Act and that he was a "strong advocate of the individual family-type farm." The FSA could not be measured by profits or interest earned, he argued, "but by how much can be achieved in family improvement at the least possible cost to both the borrower and to the Government."[93]

Hancock also tried to infuse new life into rural rehabilitation. An FSA memorandum noted that "the lending aspects of the program have constantly overshadowed the more important supervisory aspects" of rural rehabilitation, with the result that county supervisors had made "more or less a farce of supervision." The solution was a reemphasis on operating farms for profit. The FSA should also accept "only borrowers who need our type of service, who understand what the service is and who want the service."[94] Hancock and other FSA administrators believed that there was still a need for rural rehabilitation in the postwar world. FSA handbooks in 1945 instructed committeemen, "Many people are prone to forget the circumstances of thousands of our low-income farm families today because of the generally improved agricultural situation during the war period .. It would be well to remember that it was only ten years ago that about two million farm families in this country were on relief," and small farmers had received a relatively small share of the agricultural gains. If anything, new machinery and farming methods meant an even greater need for credit and supervision if low-income farmers were to increase their production and maintain their farms.[95]

Congress was not as interested in reviving rural rehabilitation, however. Even before World War II ended, Representative Harold Cooley of North Carolina introduced a bill to eliminate the FSA and replace it with a less expansive program dedicated almost exclusively to farm credit. Chairing hearings of the Select Committee of the House Committee on Agriculture, Cooley claimed that he "actively supported every appropriation which has been made for [the FSA] and the various programs which it is administrating," which proved his "friendship for the agency."[96] He (and his committee) then went on to attack the FSA for ignoring Congress and creating un-American resettlement programs and for "Tugwellian" influences, unfair and burdensome loans and supervision of clients, duplication of efforts, and general maladministration. His goal, Cooley's report noted, was to "save and make stronger the good parts of the Farm Security Administration plans for helping worthy farm people."[97]

Hancock had limited success in building a better relationship with Congress. He lasted two years as administrator, during which time the majority of the FSA's most prominent liberals had left and a large number of regional directors were replaced. Hancock resigned in the fall of 1945. Numerous individuals and groups, including the National Catholic Rural Life Conference, the CIO, the Farmers Union, and the Southern Conference for Human Welfare called for Assistant Administrator Robert Hudgens to replace Hancock. But instead, the conservative Georgian Dillard P. Lasseter, the candidate of Georgia congressman Malcolm C. Tarver (who had generally been hostile to the FSA), took over as FSA administrator.[98]

Lasseter was an extremely unpopular choice among reformers and supporters of the FSA. C. B. Baldwin called him an "unprincipled reactionary," and Farmers Union president James H. Patton called the appointment "a bitter betrayal of millions of small farmers."[99] Many others agreed. One woman wrote to Secretary of Agriculture Clinton Anderson to tell him that the appointment of Lasseter meant that he had "betrayed the F.S.A. and worked against the interests of small farmers" and urged Anderson to "cancel that appointment and appoint a man who is honestly interested in the welfare of small farmers in need of aid."[100] Another letter writer complained about the appointment of a man "so poorly qualified as Dillard Lasseter to the F.S.A., when as eminently qualified a man as R. W. Hudgens was available and was the farmers' choice."[101] USDA officials openly admitted, while claiming to respect his abilities as an administrator, that Lasseter's appointment was intended to ease relations with Congress.[102]

By this point, the FSA had begun to look into new activities and new directions.[103] Before World War II ended, it had already started to focus its efforts on aiding veterans. Henry Wallace (then vice president) had suggested that the federal government, possibly through the Farm Security Administration, should give servicemen "some priority" toward purchasing farms after the war, perhaps using government-owned land around defense industries.[104] The FSA became increasingly occupied with the issue of veterans returning to the farm following World War II. Applications for rural rehabilitation loans rose rapidly after the end of the war, close to half of them from veterans unable to find credit. While veterans had GI Bill benefits available to them, red tape and other difficulties made loans difficult to get from private lenders, especially in rural areas. In some cases, bankers in rural counties with small populations simply refused loans to veterans who were low-income farmers, even with the GI Bill guarantee, because they had not made such loans before and saw no reason to begin doing so.[105]

Having not yet killed it, the FSA's opponents tried to eliminate or refashion it. The AFBF and National Cotton Council recommended that the FSA be dismembered and its programs divided among different agencies. North Carolina's Harold Cooley introduced and held hearings on H.R. 4384, the Farmers' Home Corporation Act of 1944. It would have eliminated most of the FSA's functions save the tenant purchase program, providing government loans with a preference for veterans. But with the FSA now mostly impotent, the sense of danger dissipated; Cooley could not pass a bill for a replacement agency until 1946.[106]

The Farmers Home Administration Act of August 14, 1946, provided the authority to create the FSA's successor, the Farmers Home Administration. Dillard Lasseter continued on as administrator when the new agency began activity on November 1, 1946. The FmHA extended the coverage of the loan program to include farm owners and government-insured loans from private creditors. Preference was given to veterans with farming experience: 42 percent of the money loaned in the first fiscal year went to veterans, and the number of veteran borrowers rose rapidly. The Farmers Home Administration Act also allowed the transfer of some existing loans from the Farm Credit Administration. The remaining FSA and FCA Emergency Crop and Feed Loan offices combined into nineteen hundred FmHA county offices. The regional offices of the predecessor agencies, so objectionable to agricultural conservatives, were abolished, and their functions were transferred to state offices.[107]

On paper this constituted a considerable shift from the early goals of farm security; actually, the changes during World War II in personnel, congressional oversight, and available funds meant that the FSA had become little more than a federal credit agency well before the appropriate legislation was passed. Indeed, the FmHA, while specifically designed as a credit agency, carried over some of the functions of the FSA that had withered in the last three years of its existence: debt adjustment services, group services, tenure improvement, and agency guidance in farm and home management.[108] But overall, it marked the final defeat of the FSA's reform goals. Powerful farm-state congressmen used wartime changes to protect large agricultural interests, while at the same time the Senate and House Agricultural Committees, with the assent of the AFBF and similar groups, eliminated or altered programs that favored the less politically influential rural poor, like those of the FSA and the Bureau of Agricultural Economics.[109]

Conclusion

The United States Department of Agriculture's Economic Research Service (the successor to the Bureau of Agricultural Economics) defines a county in persistent poverty as one that has 20 percent or more of its population living in poverty over a thirty-year period. There are currently 353 such counties in the United States, and they are overwhelmingly rural (85 percent) and southern (84 percent). More than one in five southern counties is persistently poor, clustered in the Mississippi Delta, along the Black Belt, and in Appalachia. A quarter of all Americans who live in completely rural counties are in a county that qualifies as persistently in poverty.[1]

The failure of the United States to address the worst of rural poverty, much less to ease the transition into a more mechanized agriculture (which federal policies were accelerating) or to protect small farmers or help them maintain the viability of family-sized farms, is obvious. Although there have been irregular attempts to fight poverty, some aimed partly or entirely at rural Americans, the last serious efforts to reform rural life for the benefit of the poor and marginalized ended during World War II. The ultimate defeat of the Farm Security Administration was both a practical and a symbolic failure, one that still has repercussions on the United States today.

Thought the agency was ultimately dismantled, the FSA did have its accomplishments. At the time, even many of those who applauded the replacement Farmers Home Administration recognized the significance of what the FSA had done, noting that its true value lay not in a loan program, but its interest in "human beings and in the conservation of rural human resources."[2] The FSA had an impressive record in terms of the number of families it helped. By September 30, 1943, the RA and FSA had made standard rural rehabilitation loans to an estimated 695,000 families, or about one out of every nine farm operators in America. Perhaps another 334,000 families received nonstandard or emergency loans, which did not have home and farm plans and were generally much smaller. When the United States entered World War II, perhaps as many as 1.6 million farmers met the eligibility requirements for receiving rehabilitation loans—that is, they were under sixty-five years of age, not in the extremely low-income group, and had the skills to operate a family farm. Thus about two out of every five operators eligible for a standard loan, roughly, received one.[3] Overall, the FSA provided rural rehabilitation loans to over 10 percent of all U.S. farmers.[4]

Smaller, but still considerable, numbers of people took part in tenant purchase, resettlement, and other FSA programs.[5]

Several scholarly works support the view that the FSA had generally beneficial results for those involved. Looking at two northern plains counties, Michael Johnson Grant calls the rural rehabilitation program a clear success in improving family conditions and maintaining a family farm in the short, while as a long-term program for future prosperity the record was less compelling. Regarding New Deal efforts to aid the southern rural poor, Paul Mertz asserts that "rural rehabilitation efforts did the most good," though he noted that "even these were not sufficiently comprehensive." Lester M. Salamon found in the 1970s that "the long-term impact of the resettlement experiments on land retention is substantial" and even more impressive improvements in factors like civic participation and self-perception. For western resettlement projects, Brian Cannon notes that residents tended to strongly favor the goals of the resettlement project, although for many residents, remaining on the projects simply did not make economic sense.[6]

Equally important was the FSA's place as a reflection of New Deal liberalism aimed toward small farmers. The FSA was among the very few New Deal agencies primarily concerned with agricultural laborers, tenants, and small farmers; it was certainly the largest and most influential of such agencies.[7] The FSA more or less successfully implemented rural rehabilitation and other similar programs, proving it could be an effective part of a rural reform effort, though it required years of experimentation and refinement to become so. If nothing else, the intense hostility the FSA incited in those who most benefited from the status quo, especially in the South, indicates that it had a program that could threaten to overturn the existing power structure in American agriculture.

Such success and promise was not enough, however. By the time that the FSA had developed into a mature organization in the late 1930s, it had gone through a number of bureaucratic transitions. This organizational unrest and the lack of legislative foundation profoundly distorted the operations of the FSA. Even as the agency settled on rural rehabilitation and targeted loans as the best solution to rural poverty, its earlier, politically unpopular programs like the subsistence homesteads provoked loud criticism. During World War II, just as critics became most emboldened, the FSA found itself with fewer and fewer political allies. As a result, it was effectively impotent by 1943. The FSA and its most effective programs, like rural rehabilitation, thus had little part to play in the changing rural world of postwar America.

American liberalism, both in its approach to rural problems and as a whole, underwent significant changes during the Second World War. Liberalism gave up on the micromanagement of the economy that had been so evident during the New Deal. Instead, macroeconomic tools like government spending became the main avenue of controlling and influencing the economy.[8] Similarly, farm security no longer provided the solution to rural problems. Instead of improving the lives of rural people on the land, government action was more concerned with resource management and how that could improve the lives of rural Americans through things like hydropower and electrification, which would simultaneously improve America's industrial base. Keeping people on the land or protecting family farms became little more than rhetorical points. Proponents of agricultural efficiency won the day.[9]

This was by no means an inevitable development. Through most of the New Deal, American liberals had a deep suspicion of concentrated economic power as, at best, an obstacle to reform and social growth and at worst, a threat to democracy itself. But the new liberalism that emerged from World War II had an increasing confidence in the value of capitalism; for liberals, business suddenly seemed to work, both in terms of promoting the war effort and in creating almost full employment and mass consumption by the end of World War II.[10] There was potentially a role for the family-sized farm in this new agenda of promoting economic growth and a consumer economy, perhaps in bringing the middle-class consumer economy to rural America. That no such place could ever be found had to do with the specific failures of the FSA and the effectiveness of its political foes, who took control of agricultural policy during a time of great change.

The Great Depression interrupted an enormous transition in American agriculture. For decades, the trend was migration from farms and rural areas into urban industrial ones, driven by industrialization that created a demand for labor in cities and mechanization that reduced the demand in agriculture. In the 1920s, a net total of 6 million people moved from farms to cities, a development particularly evident in the South. The Depression slowed and, in some cases, reversed this trend. Some large cities even saw an overall decline in population. With the reduction in industrial employment, many people felt safer staying in rural areas.[11] However, even during the Depression, large numbers of people were leaving rural areas. One community surveyed between 1934 and 1936 found that half of the sons and three-fifths of daughters beyond school age had left the community, which was not particularly unusual given the mobility associated with tenant farming and the rural South, but about half of those had abandoned farming in favor of moving to local cities and towns.[12]

The demand for farm labor decreased every year as a result of mechanization and government policy. In the South, the most significant mechanical advance was the spindle cotton harvester. The 1930s saw the introduction of "stripper" style machines that pulled the cotton boll from the plant. These harvesters were really only useful in the dry Southwest; most of the South required a machine that could pick the lint from the bolls. John and Daniel Rust developed such a machine in the late 1930s and demonstrated them in the 1942 on Mississippi farms. Large-scale commercial production was delayed by the war but began in 1947.[13]

It took a while for mechanization to influence southern agriculture: creating a mechanical device to imitate the intricate task of picking cotton was far more difficult than, for example, one designed for harvesting wheat. It took the introduction of corporate resources, from companies like International Harvester, to create a successful commercial cotton picker.[14] The trend toward mechanization occurred faster in the rest of the country than it did in the South, which provided southerners with a picture of what would be happening in their region. Economists wondered about the "displacement" that would come to southern workers; "concern is expressed," one wrote in 1937, "as to what may happen if success is met with in the efforts to perfect a mechanical cotton picker and to extend methods of check-row planting and cross cultivation as means of decreasing chopping by hand."[15] In 1945, before mechanization took off in the South, Rupert Vance noted that "efficient practices" could allow only "half the farm people of the South" to meet the entire domestic and export demand for southern agricultural production.[16]

Sociologist Morton Rubin visited a Deep South plantation in the late 1940s and found that it operated in much the same way as it had before World War II—or World War I for that matter. The rich but perpetually-at-risk landlord ran a cotton plantation based on the unskilled labor of numerous resident tenants, sharecroppers, and laborers who lived in one- and two-room cabins, made purchases with cash or scrip at the local planter-owned commissary, and depended on credit advances to get them through the winter. The plantation had tractors and other machinery, but skilled and semiskilled labor was not particularly important: unskilled black laborers chopped and picked cotton as they had for centuries. It was not clear how long these laborers would be necessary.[17]

Between 1950 and 1970, a change occurred in American agriculture that was almost beyond comprehension for those living through it.[18] More than ever before, farming was a large-scale operation. Small farmers produced a decreasing portion of American agricultural output. As early as the mid-1940s, smaller operators found themselves left behind as they were unable to purchase equipment and machinery to keep up with bigger farms.[19] Observers at the time and since then have seen this as a change for the better. Donald Holley, for example, describes the "South's transition from labor-intensive to capital-intensive agriculture, a transition that was both inevitable and necessary for the Cotton South to achieve social and economic modernization"; this added up to a "Second Great Emancipation [that] freed the Cotton South from the plantation system and its attendant evils—cheap labor, ignorance, and Jim Crow discrimination." Holley argues, "Clearly, agencies like the Farm Security Administration (FSA) were on the wrong track with their effort to nurture the efforts of small farmers. Even on the most successful FSA projects, the farm units were too small to be economically viable. With the advent of tractors in the 1930s, leading agricultural experts realized that future cotton farms would be mechanized and therefore larger."[20]

From this perspective, the FSA was too conservative a program, trying to preserve an older style of agriculture based on small farms owned and operated by a single family. Even some FSA administrators agreed to an extent. Paul V. Maris, head of the tenant purchase program, claimed in 1944 that the FSA was in effect subsidizing a large number of clients in their decent lifestyles. He pointed to the example of a borrower in Lowndes County, Alabama, who could afford a car and a furnished home by making no payments on either of his FSA loans that year. The fact was that, especially in the low-income regions, "we have a greater population on the land than it can be made to support even when good farming practices are followed... without subsidies standards must remain very low." Simply put, there were more people farming than the land could support.[21]

Such a perspective reflected the prevailing view that the true goal of federal agricultural policy should be to increase farming efficiency and raise prices. This kind of sentiment informed the longest-lasting New Deal agricultural policies, like those of the Agricultural Adjustment Administration, whose system of benefit payments provided the capital for thousands of farmers to purchase tractors. Starting in the flatter Southwest and moving to the hillier east with each mechanical advance, each tractor did the work of several tenant or laborer families.[22] The result, in agricultural terms, was more efficient farms and a decreased need for labor.

But there are reasons to question this emphasis on increasing production as the most important measure of agricultural success. For one thing, it assumes that changes in American agriculture happened in a vacuum or were propelled purely by market and social forces. This was not the case. Government policy influenced, in ways direct and indirect, intended and not, the dramatic changes that took place in rural America in the decades following World War II. In the case of policy affecting southerners, federal policy reinforced shifts that favored the already wealthy and powerful at the expense of the poor. In historian Bruce Schulman's words, they were "policies designed not so much to uplift poor people as to enrich poor places"; as the region's upper and middle classes were pulled into the national market, much of its rural poor was left behind.[23] This meant that many moved off the farm, either to a life in rural poverty or to employment in a city. While doubtless many wanted to move (the allure of city life having long been irresistible for many young people growing up in the country), not all did. Some fought against mechanized, capital-intensive agriculture and for the traditional rural life they had grown up living, sometimes hoping to have the support of the federal government; such efforts usually proved unsuccessful.[24]

Most important, it also is not clear that production has ever really been the problem with American agriculture. It has been a long time, after all, since actually growing enough food for Americans to eat at a reasonable cost has been a serious obstacle for American farmers, and it is poverty, inequality, and structural problems, not the actual shortage of food, that most contribute to hunger in the United States and the world. As historian Paul Conkin puts it, "the greatest surplus problem in American agricultural history has involved not commodities but people."[25] This remained a problem after the end of the Second World War. As one agricultural economist described the situation in 1955, a large number of farmers "who operate continuously at low levels of output and income" were too low-income to get credit from commercial banks. This group, "nearly a third of the farmers of the United States," were "destined to continue on a bare subsistence basis in ordinary times."[26]

It is possible that the FSA, or any other agency, may never have raised the efficiency and productivity of small farmers enough to compete nationally in large-scale food production. The efficiencies of large-scale, capital-intensive, technologically advanced agribusiness meant (and continue to mean) that such processes produce at least most food stuffs cheaper. But the FSA or a similar program might have demonstrated a willingness on the part of the federal government to democratize agricultural policy or to at least provide aid to small farmers to try, in some small way, to balance out the massive amounts of support, both directly and indirectly, that the federal government provided to large farmers. More than that, one might ask, is farming as cheaply as possible necessarily the best way? The social and environmental costs of modern agriculture indicate that it is a question worth asking.

Notes

Note on Shortened Citations

To make the endnotes easier to read, citations for the following frequently cited sources and collections have been shortened throughout this book. Full documentation of these sources is available in the bibliography.

Government Documents

The yearly reports of the Resettlement Administration and the Farm Security Administration between 1936 and 1946 are listed in full in the bibliography under "Resettlement Administration," "United States Department of Agriculture, Farm Security Administration," and "United States Department of Agriculture, War Food Administration." They are cited by title throughout the footnotes.

The titles of the annual agricultural appropriations bills and hearings have been shortened and are listed by house of Congress and year, with the year the hearings were held or the bills passed in parenthesis.

"*Byrd Committee Hearings*" is the shortened citation for U.S. Congress, Joint Committee on Reduction of Nonessential Federal Expenditures, *Reduction of Nonessential Federal Expenditures, Part II.* 77th Cong., 1st–2nd sess., December 1, 1941, January 23, 1942, and February 3, 1942; and U.S. Congress, Joint Committee on Reduction of Nonessential Federal Expenditures, *Reduction of Nonessential Federal Expenditures, Part III,* 77th Cong., 2nd sess., February 6, 10, 13, and 27, 1942.

"FDR, *PPA*" is the shortened citation for Franklin D. Roosevelt, *The Public Papers and Addresses of Franklin D. Roosevelt,* 13 vols., ed. Samuel I. Rosenman (New York: Random House, 1938–1950).

"House, *Activities of the FSA*" is the shortened citation for U.S. Congress, House, *Activities of the Farm Security Administration, Report of Select Committee of the House Committee On Agriculture to Investigate the Activities of the Farm Security Administration,* 78th Cong., 2nd sess., 1944, H.R. 1430.

"House, *Farm Labor Program, 1943*" is the shortened citation for U.S. Congress, House, *Farm Labor Program, 1943, Hearings before the Subcommittee of the Committee on Appropriations, on the Appropriation for the Farm Labor Program, Calendar Year 1943,* 78th Cong., 1st sess., 1943.

"House, *Farm Security Hearings*" is the shortened citation for U.S. Congress, House, *Hearings Before the Select Committee of the House Committee on Agriculture, to Investigate the Activities of the Farm Security Administration,* 78th Cong., 1st sess., 1943–1944.

"House, *Farm Tenancy Hearing*" is the shortened citation for U.S. Congress, House, *Farm Tenancy, Hearing before the Committee on Agriculture on H.R. 8.* 75th Cong., 1st sess., 1937.

"House, *Payments in Lieu of Taxes on Resettlement Projects*" is the shortened citation for U.S. Congress, House, *Payments in Lieu of Taxes on Resettlement Projects, Hearings before a Subcommittee of the Committee on Ways and Means.* 74th Cong., 2nd sess., 1936.

"Senate, *Resettlement Administration Program*" is the shortened citation for U.S. Congress, Senate, *Resettlement Administration Program*, 74th Cong., 2nd sess., 1936, Doc. 213.

Archival Sources

"Alabama Relief Administration Papers" is the shortened citation for Alabama Relief Administration, State Publications, 1932–1935, SG014348, Alabama Department of Archives and History, Montgomery, Alabama.

"Bankhead Papers" is the shortened citation for John H. Bankhead Jr., Papers, LPR53, Alabama Department of Archives and History, Montgomery, Alabama.

NAL refers to documents and publications currently housed at the National Agricultural Library, Beltsville, Maryland. Many of these documents, such as reports and brochures, lack other clear information regarding publication and source.

The following collections from the Records of the Farmers Home Administration, Record Group 96, held at the National Archives at College Park, Maryland, have been shortened in the endnotes as follows, with the current archival entry number for each collection in parenthesis.

"FSA Historical Records, RG 96, NACP" refers to Records Relating to the History of the Farm Security Administration (Entry 19).

"General Administrative Records, RG 96, NACP" refers to General Administrative Records, 1943–56 (Entry 48).

"General Correspondence, Washington Office, RG 96, NACP" refers to General Correspondence Maintained in the Washington Office, 1935–38 (Entry 1).

"Project Records, RG 96, NACP" refers to Project Records, 1935–1940 (Entry 71).

"Rural Rehabilitation, General Correspondence, RG 96, NACP" refers to Rural Rehabilitation Division, General Correspondence, 1935–38 (Entry 5).

The following collections from the Records of the Farmers Home Administration, Record Group 96, held at the National Archives Southeast Region at Morrow, Georgia, have been shortened in the endnotes (the current archival entry number for each collection is in parenthesis):

"Farm Ownership Files, RG 96, NARASE" refers to Records of the Rural Rehabilitation Division, Region 5, Montgomery, Alabama, Farm Ownership Case Files, 1938–1946 (Entry 94).

"Office of the Director, General Correspondence, RG 96, NARASE" refers to Records of the Office of the Director, General Correspondence, 1934–1942 (Entry 93).

"Project Construction Files, RG 96, NARASE" refers to Project Construction Files, 1938–1941 (Entry 97).

"Rural Rehabilitation Loan Files, RG 96, NARASE," refers to Rural Rehabilitation Loan Case Files, 1934–1944 (Entry 133).

The following collections from the Records of the Office of the Secretary of Agriculture, Record Group 16, held at the National Archives at College Park, Maryland, have been shortened in the endnotes (the current archival entry number for each collection is in parenthesis):

"Office of the Secretary, General Correspondence, RG 16, NACP" refers to Office of the Secretary, General Correspondence, 1906–1975 (Entry 17).

"Office of the Solicitor, Resettlement Cases, RG 16, NACP" refers to Records of the Office of the Solicitor, Records of Resettlement Administration Cases, 1935–1937 (Entry 113).

"Office of the Solicitor, Rural Rehabilitation, RG 16, NACP" refers to Records of the Office of the Solicitor, Records of Rural Rehabilitation Cases, 1937–1943 (Entry 101).

Preface

1. Mattie Cole Stanfield, *Palmerdale, Alabama: 50-year History, 1934–1984* [Palmerdale, AL: self-published, 1984], 2.
2. Stanfield, *Palmerdale, Alabama*, 4–8, 14.

Introduction

1. "Roosevelt Begins Providing Jobless with Land to Work," *New York Times*, May 28, 1932.
2. This variety of approaches to reform matched the first years of the New Deal as a whole; see Alan Brinkley, *The End of Reform: New Deal Liberalism in Recession and War* (New York: Random House, 1995), 4–6.
3. Anthony Badger, *The New Deal: The Depression Years, 1933–1940* (New York: Hill and Wang, 1989; repr., Chicago: Ivan R. Dee, 2002), 6. See also David M. Kennedy, *Freedom from Fear: The American People in Depression and War, 1929–1945* (New York: Oxford University Press, 2005), 153–54.
4. For example, government scientists developed a hybrid seed corn that improved yields by about 20 percent at the same time that the government was trying to reduce corn acreage. Donald C. Blaisdell, *Government and Agriculture: The Growth of Federal Farm Aid* (New York: Farrar and Rinehart, 1940), 124.
5. John H. Caufield, "Dalworthington Families Combine Rural and City Life at Small Expense," *Dallas Morning News*, May 14, 1938.
6. See Numan V. Bartley, *The New South, 1945–1980* (Baton Rouge: Louisiana State University Press, 1995), 122–31; Pete Daniel, *Breaking the Land: The Transformation of Cotton, Tobacco, and Rice Cultures since 1880* (Urbana: University of Illinois Press, 1985), 239–298; Gilbert C. Fite, *Cotton Fields No More: Southern Agriculture, 1865–1980* (Lexington: University Press of Kentucky, 1984), 180–225; and Bruce J. Schulman, *From Cotton Belt to Sunbelt: Federal Policy, Economic Development, and the Transformation of the South, 1938–1980* (Durham, NC: Duke University, 1994), 88–173.
7. Carol A. Horton, *Race and the Making of American Liberalism* (New York: Oxford University Press, 2005), 121–30. See also Brinkley, *End of Reform*, 265–70. Gail Radford describes how New Deal housing programs came to most benefit prosperous homeowners in *Modern Housing for America: Policy Struggles in the New Deal Era* (Chicago: University of Chicago Press, 1996).

8. Joseph W. Eaton, *Exploring Tomorrow's Agriculture: Co-Operative Group Farming—A Practical Program of Rural Rehabilitation* (New York: Harper and Brothers, 1943), 3.
9. House, *Farm Security Hearings, Part I,* 193, 194.
10. Sidney Baldwin, *Poverty and Politics: The Rise and Decline of the Farm Security Administration* (Chapel Hill: University of North Carolina Press, 1963).
11. Paul E. Mertz, *New Deal Policy and Southern Rural Poverty* (Baton Rouge: Louisiana State University Press, 1978); Jess Gilbert, "Eastern Urban Liberals and Midwestern Agrarian Intellectuals: Two Group Portraits of Progressives in the New Deal," *Agricultural History* 74.2 (2000): 162–80; Jess Gilbert and Carolyn C. Howe, "Beyond 'State vs. Society': Theories of the State and New Deal Agricultural Policies," *American Sociological Review* 56.2 (April 1991): 204–20; Jess Gilbert, "A Usable Past: New Dealers Henry A. Wallace and M. L. Wilson Reclaim the American Agrarian Tradition," in *Rationality and the Liberal Spirit: A Festschrift Honoring Ira Lee Morgan,* ed. Centenary College Department of English ([Shreveport, La.]: Centenary College of Louisiana, 1997), 134–42; Jess Gilbert and Alice O' Connor, "Leaving the Land Behind: Struggles for Land Reform in U.S. Federal Policy, 1933–1965," in *Who Owns America? Social Conflict over Property Rights,* ed. Harvey M. Jacobs (Madison: University of Wisconsin Press, 1998), 114-30. See also Jane Adams and D. Gorton, "This Land Ain't My Land: The Eviction of Sharecroppers by the Farm Security Administration," *Agricultural History* 83.3 (2009): 323–51; Lee J. Alston and Joseph P. Ferrie, *Southern Paternalism and the American Welfare State: Economics, Politics, and Institutions in the South, 1865–1965* (Cambridge, UK: Cambridge University Press, 1999); Mary Summers, "The New Deal Farm Programs: Looking for Reconstruction in American Agriculture," *Agricultural History* 74.2 (2000): 241–57.
12. Michael Johnston Grant, *Down and Out on the Family Farm: Rural Rehabilitation in the Great Plains, 1929–1945* (Lincoln: University of Nebraska Press, 2002); Jarod Roll, *Spirit of Rebellion: Labor and Religion in the New Cotton South* (Urbana: University of Illinois Press, 2010). See also Michael R. Grey, *New Deal Medicine: The Rural Health Programs of the Farm Security Administration* (Baltimore: Johns Hopkins University Press, 1996).
13. Brian Q. Cannon, *Remaking the Agrarian Dream: New Deal Rural Resettlement in the Mountain West* (Albuquerque: University of New Mexico Press, 1996); Paul K. Conkin, *Tomorrow a New World: The New Deal Community Program* (Ithaca, NY: Cornell University Press, 1959); Donald Holley, *Uncle Sam's Farmers: The New Deal Communities in the Lower Mississippi Valley* (Urbana: University of Illinois Press, 1975); Sara M. Gregg, *Managing the Mountains: Land Use Planning, the New Deal, and the Creation of a Federal Landscape in Appalachia* (New Haven, CT: Yale University Press, 2010); Neil M. Maher, *Nature's New Deal: The Civilian Conservation Corps and the Roots of the American Environmental Movement* (New York: Oxford University Press, 2008); Sarah T. Phillips, *This Land, This Nation: Conservation, Rural America, and the New Deal* (Cambridge, UK: Cambridge University Press, 2007). See also Joseph L. Arnold, *The New Deal in the Suburbs: A History of the Greenbelt Town Program, 1935–1954* (Columbus: Ohio State University Press, 1971); Edward C. Banfield, *Government Project* (Glencoe, IL: Free Press, 1951); and Cathy D. Knepper, *Greenbelt, Maryland: A Living Legacy of the New Deal* (Baltimore: Johns Hopkins University Press, 2001).
14. Howard A. Turner, "Farm Tenancy Distribution and Trends in the United States," *Law and Contemporary Problems* 4.4 (1937): 426. See also Blaisdell, *Government and Agriculture,* 2–27; and Kennedy, *Freedom from Fear,* 195. On tenancy on the northern plains prior to the Great Depression, see Grant, *Down and Out on the Family Farm,* 36–59.

15. T. J. Woofter Jr., *Landlord and Tenant on the Cotton Plantation* (Washington, DC: Works Progress Administration, Division of Social Research, 1936), 62–63.
16. James Agee and Walker Evans, *Let Us Now Praise Famous Men* (Boston: Houghton Mifflin Company, 1939; 1960), 118. See also L. C. Gray, "Our Major Agricultural Land Use Problems and Suggested Lines of Action," in *Farmers in a Changing World,* by U.S. Department of Agriculture (Washington, DC: U.S. Government Printing Office, 1940), 407. On how those who did work hard might see any gains go to the benefit of the property's owners, not for themselves, see Theodore Rosengarten, *All God's Dangers: The Life of Nate Shaw* (New York: Alfred A. Knopf, 1974), 146. The connection between agricultural practices and social problems had been clear for years, and people like George Washington Carver hoped that better use of the land might provide economic independence for the downtrodden black tenant; see Mark D. Hersey, *My Work Is That of Conservation: An Environmental Biography of George Washington Carver* (Athens: University of Georgia Press, 2011).
17. Works Progress Administration, *Farm Tenure Statistics in Southeastern States* [Birmingham, AL: Works Progress Administration, 1938], 30–35; Works Progress Administration, *Farm Tenants in the Southeast, 1920–1935* [Birmingham, AL: Works Progress Administration, 1938], 5–6. A statistically insignificant number (about 11,400 in 1930) were farm managers, most prominently in the black belt and citrus and vegetable regions of Florida. Relatively few southern farmers (about 238,000) were cash tenants, the highest grade of tenant. U.S. Census Bureau, *15th Census of Agriculture, 1930* (Washington, DC: Government Printing Office, 1932), vol. 3, *Type of Farm,* pt. 2, *The Southern States,* 31. On the differences between sharecroppers, tenants, and laborers, see John D. Black and R. H. Allen, "The Growth of Farm Tenancy in the United States," *Quarterly Journal of Economics* 51.3 (1937): 406–7; Jack Temple Kirby, *Rural Worlds Lost: The American South, 1920–1960* (Baton Rouge: Louisiana State University Press, 1987), 140–42; and Woofter, *Landlord and Tenant on the Cotton Plantation,* 10. A good description of the origins of sharecropping and southern tenancy can be found in Donald H. Holley, *The Second Great Emancipation: The Mechanical Cotton Picker, Black Migration, and How They Shaped the Modern South* (Fayetteville: University of Arkansas Press, 2000), 21–28. On the emergence of sharecropping and tenancy as a market process, see Gavin Wright, *Old South, New South: Revolutions in the Southern Economy since the Civil War* (New York: Basic Books, Inc., 1986), 84–106. On patron-client relations in the South as an outgrowth of premechanized agriculture and the resulting opposition among the southern elite to competing government welfare, see Lee J. Alston and Joseph P. Ferrie, *Southern Paternalism and the American Welfare State: Economics, Politics, and Institutions in the South, 1865–1965* (Cambridge, UK: Cambridge University Press, 1999).
18. A. B. Book, "A Note on the Legal Status of Share-Tenants and Share-Croppers in the South," *Law and Contemporary Problems* 4.4 (1937): 541. Some scholars do argue for the efficiency of sharecropping; see, for example, Martin A. Garrett Jr. and Zhenhu Xu, "The Efficiency of Sharecropping: Evidence from the Postbellum South," *Southern Economic Journal* 69.3 (2003): 578–95.
19. Charles S. Johnson, Edwin R. Embree, and W. W. Alexander, *The Collapse of Cotton Tenancy: Summary of Field Studies and Statistical Surveys, 1933–1935* (Chapel Hill: University of North Carolina Press, 1935), 4–5.
20. E. C. Branson, "Farm Tenancy in the Cotton Belt: How Farm Tenants Live," *Journal of Social Forces* 1.3 (1923): 213. See also David Eugene Conrad, *The Forgotten Farmers: The*

Story of Sharecroppers in the New Deal (Urbana: University of Illinois Press, 1965), 2; Johnson, Embree, and Alexander, *Collapse of Cotton Tenancy,* 10.
21. John H. Bankhead to J. L. Edwards, [early December 1936?], Folder 1, Box 5, Bankhead Papers.
22. Rosengarten, *All God's Dangers,* 33.
23. Paul K. Conkin, *A Revolution Down on the Farm: The Transformation of American Agriculture Since 1929* (Lexington: University of Kentucky Press, 2008), 4–19. See also Robert E. Zabawa and Sarah T. Warren, "From Company to Community: Agricultural Community Development in Macon County, Alabama, 1881 to the New Deal," *Agricultural History* 72.2 (1998): 460–61, 471–74.
24. Paul W. Terry and Verner M. Sims, *They Live on the Land: Life in an Open-Country Southern Community* (Tuscaloosa: University of Alabama Press, 1940; 1993), 46.
25. Arthur F. Raper and Ira de Reid, *Sharecroppers All* (Chapel Hill: University of North Carolina Press, 1941), 21; Conkin, *Revolution Down on the Farm,* 49.
26. Margaret Jarman Hagood, *Mothers of the South: Portraiture of the White Tenant Farm Woman* (Chapel Hill: University of North Carolina Press, 1939; repr., New York: Greenwood Press, 1969), 77–91; Melissa Walker, *All We Knew Was to Farm: Rural Women in the Upcountry South, 1919–1941* (Baltimore: Johns Hopkins University Press, 2000), 20–24, 21 (quote). On women's economic role in the rural South, see also Lu Ann Jones, *Mama Learned Us to Work: Farm Women in the New South* (Chapel Hill: University of North Carolina Press, 2002), esp. 49–79.
27. Woofter, *Landlord and Tenant,* 83–86. Subtracting advances for sustenance throughout the year, the cash income average was only $122 for croppers and $202 for tenants. Tenant income was also, as in the rest of the South, irregular: two-thirds of the yearly income came after selling crops. See also Carl C. Taylor, Helen W. Wheeler, and E. L. Kirkpatrick, *Disadvantaged Classes in American Agriculture,* Social Research Report No. 8 (Washington, DC: United States Department of Agriculture, Farm Security Administration, and Bureau of Agricultural Economics Cooperating, 1938), 8–9. On Depression-era rural poverty in the North, one first-hand account is Louise V. Armstrong, *We Too Are the People* (Boston: Little, Brown, 1938; repr., New York: Arno Press, 1971). As Gavin Wright points out in *Old South, New South* (202–3), even considering the difficulties of making such a comparison accurately, real wages in the Deep South were reported to be no higher in 1929 than they had been in 1890.
28. Woofter, *Landlord and Tenant,* 9, 49–62; Raper and Reid, *Sharecroppers All,* 38–39; Johnson, Embree, and Alexander, *Collapse of Cotton Tenancy,* 26–29; Fite, *Cotton Fields No More,* 23–24.
29. Rosengarten, *All God's Dangers,* 33.
30. William Alexander Percy, *Lanterns on the Levee: Recollections of a Planter's Son* (New York: Alfred A. Knopf, 1941), 282.
31. Erskine Caldwell and Margaret Bourke-White, *You Have Seen Their Faces* (New York: Viking Press, 1937), 51–54. The story of the landlord who had to refigure an account upon learning of an extra bale of cotton, in order to maintain a renter's debt, reached folkloric status in the post–Civil War South. See Johnson, Embree, and Alexander, *Collapse of Cotton Tenancy,* 8–11.
32. Woofter, *Landlord and Tenant,* 11; H. L. Mitchell, *Mean Things Happening in This Land: The Life and Times of H. L. Mitchell, Co-founder of the Southern Tenant Farmers Union* (Montclair, NJ: Allanheld, Osmun, 1979; repr., Norman: University of Oklahoma Press,

2008), 22–23; Charles S. Johnson, *Shadow of the Plantation* (Chicago: University of Chicago Press, 1934), 120–23; Raper and Reid, *Sharecroppers All*, 26.
33. Hersey, *My Work Is That of Conservation*, 202–10. This kind of mistreatment contributed to a distaste for farm life among many black former farmers; see Melissa Walker, *Southern Farmers and Their Stories: Memory and Meaning in Oral History* (Lexington: University of Kentucky Press, 2006), 156–64.
34. Ben Robertson, *Red Hills and Cotton: An Upcountry Memory* (New York: Alfred A. Knopf, 1942), 82. See also Johnson, Embree, and Alexander, *Collapse of Cotton Tenancy*, 30–31.
35. Mitchell, *Mean Things Happening in This Land*, 142–43; Robert W. Hudgens, "Rural Problems—The Present Situation in the South" [March 9, 1937], p. 2, Folder "161-Speeches," Box 11, Rural Rehabilitation General Correspondence, RG 96, NACP. See also Edwin R. Embree, "Southern Farm Tenancy: The Way Out of Its Evils," *Survey Graphic* 25.3 (1936): 152; House, *Farm Tenancy Hearing*, 67
36. National Resources Committee, *Farm Tenancy: The Report of the Special Committee on Farm Tenancy* (Washington, DC: U.S. Government Printing Office, 1937), 7–8; Johnson, Embree, and Alexander, *Collapse of Cotton Tenancy*, 15. See also Kirby, *Rural Worlds Lost*, 174–81.
37. Hagood, *Mothers of the South*, 9–10.
38. Morris Llewellyn Cooke, "Electricity Goes to the Country," *Survey Graphic* 25.9 (1936): 510; *15th Census of Agriculture, 1930*, vol. 2, *Reports by States*, part 2, *The Southern States*, 56–57.
39. Woofter, *Landlord and Tenant*, 101, 6.
40. Louis M. Kyriakoudes, "Lookin' for Better All the Time: Rural Migration and Urbanization in the South, 1900–1950," in *African American Life in the Rural South: 1900–1950*, ed. R. Douglas Hurt (Columbia: University of Missouri Press, 2003), 14–18. On southern out-migration, see Wright, *Old South, New South*, 200–207. On efforts to stop migration, see Pete Daniel, *The Shadow of Slavery: Peonage in the South, 1901–1969* (Urbana: University of Illinois Press, 1972), especially 19–42 and 170–92; Cartern Godwin Woodson, *The Rural Negro* (Washington, DC: Association for the Study of Negro life and History, 1930), 67–88.
41. Woofter, *Landlord and Tenant*, 125–28. See also Karl Shafer, *A Basis for Social Planning in Coffee County, Alabama*, Social Research Report, No. 6 (Washington, DC: U.S. Department of Farm Agriculture, the Farm Security Administration, and the Bureau of Agricultural Economics, Cooperating, 1937), 28–29.
42. Terry and Sims, *They Live on the Land*, 189–95, 210–11.
43. Joseph Gaer, *Toward Farm Security: The Problem of Rural Poverty and the Work of the Farm Security Administration* (Washington, DC: Government Printing Office, 1941), 33.
44. Johnson, Embree, and Alexander, *Collapse of Cotton Tenancy*, 16–17; Woofter, *Landlord and Tenant*, 101–3; Johnson, *Shadow of the Plantation*, 100–102; Rupert B. Vance, *Human Geography of the South: A Study in Regional Resources and Human Adequacy* (Chapel Hill: University of North Carolina Press, 1935), 423–33.
45. Kenneth J. Bindas, *Remembering the Great Depression in the Rural South* (Gainesville: University of Florida Press, 2007), 16.
46. Kirby, *Rural Worlds Lost*, 188–90.
47. Vance, *Human Geography of the South*, 376; O. V. Wells, "Agriculture Today: An Appraisal of the Agricultural Problem," in *Farmers in a Changing World*, by United States Department of Agriculture (Washington, DC: Government Printing Office, 1940), 388;

National Emergency Council, *Report on Economic Conditions of the South* (Washington, DC: U.S. Government Printing Office, 1938), 27–30; Raper and Reid, *Sharecroppers All,* 21.
48. Mertz, *New Deal Policy and Southern Rural Poverty,* 13; National Emergency Council, *Report on Economic Conditions of the South,* 29–30. See also Woofter, *Landlord and Tenant,* 97–98.
49. Hagood, *Mothers of the South,* 108–27; Johnson, *Shadow of the Plantation,* 192–96; Terry and Sims, *They Live on the Land,* 110–27.
50. Terry and Sims, *They Live on the Land,* 74–78. See also National Emergency Council, *Report on Economic Conditions of the South,* 23. On the poll tax, see V. O. Key, *Southern Politics in State and Nation* (New York: Alfred A. Knopf, 1949; repr., Knoxville: University of Tennessee Press, 1984), 578–618. Key estimated that the poll tax eliminated 5–10 percent of white voters in most southern states.
51. Key, *Southern Politics in State and Nation,* 504.

Chapter 1

1. "Field Notes," *Rural Rehabilitation* 1.1 (1934): 16; Alabama Relief Administration, "Rural Rehabilitation Program for 1934," p. 7, Folder 4, Alabama Relief Administration Papers; "The Steer Aids in Farm Relief," *New York Times,* January 20, 1935.
2. Alabama Relief Administration, "Rural Rehabilitation Program for 1934," p. 5, Folder 4, Alabama Relief Administration Papers.
3. Terry and Sims, *They Live on the Land,* 45. On the importance of mules in southern agriculture, see Kirby, *Rural Worlds Lost,* 197–204.
4. Lorena Hickock to Harry L. Hopkins, April 7, 1934, in *One Third of a Nation: Lorena Hickock Reports on the Great Depression,* ed. Richard Lowitt and Maurine Beasley (Urbana: University of Illinois Press, 1981), 210–11.
5. McConnell, *The Decline of Agrarian Democracy,* 85–86. On the development of rural rehabilitation in the northern plains, see Grant, *Down and Out on the Family Farm,* 60–97.
6. James G. Maddox, "The Farm Security Administration" (PhD diss., Harvard University, 1950), 157–58. This statistic refers more specifically to those who would become rural rehabilitation clients. At the other end of the scale were borrowers in the West (what would become Resettlement Administration Regions IX, X, and XI), who had generally the smallest families, most education, and highest incomes, with the Northeast (Region I) close behind. The Midwest, Lake States, and Plains were somewhere in between.
7. Howard B. Myers, "Relief in the Rural South," *Southern Economic Journal* 3.3 (1937): 283.
8. W. J. Spillman, "The Agricultural Ladder," *American Economic Review* 9.1 Supplement (1919): 170; Olaf F. Larson and Julie N. Zimmerman, *Sociology in Agriculture: The Galpin-Taylor Years in the U.S. Department of Agriculture, 1919–1953* (University Park: Pennsylvania State University Press, 2003), 149; Edward L. Ayers, *The Promise of the New South* (New York: Oxford University Press, 1992, 2007), 195–200; Wright, *Old South, New South,* 99–107.
9. E. A. Schuler, *Social Status and Farm Tenure—Attitudes and Social Conditions of Corn Belt and Cotton Belt Farmers,* Social Research Report No. 4 (Washington, DC: United States Department of Agriculture, the Farm Security Administration, and the Bureau of Agricultural Economics, Cooperating, 1938), 108–12; Josephine C. Brown, "Rural Families on Relief," *Annals of the Academy of Political and Social Science* 176 (1934): 90. See also William T. Ham, "The Status of Agricultural Labor," *Law and Contemporary Problems* 4.4 (1937): 559.

10. David B. Danbom, *The Resisted Revolution: Urban America and the Industrialization of Agriculture, 1900–1930* (Ames: Iowa State University Press, 1979), 24–74; Dona Brown, *Back to the Land: The Enduring Dream of Self-Sufficiency in Modern America* (Madison: University of Wisconsin Press, 2011), 23–26; 79–105; Gene Wunderlich, *American Country Life: A Legacy* (Lanham, MD: University Press of America, 2003), 25–55; Daniel T. Rodgers, *Atlantic Crossings: Social Politics in a Progressive Age* (Cambridge, MA: Belknap Press/Harvard University Press, 1998), 324–26, 330–35.

11. Conkin, *Revolution Down on the Farm*, 19–25; Everett E. Edwards, "American Agriculture—The First 300 Years," in *Farmers in a Changing World*, by U.S. Department of Agriculture (Washington, DC: U.S. Government Printing Office, 1940), 247–56; William J. Block, *The Separation of the Farm Bureau and the Extension Service: Political Issue in a Federal System* (Urbana: University of Illinois, 1960); Gladys Baker, *The County Agent* (Chicago: University of Chicago Press, 1939), 1–94; Walker, *Southern Farmers and Their Stories*, 21–22; Jones, *Mama Learned Us to Work*, 107–38; M. S. McDowell, "What the Agricultural Extension Service Has Done for Agriculture," *Annals of the American Academy of Political and Social Science* 142 (1929): 251–55.

12. H. G. Seyforth, "Helping Low-Income Farmers," *Extension Service Review* 8.3 (1937): 35; M. L. Wilson, "Problem of Poverty in Agriculture," *Journal of Farm Economics* 22.1 (1940): 10–11; Lawrence Westbrook, "Relief Supports Live-at-Home Program," *Extension Service Review* 5.5 (1934): 67.

13. Westbrook, "Relief Supports Live-at-Home Program," 67.

14. Arthur L. Deering, "Farm and Home Visits," *Extension Service Review* 1.8 (1930): 116.

15. T. Roy Reid, "Public Assistance to Low-Income Farmers of the South," *Journal of Farm Economics* 21.1 (1939): 189–90; John L. Shover, *First Majority—Last Minority: The Transforming of Rural life in America* (De Kalb: Northern Illinois University Press, 1976), 232; Gilbert and Howe, "Beyond 'State vs. Society,'" 204–20; Debra A. Reid, *Reaping a Greater Harvest: African Americans, the Extension Service, and Rural Reform in Jim Crow Texas* (College Station: Texas A&M University Press, 2007). On postwar racism in the Extension Service and other federal agricultural agencies, see Pete Daniel, *Dispossession: Discrimination against African American Farmers in the Age of Civil Rights* (Chapel Hill: University of North Carolina Press, 2013).

16. James G. Maddox, "An Historical Review of the Nation's Efforts to Cope with Rural Poverty," *American Journal of Agricultural Economics* 50.5 (1968): 1355; Grant McConnell, *The Decline of Agrarian Democracy* (Berkeley: University of California Press, 1953), 84. On the politics of parity and the man behind it, see Gilbert C. Fite, *George N. Peek and the Fight for Farm Parity* (Norman: University of Oklahoma Press, 1954). Even unorthodox thinkers like Rexford Tugwell agreed that the best way to end the Depression was to lift the prices of products that had flexible costs (like agriculture) and increase the production of those with rigid cost (like industry). See Rexford G. Tugwell, "The Road to Economic Recovery," *Extension Service Review* 5.3 (1934): 33–34.

17. Kennedy, *Freedom from Fear*, 199–202. On federal agricultural policy and the response to the agricultural crisis under Herbert Hoover, see David E. Hamilton, *From New Day to New Deal: American Farm Policy from Hoover to Roosevelt, 1928–1933* (Chapel Hill: University of North Carolina Press, 1991). The role of the AAA in worsening the plight of tenants has been carefully studied. See Conrad, *Forgotten Farmers*, and Donald H. Grubbs, *Cry from the Cotton: The Southern Tenant Farmers' Union and the New Deal* (Chapel Hill: University of North Carolina Press, 1971). See also Richard S. Kirkendall, *Social Scientists and Farm Politics in the Age of Roosevelt* (Columbia: University of

Missouri Press, 1966), 30–148; Theodore Saloutos, *The American Farmer and the New Deal* (Ames: Iowa State University Press, 1982), 98–109; Mertz, *New Deal Policy and Southern Rural Poverty*, 20–44; Kirby, *Rural Worlds Lost*, 60–67; Daniel, *Breaking the Land*, 94–109; and Holley, *Second Great Emancipation*, 59–74.

18. Douglass V. Brown, Edward Chamberlin, et al., *The Economics of the Recovery Program* (New York: Whittlesey House, 1934), 139–59; Blaisdell, *Government and Agriculture*, 39–41; Saloutos, *The American Farmer and the New Deal*, 34–39.

19. Conkin, *A Revolution Down on the Farm*, 63–66. On the response by the leaders of the three major farm organizations to the plan (which they generally opposed, although they could not get united behind any of their own plans), see William R. Johnson, "National Farm Organizations and the Reshaping of Agricultural Policy in 1932," *Agricultural History* 37.1 (1963): 35–39.

20. Edwin G. Nourse, Joseph S. Davis, and John D. Black, *Three Years of the Agricultural Adjustment Administration* (Washington, DC: Brookings Institute, 1937), 53–54, 78–81; Henry I. Richards, *Cotton and the AAA* (Washington, DC: Brookings Institute, 1936), 74–79; Roger Biles, *The South and the New Deal* (Lexington: University Press of Kentucky, 1994), 40–41; Kennedy, *Freedom from Fear*, 206; James MacGregor Burns, *Roosevelt: The Lion and the Fox* (San Diego: Harcourt Brace, 1956), 1:193–94, 1:351; Blaisdell, *Government and Agriculture*, 169–71; George B. Tindall, *The Emergence of the New South, 1913–1945*, vol. 10 of *A History of the South* (Baton Rouge: Louisiana State University Press, 1967), 394. For a fuller discussion of the vote on and debates about the AAA, see Conrad, *Forgotten Farmers*, 22–36.

21. Kirkendall, *Social Scientists and Farm Politics*, 65–66; Saloutos, *American Farmer and the New Deal*, 58. Saloutos also examines staffing the AAA and some of the difficulties thereof (50–65). On whether it was unrealistic to expect the AAA to do much for small farmers, see Saloutos, *American Farmer and the New Deal*, 261–66; Theodore Saloutos, "New Deal Agricultural Policy: An Evaluation," *Journal of American History* 61.2 (1974): 403–4, 09–10; and Nourse, Davis, and Black, *Three Years of the Agricultural Adjustment Administration*, 345. For the counterargument, that the government inserted itself into the situation and changed the relationship between tenants and landlords to the latter's favor, see Daniel, *Breaking the Land*, 101.

22. Alger Hiss, "Memorandum for Mr. Jones," August 11, 1933, and Alger Hiss, "Memorandum to Mr. Frank," August 16, 1933, both in Folder "Department of Agriculture—Resettlement Administration. 1933–35," Box 31, Baldwin Papers; Saloutos, *American Farmer and the New Deal*, 88. In "The New Deal Farm Programs" (241–57), Mary Summers argues that it was not New Dealers who institutionalized conservative farm programs but rather their opponents. But whatever the intentions of its creators and administrators, AAA was certainly a part of the New Deal.

23. Chester C. Davis, "Production-Control Measures Under Adjustment Act Are Offered to Farmers," *Extension Service Review* 4.4 (1933): 49–50; Conrad, *Forgotten Farmers*, 52–56; Harold Hoffsommer, "The AAA and the Cropper," *Social Forces* 13.4 (1935): 496.

24. William R. Amberson, "The New Deal for Share-Croppers," *The Nation*, February 13, 1935, 185.

25. Richards, *Cotton and the AAA*, 141–42. Richards argues that to a degree, the way that the contracts so favored landlords discouraged the displacement of tenants or reduction of their status: they were getting so little already that there was little reason to try to remove them. Various examples of just how little a tenant could expect to receive can be found in

Grubbs, *Cry from the Cotton*, 19–20; Hoffsommer, "AAA and the Cropper," 498; Howard Kester, *Revolt among the Sharecroppers* (New York: Covici, Friede, 1936; repr., Knoxville; University of Tennessee Press, 1997), 29; and Nourse, Davis, and Black, *Three Years of the Agricultural Adjustment Administration*, 95–100.

26. Conrad, *Forgotten Farmers*, 64–82; Gordon W. Blackwell, "The Displaced Tenant Farm Family in North Carolina," *Social Forces* 13.1 (1934): 67–69; Johnson, Embree, and Alexander, *Collapse of Cotton Tenancy*, 51–56.

27. Amberson, "New Deal for Share-Croppers," 186; Biles, *South and the New Deal*, 43–44; Hoffsommer, "AAA and the Cropper," 497; Conrad, *Forgotten Farmer*, 40–42; Johnson, Embree, and Alexander, *Collapse of Cotton Tenancy*, 51–56. See also Daniel, *Breaking the Land*, 92–105. On how local disputes played out in Texas, which was also further along in the process of mechanization than most of the cotton South, see Neil Foley, *The White Scourge: Mexicans, Blacks, and Poor Whites in Texas Cotton Culture* (Berkeley: University of California Press, 1997), 163–82.

28. D. P. Trent, "Memorandum for the Secretary," July 2, 1934, Folder "Tenancy," Box 2081, Office of the Secretary, General Correspondence, RG 16, NACP.

29. Conrad, *Forgotten Farmers*, 66, 209. See also Kester, *Revolt Among the Sharecroppers*, 27–28.

30. Kennedy, *Freedom from Fear*, 209–12; Reid, *No Greater Harvest*, 116–31; John P. Davis, "A Survey of the Problems of the Negro Under the New Deal," *The Journal of Negro Education* 5.1 (1936): 6; Sherwood Eddy to Henry Wallace, March 23, 1936, Folder "Tenancy," Box 2439, Office of the Secretary, General Correspondence, RG 16, NACP; Conrad, *Forgotten Farmers*, 154–76. As Jason Manthorne points out, STFU members were less likely to oppose the AAA than were its leaders: the rank and file tended to have a relatively traditional view of land ownership and agriculture, and they wanted to use the organization to improve their position vis-à-vis their landlords or to increase their own chances for land ownership. Jason Manthorne, "The View from the Cotton: Reconsidering the Southern Tenant Farmers' Union," *Agricultural History* 84.1 (2010): 20–45.

31. Tindall, *Emergence of the New South*, 395–96; Nourse, Davis, and Black, *Three Years of the Agricultural Adjustment Administration*, 325.

32. Kirkendall, *Social Scientists and Farm Politics*, 100. On the purge, see Raymond Gram Swing, "The Purge at the AAA," *The Nation*, February 20, 1935, 216–17; William E. Leuchtenburg, *Franklin D. Roosevelt and the New Deal, 1932–1940* (New York: Harper and Row, 1963), 139; Russell Lord, *The Wallaces of Iowa* (Boston: Houghton Mifflin, 1947), 393–409; Conrad, *Forgotten Farmers*, 56–58, 140–50; Kirkendall, *Social Scientists and Farm Politics*, 101–5; Saloutos, *American Farmer and the New Deal*, 110–123; "The Reminiscences of Howard R. Tolley," Oral History Research Office, Columbia University, 1975 (microfiche), 316–31; John C. Culver and John Hyde, *American Dreamer: The Life and Times of Henry A. Wallace* (New York: W. W. Norton, 2000), 154–57.

33. Bindas, *Remembering the Great Depression*, 58–59.

34. Wayne D. Rasmussen, "The New Deal Farm Programs: What They Were and Why They Survived," *American Journal of Agricultural Economics* 65.5 (1983): 1158; Tindall, *Emergence of the New South*, 404.

35. Daniel, *Breaking the Land*, 105. See also Holley, *Second Great Emancipation*, 59–68. As Holley points out, efforts to fix the AAA also contributed to pushing tenants and sharecroppers off the land, as pro-tenant policies only encouraged a shift to wage labor and then mechanization.

36. E. L. Morgan, "National Policy and Rural Public Welfare," *Rural Sociology* 1.1 (1936): 8–12.
37. Robert P. Post, "Relief Policies," *New York Times*, March 4, 1934.
38. Federal Emergency Relief Administration, *Purposes and Activities of the Federal Emergency Relief Administration* (Washington, DC: U.S. Government Printing Office, 1935), 4 (quote); Olaf F. Larson, *Ten Years of Rural Rehabilitation in the United States* (Washington, DC: Bureau of Agricultural Economics, U.S. Department of Agriculture, 1947), 19.
39. David R. Williams, "Rural Industrial Communities," *Rural Rehabilitation* 1.1 (1934), 4.
40. "The Spiritual Side of Rehabilitation," *Rural Rehabilitation* 1.2 (1935): 27.
41. Mertz, *New Deal Policy and Southern Rural Poverty*, 46–47; Harry L. Hopkins, *Spending to Save: The Complete Story of Relief* (New York: W. W. Norton, 1936), 140–42.
42. Katie Louchheim, ed., *The Making of the New Deal: The Insiders Speak* (Cambridge: Harvard University Press, 1983), 179.
43. Raper and Reid, *Sharecroppers All*, 25; Johnson, Embree, and Alexander, *Collapse of Cotton Tenancy*, 57–59; Woofter, *Landlord and Tenant*, 150–53; Robin D. G.. Kelley, *Hammer and Hoe: Alabama Communists during the Great Depression* (Chapel Hill: University of North Carolina Press, 1990), 54–55.
44. Gove Hambidge, "Farmers in a Changing World—A Summary," in *Farmers in a Changing World*, by U.S. Department of Agriculture (Washington, DC: Government Printing Office, 1940), 3; Brown, "Rural Families on Relief," 91–92; Lorena Hickock to Eleanor Roosevelt, June 6, 1934, *One Third of a Nation*, 272.
45. Mertz, *New Deal Policy and Rural Southern Poverty*, 47–67; USDA, "History of the Farm Security Administration," October 10, 1940, pp. 2–3, Folder "85-160 Public Relations, July thru December, 1940," Box 14, Office of the Director, General Correspondence, RG 96, NARASE.
46. Alabama Relief Administration, "Rural Rehabilitation Program for 1934," pp. 2–13, Folder 4, Alabama Relief Administration Papers. Robert Hudgens, who would become important in the FSA, was involved in a rural rehabilitation program in South Carolina. "The Reminiscences of Robert W. Hudgens," Oral History Research Office, Columbia University, 1957 (microfiche), 116–22.
47. Holley, *Uncle Sam's Farmers*, 25; Mertz, *New Deal Policy and Southern Rural Poverty*, 67; Aubrey Williams, "The New Relief Program: Three Great Aims," *New York Times*, April 1, 1934. On the shift from the CWA to FERA and the WPA, see Bonnie Fox Schwartz, *The Civil Works Administration, 1933–1934: The Business of Emergency Employment in the New Deal* (Princeton, NJ: Princeton University Press, 1984), 213–59.
48. Hopkins, *Spending to Save*, 143.
49. Maddox, "Farm Security Administration," 10–11; Baldwin, *Poverty and Politics*, 62–63; Lawrence Westbrook, "The Program of Rural Rehabilitation of the FERA," *Journal of Farm Economics* 17.1 (1935): 91–93; USDA, "History of the Farm Security Administration," October 10, 1940, p. 3, Folder "85-160 Public Relations, July thru December, 1940," Box 14, Office of the Director, General Correspondence, RG 96, NARASE.
50. Larson, *Ten Years of Rural Rehabilitation*, 29–30.
51. Westbrook, "Program of Rural Rehabilitation," 89.
52. Paul V. Maris, "Rehabilitation Policies," *Rural Rehabilitation* 1.1 (1934), 21. On the New Deal's early emphasis on economizing as well as battling the Depression, see Adam Cohen, *Nothing to Fear: FDR's Inner Circle and the Hundred Days That Created Modern America* (New York: Penguin Books, 2009), 84–108.

53. "Making Tax Payers out of Tax Consumers," *Rural Rehabilitation* 1.2 (1935): 4.
54. Westbrook, "Program of Rural Rehabilitation," 97.
55. W. T. Frazier, "Farm Security Administration," vol. 2, n.d., Box 3, FSA Historical Record, RG 96, NACP; Larson, *Ten Years of Rural Rehabilitation*, 62–65. For an early example of how decentralized structure could hinder rehabilitation efforts: South Carolina saw its Rural Rehabilitation Advisory Committee limit rehabilitation loans to $500, compared to averages in other states between $2,500 and $3,000. Jack Irby Hayes Jr., *South Carolina and the New Deal* (Columbia: University of South Carolina Press, 2001), 132.
56. Paul V. Maris, "Policy Interpretations," *Rural Rehabilitation* 1.2 (1935), 13.
57. Larson, *Ten Years of Rural Rehabilitation*, 154.
58. FERA, "Note [Form FRR-20]," *Rural Rehabilitation Forms, Section II, Forms FRR 20–27*, p. 7, NAL ; F. E. Patch to Paul V. Maris, April 22, 1936, Folder "RR-101-3 P Reports," Box 6, Rural Rehabilitation General Correspondence, RG 96, NACP; Maddox, "Farm Security Administration," 200–202.
59. "80,000 Families End Direct Relief," *New York Times*, October 6, 1934.
60. Lorena Hickock to Harry L. Hopkins, June 7, 1934, *One Third of a Nation*, 272–75, 274 (quote); Larson, *Ten Years of Rural Rehabilitation*, 62–65, 153–54; Monroe Oppenheimer, "The Development of the Rural Rehabilitation Loan Program," *Law and Contemporary Problems* 4.4 (1937): 477–78. On farm supervisors on the Great Plains, see Grant, *Down and Out on the Family Farm*, 110–17.
61. Larson, *Ten Years of Rural Rehabilitation*, 32–33; Oppenheimer, "Development of the Rural Rehabilitation Loan Program," 475–76; Berta Asch and A. R. Magnus, *Farmers on Relief and Rehabilitation* (Washington, DC: Government Printing Office, 1937), 17–19.
62. Larson, *Ten Years of Rural Rehabilitation*, 155; Theodore E. Whiting, dir., *Final Statistical Report of the Federal Emergency Relief Administration* (Washington, DC: U.S. Government Printing Office, 1942), 67, 233–34.
63. "Rehabilitation Makes Progress in Alabama," *Extension Service Review* 5.7 (1934): 104; Lorena Hickock to Harry L. Hopkins, June 7, 1934, *One Third of a Nation*, 272–73.
64. FERA, "Application for Rural Rehabilitation" and "Family Rehabilitation Plan," *Rural Rehabilitation Forms—Section I, Forms FRR 1, 2, and 3*, NAL.
65. John B. Holt, *An Analysis of Methods and Criteria Used in Selecting Families for Colonization Projects*, Social Research Report, no. 1 (Washington, DC: U.S. Department of Agriculture, the Farm Security Administration, and the Bureau of Agricultural Economics, Cooperating, 1937), 39–43.
66. Woofter, *Landlord and Tenant*, 174.
67. "Reminiscences of Robert W. Hudgens," 188, 262–64; see also Laurence Hewes, *Boxcar in the Sand* (New York: Alfred A. Knopf, 1957), 73–74.
68. Julian Harris, "Pioneer Spirit Is Awakened in Georgians by Amazing Rural Rehabilitation Plan," *Atlanta Constitution*, March 10, 1935.
69. Brown, "Rural Families on Relief," 94.
70. Lorena Hickock to Harry L. Hopkins, April 7, 1934, *One Third of a Nation*, 210 (quote); Maddox, "Farm Security Administration," 18; Paul V. Maris, "Rehabilitation Policies," *Rural Rehabilitation* 1.1 (1934); 21.
71. Whiting, *Final Statistical Report of the Federal Emergency Relief Administration*, 65; USDA, "History of the Farm Security Administration," October 10, 1940, Folder "85–160 Public Relations, July thru December, 1940," Box 14, Office of the Director, General Correspondence, RG 96, NARASE.

72. Larson, *Ten Years of Rural Rehabilitation*, 84.
73. Asch and Magnus, *Farmers on Relief and Rehabilitation*, 28–29.
74. Clyde Cruse, "Forty Acres and a Home and It's Mine!" *Birmingham Post*, October 30, 1935.
75. Gaer, *Toward Farm Security*, 52; Brown, "Rural Families on Relief," 90–91; Kevin J. Cahill, "Fertilizing the Weeds: The Rural Rehabilitation Program in West Virginia," *Journal of Appalachian Studies* 4.2 (1998): 288–89; Maddox, "Farm Security Administration," 17–18.
76. Westbrook, "Program of Rural Rehabilitation," 95–99.
77. Lawrence Westbrook, *Rural-Industrial Communities for Stranded Families: An Outline of Suggested Procedure for the Guidance of State and County Public Relief Administration and Cooperating Public Agencies* (Washington, DC: Federal Emergency Relief Administration, Division of Rural Rehabilitation and Stranded Populations, [1934?]), 1–4.
78. "U.S. Speeding Plans to Form Relief Towns," *Washington Post*, April 24, 1934.
79. "Irwinville Farms Project," pp. 7–11, Folder "Farm Security 3 Projects Jan-April 21," Box 77, Office of the Secretary, General Correspondence, RG 16, NACP.
80. Westbrook, "Program of Rural Rehabilitation," 99–100; W. T. Frazier, "Farm Security Administration," vol 1, n.d., Box 3, FSA Historical Record, RG 96, NACP. A good discussion of the project can be found in Shafer, *A Basis for Social Planning in Coffee County, Alabama*.
81. Paul H. Johnstone, "Memorandum for Mr. M. L. Wilson," March 8, 1939, p. 1, Folder "Farm Security 2 (Projects)," Box 3019, Office of the Secretary, General Correspondence, RG 16, NACP. Recent works touching on southeast Alabama in general or Coffee County in particular in the early twentieth century include Hersey, *My Work Is That of Conservation*, and James C. Geisen, *Boll Weevil Blues: Cotton, Myth, and Power in the American South* (Chicago: University of Chicago Press, 2011).
82. R. W. Hudgens to R. G. Tugwell, November 3, 1936, Folder "AD-AL-17 (060) Coffee Farms," Box 52, Project Records, RG 96, NACP; Ernie Pyle, *Home Country* (New York: William Sloane Associates, 1947), 365.
83. "Agricultural and Economic Brief of Coffee County Farms, RF-AL-17," Folder "Coffee County, RF-AL-17," Box 67, Project Records, RG 96, NACP; Resettlement Administration, "Request for approval of final plans and authorization of funds to complete the development of the Coffee County Project, Coffee County, Alabama," p. 2, in "Final Budget: Coffee County Homesteads, Coffee Co., Alabama, RF-AL-17," Box 62, Project Records, RG 96, NACP; W. L. McArthur to J. B. Kasten, November 5, 1936, Folder "AD-AL-17 (060) Coffee Farms," Box 52, Project Records, RG 96, NACP.
84. USDA, "Skyline Farms," August 20, 1940, and USDA, "Skyline Farms," n.d., p. 2 (quote), both in Folder "85-1-16-60 Projects," Box 18, Project Construction Files, RG 96, NARASE. On Skyline Farms, see David Campbell and David Coombs, "Skyline Farms: A Case Study of Community Development and Rural Rehabilitation," *Appalachian Journal* 10.3 (1983): 244–54.
85. "Two Units Push Stranded Aid, Relief Plans," *Washington Post*, March 4, 1934. Harriman did, however, support the similar work of the Division of Subsistence Homesteads, apparently because he saw it as the beginning of a long-term process that would promote private ownership of land and industry without too much burden on the government. "Harriman Praises Homestead Project as Key to Stability," *Washington Post*, July 30, 1934; Henry I. Harriman, "A New Pattern for Industrial America," *New York Times*, September 23, 1934.
86. Rexford G. Tugwell, "The Place of Government in a National Land Program," *Journal of Farm Economics* 16.1 (1934): 58–63; Arthur M. Hyde, "Needed—A Land Policy,"

Extension Service Review 2.12 (1931): 177–78; "Recommendations of Land-Utilization Conference," *Extension Service Review* 3.1 (1932): 8–11; L. C. Gray, "The Social and Economic Implications of the National Land Program," *Journal of Farm Economics* 18.2 (1936): 257. Gray's emphasis on land use reform led some RA employees to complain that he was "primarily interested in land, not in people." Hewes, *Boxcar in the Sand,* 81. On the background of conservationism prior to the New Deal, see Phillips, *This Land, This Nation,* 21–59; on reform sentiment between the turn of the century and the New Deal, see Gregg, *Managing the Mountains,* 76–101.

87. G. W. Forster, "Progress and Problems in Agricultural Planning in the Southern States," *Journal of Farm Economics* 18.1 (1936): 92; Woofter, *Landlord and Tenant on the Cotton Plantation,* 43–44; Owen P. White, "All Washed Up," *Collier's,* September 29, 1934, 12–13; H. H. Bennett, "Our Soil Can Be Saved," in *Farmers in a Changing World,* by United States Department of Agriculture (Washington, DC: Government Printing Office, 1940), 430. Providence Gully is today part of a conservation area known as Providence Canyon State Park. See Paul S. Sutter, "What Gullies Mean: Georgia's 'Little Grand Canyon' and Southern Environmental History," *Journal of Southern History* 76.3 (2010): 579–616. As Chris Rasmussen points out, whatever the actual relationship, tenancy and erosion had become powerfully linked in the public mind during the Great Depression. Chris Rasmussen, "'Never a Landlord for the Good of the Land': Farm Tenancy, Soil Conservation, and the New Deal in Iowa," *Agricultural History* 73.1 (1999): 70–95.
88. Turner, "Farm Tenancy Distribution and Trends in the United States," 427.
89. National Resources Committee, *Farm Tenancy,* 7.
90. Maddox, "Farm Security Administration," 56–57.
91. Rexford G. Tugwell, "The Place of Government in a National Land Program," *Journal of Farm Economics* 16.1 (1934): 58–63.
92. Maddox, "Farm Security Administration," 151.
93. Nat T. Frame, "Larger Objectives in Extension Work," *Extension Service Review* 1.5 (1930): 67 (quote); Myers, "Relief in the Rural South," 282. Myers also provides a brief history of non-RA relief between the start of the FERA program and 1937.

Chapter 2

1. Henry Edison Williams, "Roosevelt Expects Farm-Factory Plan to Cut Down Doles," *Christian Science Monitor* February 5, 1934.
2. Holley, *Uncle Sam's Farmers,* 53; Gregg, *Managing the Mountains,* 187–90.
3. Brown, *Back to the Land,* 23–32, and, on the extent to which New Dealers built on existing back to the land ideology, 150–54; Danbom, *The Resisted Revolution,* 36–41. See also Stanford J. Layton, *To No Privileged Class: The Rationalization of Homesteading and Rural Life in the Early Twentieth-Century American West* (Salt Lake City: Charles Redd Center for Western Studies, Brigham Young University, 1988), 37; Paul H. Johnstone and Dorothy C. Goodwin, "The Back-to-the-Land Movement," in *A Place on Earth: A Critical Appraisal of Subsistence Homesteads,* ed. Russell Lord and Paul H. Johnstone (Washington, DC: U.S. Department of Agriculture, Bureau of Agricultural Economics, 1942), 11–13. For a summary of back-to-the-landism in America prior to the New Deal, see Conkin, *Tomorrow a New World,* 11–36; Holley, *Uncle Sam's Farmers,* 22. The debate over how to best combine the advantages of urban modernity and rural life was felt in other areas, including conservation and rural electrification; see Phillips, *This Land, This Nation,* 25–46.

4. J. Blaine Gwin, "Subsistence Homesteads," *Social Forces* 12.4 (May 1934): 522. The idealization of suburban living as a mix of the best of rural and urban life had begun in the early twentieth century and especially the 1920s; for an overview, see Brown, *Back to the Land*, 106-37.
5. Philip M. Glick, "The Federal Subsistence Homesteads Program," *Yale Law Journal* 44.8 (1935): 1327.
6. FDR, *PPA*, 1:516-17. For similar sentiment as candidate and governor, see FDR, *PPA*, 1:142-43, 699-700, as well as 701-2, 813-15, where he discusses his concerns about the farm mortgage problem, which endangered a farmer's position on the land. See also Leuchtenburg, *Franklin D. Roosevelt and the New Deal*, 136. On governor and candidate Roosevelt and land use reform, see Phillips, *This Land, This Nation*, 59-72.
7. "Text of the Inaugural Address; President for Vigorous Action," *New York Times*, March 5, 1933.
8. FDR, *PPA*, 1:507 (quote), 511; Rexford G. Tugwell, *The Brains Trust* (New York: The Viking Press, 1968), 178-79; Franklin D. Roosevelt, "Back to the Land," *Review of Reviews*, October, 1931, 63-64.
9. FDR, *PPA*, 1:116.
10. "Roosevelt Begins Providing Jobless With Land to Work," *New York Times*, May 28, 1932.
11. FDR, *PPA*, 1:487.
12. M. L. Wilson, "Problem of Poverty in Agriculture," *Journal of Farm Economics* 22.1 (1940): 26 (quote); The Unofficial Observer, "'M.L.' Another New Deal Natural," and "Perfect Prefecture of New Deal," *Washington Post*, March 2, 3, 1934; Johnstone and Goodwin, "The Back-to-the-Land Movement," 20-21; Gwin, "Subsistence Homesteads," 523-24. See also Kirkendall, *Social Scientists and Farm Politics in the Age of Roosevelt*, 11-15; 22-49. Wilson also worked with very large units, as seen in his Fairway Farms project—in all cases, the goal was to fit farmers to the land on which they worked. See Phillips, *This Land, This Nation*, 42-43.
13. Frances Durham, "Homesteads Sought for State by Native," *Birmingham News*, February 11, 1934.
14. Kennedy, *Freedom from Fear*, 123-24.
15. Rexford G. Tugwell, *In Search of Roosevelt* (Cambridge, MA: Harvard University Press, 1972), 123.
16. Rexford G. Tugwell, "National Significance of Recent Trends in Farm Population," *Social Forces* 14.1 (1935): 1.
17. "A Back-to-Farm Plan That Works in Georgia," *New York Times*, July 23, 1933. See also the similar theory of "decentration," which held that breaking up large industrial units into smaller ones scattered across the countryside would protect against economic depression and promote craft pride among more independent workers. Rader Winget, "Smaller Units in Industry Planned," *Birmingham News*, December 17, 1933.
18. "Subsistence Homesteads," *Dallas Morning News*, August 1, 1934.
19. S. 69, 73rd Cong., 1st sess. (introduced March 10, 1933), and S. 1503, 73rd Cong., 1st sess. (introduced April 27, 1933), both in Box 10, Bankhead Papers; Conkin, *Tomorrow a New World*, 87-89; Holley, *Uncle Sam's Farmers*, 26-27.
20. T. J. Woofter, "Southern Population and Social Planning," *Social Forces* 14.1 (1935): 20.
21. National Industrial Recovery Act, H.R. 5755, 73rd Cong., 1st sess.
22. FDR, Executive Order 6209, 1933.
23. Paul H. Johnstone and Dorothy C. Goodwin, "The Administration of the Subsistence-Homesteads Program," in *A Place on Earth: A Critical Appraisal of Subsistence*

Homesteads, ed. Russell Lord and Paul H. Johnstone (Washington, DC: U.S. Department of Agriculture, Bureau of Agricultural Economics, 1942), 39.
24. Ray Rutledge, "The Development of the Bankhead Farmsteads Community" (M.A. thesis, Auburn University, 1951), 15-18.
25. Leonard A. Salter Jr., "Research and Subsistence Homesteads," *Rural Sociology* 2.2 (1937): 208.
26. John H. Bankhead Jr. to H. H. MacCormack, December 11, 1934, Folder 5, Box 4, Bankhead Papers.
27. Carl C. Taylor, "Research Needed as Guidance to the Subsistence Homesteads Program," *Journal of Farm Economics* 16.2 (1934): 310. See also Johnstone and Goodwin, "The Administration of the Subsistence-Homesteads Program," 35-36. Taylor later defined five main divisions among supporters of the subsistence homesteads: back-to-the-landers, those who wanted a low-cost housing program, those who wanted to decentralize industry, those promoting "a recrudescence of the simple handicraft life," and those who simply thought it was being done wrong. Carl C. Taylor, "Social and Economic Significance of the Subsistence Homesteads Program," *Journal of Farm Economics* 17.4 (1935): 722.
28. John H. Bankhead Jr., "Back to the Farm," Folder 2, Box 13, Bankhead Papers; M. L. Wilson, "The Subsistence Homestead Program," in *Proceedings of the Institute of Public Affairs* (Athens: University of Georgia Press, 1934), 166; Gwin, "Subsistence Homesteads," 522. See also Paul W. Wager, *One Foot on the Soil: A Study of Subsistence Homesteads in Alabama* (Tuscaloosa: Bureau of Public Administration, University of Alabama, 1945), 1; Paul H. Johnstone, Introduction to Lord and Johnstone, *A Place on Earth,* 3. On the negativity of Depression-era back-to-the-land thinking, see Brown, *Back to the Land,* 143-47.
29. The Unofficial Observer, "'M.L.' Another New Deal Natural," and "Perfect Prefecture of New Deal" *Washington Post,* March 2, 3, 1934.
30. Wilson, "Subsistence Homestead Program," 163.
31. M. L. Wilson, "The Place of Subsistence Homesteads in Our National Economy," *Journal of Farm Economics* 16.1 (1934): 74-76.
32. "Reminiscences of Howard R. Tolley," 422-23; Johnstone and Goodwin, "The Administration of the Subsistence-Homesteads Program," 41, 44-45.
33. U.S. Department of Agriculture, Soil Conservation Service, *The Preparation of the Standard State Soil Conservation Districts Law: An Interview with Philip M. Glick* (Washington, DC: U.S. Government Printing Office, 1990), 6, 15; Conkin, *Tomorrow a New World,* 96. See also Gilbert, "A Usable Past," 134-142; Jess Gilbert, "Low Modernism and the Agrarian New Deal," in *Fighting for the Farm: Rural American Transformed,* ed. Jane Adams (Philadelphia: University of Pennsylvania Press, 2003), 129-31.
34. "Federal Funds for Farms Only Drop in Bucket," *Christian Science Monitor,* September 29, 1933.
35. Glick, "Federal Subsistence Homesteads Program," 1332-33; "Dr. Wirt's Testimony on His 'Brain Trust' Revolt Charges," *Washington Post,* April 11, 1934.
36. Gwin, "Subsistence Homesteads," 524; Wilson, "Subsistence Homestead Program," 172-74; Conkin, *Tomorrow a New World,* 104-5.
37. Glick, "Federal Subsistence Homesteads Program," 1334-35; Maddox, "Farm Security Administration," 23-24; Conkin, *Tomorrow a New World,* 105-7. Many of these still exist among the records of the Farmers Home Administration housed in the National Archives, Record Group 96, National Archives at College Park, Maryland.

38. John H. Bankhead Jr. to Charles E. Pynchon, July 26, 1934, Bankhead Papers. On similar projects in the Appalachian Mountains of Vermont and Virginia, see Gregg, *Managing the Mountains*, 190–212.
39. "Subsistence Garden Homestead Project in Birmingham, Alabama," Folder "Alabama No. 3 Birmingham 310," Project Records, RG 96, NACP. Supporters and administrators in other projects fueled similarly unrealistic ideas about the resettlement projects; see, for example, Cannon, *Remaking the Agrarian Dream*, 25–36.
40. Walter M. Kollmorgen, "The Subsistence-Homesteads Near Birmingham," in Lord and Johnstone, *A Place on Earth*, 66.
41. Holt, *Analysis of Methods and Criteria*, 36–38.
42. Kollmorgen, "Subsistence Homesteads Near Birmingham," 69.
43. "Economic and Social Merits of Proposed Project," p. 1, in "200-Plans-Project Book AL-3," Box 10, Project Records, RG 96, NACP.
44. Schedule XII in "Subsistence Garden Homestead Project in Birmingham, Alabama," Folder "Alabama No. 3 Birmingham 310," Project Records, RG 96, NACP.
45. Wilson, "Subsistence Homestead Program," 173.
46. Erskine Ramsay to Oscar M. Dugger, April 19, 1934, Folder "843 Birmingham General," Box 15, Project Records, RG 96, NACP. Something similar happened at the Tupelo Homesteads; industrial residents there had incomes exceeding the Mississippi average, and the first homesteaders had incomes in the $1,300–$1,900 range. Fred C. Smith, "The Tupelo Homesteads: A New Deal Agrarian Experimentation," *Journal of Mississippi History* 68.2 (2006): 103–4.
47. Wager, *One Foot on the Soil*, 151–52.
48. Glick, "Federal Subsistence Homesteads Program," 1325–26; C. L. Burton, "Schedule of Preliminary Information for Proposed Jasper Subsistence Homesteads Unit," December 30, 1933, pp. 1–3, Folder "200-Preliminary Plans AL-12," Box 31, Project Records, RG 96, NACP.
49. Oscar M. Dugger, "Jasper Homesteads Jasper, Alabama Project #34," December 5, 1934, Folder "200-Preliminary Plans AL-12," Box 29, Project Records, RG 96, NACP.
50. Wager, *One Foot on the Soil*, 190–91.
51. Kollmorgen, "Subsistence Homesteads Near Birmingham," 70.
52. Kollmorgen, "Subsistence Homesteads Near Birmingham," 67–68; Wager, *One Foot on the Soil*, 18–19.
53. Clarence E. Pickett, *For More than Bread: An Autobiographical Account of Twenty-Two Year's Works with the American Friends Service Committee* (Boston: Little, Brown, 1953), 53–54.
54. Johnstone and Goodwin, "Administration of the Subsistence-Homesteads Program," 46–47; Baldwin, *Poverty and Politics*, 73, Conkin, *Tomorrow a New World*, 118–21; Smith, "Tupelo Homesteads," 90–92.
55. Harold Ickes, *The Secret Diaries of Harold L. Ickes*, vol. 1, *The First Thousands Days, 1933–1936* (New York: Simon and Schuster, 1953), 160.
56. Johnstone and Goodwin, "The Administration of the Subsistence-Homesteads Program," 48.
57. Ickes, *Secret Diaries*, 1:218; Johnstone and Goodwin, "Administration of the Subsistence-Homesteads Program," 46; Conkin, *Tomorrow a New World*, 121–22.
58. John H. Bankhead Jr. to Hugh MacRae, February 14, 1935, Folder 6, Box 4, Bankhead Papers. Bankhead received voluminous correspondence asking for help with some aspect of the work of the Division of Subsistence Homesteads in Alabama.

59. Bruce L. Melvin, "Emergency and Permanent Legislation with Special Reference to the History of Subsistence Homesteads," *American Sociological Review* 1.4 (1936): 626.
60. "Reminiscences of Robert W. Hudgens," 142–47.
61. John H. Bankhead Jr. to Charles E. Pynchon, July 26, 1934, Folder 5, Box 4, Bankhead Papers.
62. Pickett, *For More than Bread*, 44.
63. "Pynchon Takes Over Federal Homestead Job," *Washington Post*, May 27, 1934; Ickes, *Secret Diaries*, 1:206, 219; Maddox, "Farm Security Administration," 24–25.
64. "Progress Made in Homesteading," *New York Times*, September 16, 1934.
65. Stanley W. Brown and Virgil E. Baugh, *Preliminary Inventory of the Records of the Farmers Home Administration (Record Group 96)* (Washington, DC: National Archives, 1959), 1, 18. See also Johnstone and Goodwin, "Administration of the Subsistence-Homesteads Program," 49–51.
66. Henry A. Wallace to FDR, March 19, 1937, and Will W. Alexander, "Memorandum to the Secretary, March 17, 1937," both in Folder "Resettlement 2 (Projects) (Jan.–Jul)," Box 2630, Office of the Secretary, General Correspondence, RG 16, NACP; House, *Farm Tenancy Hearing*, 76, 84, 95–96.
67. "Government to Call Bids Soon," *Birmingham News*, August 10, 1934. The Jasper Homesteads, expected to be constructed at about the same time, did not select a contractor until January 1935. "First Homestead Units at Jasper Will Be Erected," *Birmingham News*, January 20, 1935.
68. Harold Helfer, "'Tomorrow's Town' Is Full of Homes that Jack Built," *Birmingham Post*, January 12, 1935.
69. "Model Colony Work Started," *Birmingham Post*, February 11, 1935; "First Homesteads Near Completion," *Birmingham Post*, October 10, 1935; "Applications for Homesteads Set," *Birmingham News*, April 30, 1935.
70. "Utopia Comes True to Many Homesteaders," *Birmingham Post*, July 6, 1935.
71. "Workers Called for Homestead," *Birmingham News*, January 21, 1934; D. H. Greer to J. H. Jenkins, April 17, 1934, Folder "843 Birmingham, General," Box 15, Project Records, RG 96, NACP; Albert Maverick Jr. to John H. Bankhead Jr., June 8, 1935, Folder 7, Box 4, Bankhead Papers.
72. Banfield, *Government Project*, 148. Casa Grande is also discussed throughout Cannon, *Remaking the Agrarian Dream*.
73. "Tugwell Dream Fades," *Birmingham Age-Herald*, May 3, 1937.
74. Ickes, *Secret Diaries*, 1:227; Conkin, *Tomorrow a New World*, 142–43; Baldwin, *Poverty and Politics*, 74–75; John H. Bankhead Jr. to Thomas W. Gilmer, December 26, 1934, Folder 5, Box 4, Bankhead Papers.
75. Conkin, *Tomorrow a New World*, 129–30.
76. Melvin, "Emergency and Permanent Legislation," 623.
77. Leon Dure Jr., "Housing Plan to Be Pushed by President," *Washington Post*, October 18, 1934.
78. Warner Ogden, "Paradise Lost: or the Sad Story of Another Utopia That Failed," *Wall Street Journal*, February 15, 1950.
79. Glick, "Federal Subsistence Homesteads Program," 1345–47 (on the question of whether the federal or state government exercised civil and criminal jurisdictions over the homesteaders, see 1360–67); Melvin, "Emergency and Permanent Legislation," 627. This would also be a criticism of FERA's successor, the Resettlement Administration, which would eventually make payments to local governments for citizens living in nontaxable

projects; see House, *Payments in Lieu of Taxes on Resettlement Projects*, 19–20; Brown and Baugh, *Preliminary Inventory*, 23; House, *Farm Tenancy Hearing*, 105–11; Gray, "Social and Economic Implications of the National Land Program," 266.
80. Conkin, *Tomorrow a New World*, 130; Baldwin, *Poverty and Politics*, 73.
81. Wilson, "Subsistence Homestead Program," 171.
82. Ickes, *Secret Diaries*, 1:129. An extremely critical view can be found in C. J. Maloney, *Back to the Land: Arthurdale, FDR's New Deal, and the Cost of Economic Planning* (Hoboken, NJ: John Wiley & Sons, 2011).
83. Calvert L. Estill, "Blunders at Arthurdale," *Washington Post*, August 12, 1934; Pickett, *For More than Bread*, 55–56; "Answers Attacks on Reedsville Plan," *New York Times*, October 4, 1934.
84. Conkin, *Tomorrow a New World*, 128–30. Glick, the former general counsel of the Federal Division of Subsistence Homesteads, believed the Comptroller General was incorrect about this decision, arguing that similar setups had been found acceptable in the past and that the "almost frankly experimental" nature of Section 208 indicated that it was better to err on the side of a more liberal grant of charter powers. See Glick, "Federal Subsistence Homesteads Program," 1337–39. The decision almost ended several projects underway; see Gregg, *Managing the Mountains*, 190–91.
85. Kennedy, *Freedom from Fear*, 216–27; Brinkley, *End of Reform*, 74–75.
86. Kennedy, *Freedom from Fear*, 243–48; Badger, *The New Deal*, 95–97.
87. Lillian Perrine Davis, "Relief and the Sharecropper," *Survey Graphic* 25.1 (1936): 22; Harold Hoffsommer, *Landlord-Tenant Relations and Relief in Alabama* (Washington, DC: Division of Research, Statistics and Finance, Federal Emergency Relief Administration), i; "Subsistence Plans Ask $500,000,000," *New York Times*, December 10, 1934.
88. R. L. Duffus, "Vast Drive 'To End Relief' Gets Under Way," *New York Times*, April 14, 1935.
89. Myers, "Relief in the Rural South," 284–91.

Chapter 3

1. Knott, "Modern Gulliver Helps the Little People," *Dallas Morning News*, November 17, 1936.
2. Summers, "New Deal Farm Programs," 249–250. Despite the notion that the court-packing plan represented the end of meaningful New Deal legislation, the Roosevelt Administration continued to fight over its influence on the southern economy into World War II, and federal policy contributed to the economic changes following the war; see Schulman, *From Cotton Belt to Sunbelt*, esp. 1–87.
3. Kennedy, *Freedom from Fear*, 249–54; Leuchtenburg, *Franklin D. Roosevelt and the New Deal*, 123–24; Blaisdell, *Government and Agriculture*, 150–51; Rebecca Farnham and Irene Link, *Effects of the Works Program on Rural Relief: A Survey of Rural Relief Cases Closed in Seven States, July through November 1935* (Washington, DC: U.S. Government Printing Office, 1938), 96.
4. Culver and Hyde, *American Dreamer*, 156.
5. Maddox, "Historical Review of the Nation's Efforts to Cope with Rural Poverty," 1356–57; Maddox, "Farm Security Administration," 26–27; Tugwell, "The Resettlement Idea," 159; House, *Payments in Lieu of Taxes on Resettlement Projects*, 8–12; Grant, *Down and Out on the Family Farm*, 86–88.
6. Gregg, *Managing the Mountains*, 7. Bureaucratic transformations reflected and influenced ideological changes in other New Deal agencies; see Maher, *Nature's New Deal*, esp. 45–48.

7. FDR, Executive Order 7027, 1935; FDR, Executive Order 7200, 1935. See also Senate, *Resettlement Administration Program*, 1–2; "Text of Executive Order," *New York Times*, May 2, 1935; and Leon Udre Jr., "Senate Orders Inquiry of Hopkins' Demands Upon States on Relief," *Washington Post*, May 2, 1935.
8. "A Staggering Task," *Washington Post*, May 5, 1935.
9. Maddox, "Farm Security Administration," 32.
10. Hewes, *Boxcar in the Sand*, 68.
11. *First Annual Report, Resettlement Administration*, 16. The number depends on how one counts projects. Not even RA administrators were quite sure what they had picked up.
12. FDR, Executive Order 7028, 1935; FDR, Executive Order 7041, 1935; House, *Activities of the FSA*, 64–67; House, *Farm Security Hearings, Part 1*, 116.
13. W. T. Frazier, "Farm Security Administration," vol. 1, n.d., Box 3, FSA Historical Record, RG 96, NACP; Maddox, "Farm Security Administration," 95–96; Oppenheimer, "Development of the Rural Rehabilitation Loan Program," 480–82. The process, in short: the various state corporations gave the Resettlement Administration management powers for an interim period, before the transfer became official. This required both a resolution by the board of directors for each state corporation and approval by each state attorney general, since funds had technically become state money. Thereafter, the states transferred their rural rehabilitation programs to the RA. The funds were not directly transferred to the Treasury but rather were placed in a trust fund to ensure that the state involved would get those rural rehabilitation expenditures by the federal government.
14. House, *Farm Security Hearings, Part I*, 117; *First Annual Report, Resettlement Administration*, 16; "Reminiscences of Howard R. Tolley," 339–40; Henry A. Wallace to Louis Brownlow, February 3, 1937, Folder "Resettlement," Box 2629, Office of the Secretary, General Correspondence, RG 16, NACP.
15. *First Annual Report, Resettlement Administration*, 16.
16. Will W. Alexander, "Memorandum for Mr. M. L. Wilson," November 19, 1937, pp. 1–4, Folder "Resettlement 1 (Rural Rehabilitation)," Box 2629, Office of the Secretary, General Correspondence, RG 16, NACP.
17. Sidney Olson, "RA Employees Paid $181,000 Extra in Error in Early Days," *Washington Post*, August 21, 1936.
18. John A. Overholt, et al., "Report and Recommendations on key positions in the Resettlement Administration in two parts: Part I as of July 1, 1935, Part II from July 1, 1935 to March 1, 1936," p. 2, 19, Folder "Dept. of Agriculture—Resettlement Adm.—Report and Recommendations on Key Positions. 1935–1936." Box 31, Baldwin Papers.
19. "Tugwell to Drop 8,000 from Staff," *New York Times*, July 31, 1935; Senate, *Agricultural Appropriation Bill for 1938* (1937), 423.
20. Maddox, "Farm Security Administration," 28–31.
21. "Biographical Sketches of Brown, Black, Alexander, and Gray," n.d., Folder "Tenancy 1 (Farmer's Home Corporation)," Box 2663, Office of the Secretary, General Correspondence, RG 16, NACP; Calvin Benham Baldwin, "Personnel Questionnaire" November 4, 1939, Folder "Dept. of Agriculture—Farm Security Administration—Personnel, 1933–1943," Box 31, Baldwin Papers; Overholt, et al, "Report and Recommendations on key positions," 27–32, 49–52, 106–7; Baldwin, *Poverty and Politics*, 95–98; Conkin, *Tomorrow a New World*, 146–55. Many of these assistants, most importantly C. B. Baldwin but also others like Laurence Hewes, entered into a sort of pupil-mentor relationship with Tugwell. Hewes, *Boxcar in the Sand*, 53; "The

Reminiscences of Will W. Alexander," Oral History Research Office, Columbia University, 1972 (microfiche), 470

22. Baldwin, *Poverty and Politics*, 95, 244; Bernard Sternsher, *Rexford Tugwell and the New Deal* (New Brunswick, NJ: Rutgers University Press, 1964), 263.

23. R. G. Tugwell, Memorandum, "Functional organization of the Resettlement Administration," June 11, 1935, Folder "(AD-101. And. Gen.) Organization Charts, Etc," Box 26, General Correspondence, Washington Office, RG 96, NACP; Gaer, *Toward Farm Security*, 129–30; "Regional Offices of the Resettlement Administration," [n.d.], Folder "Department of Agriculture—Resettlement Administration. 1933–35," Box 31, Baldwin Papers; Maddox, "Farm Security Administration," 32–33; R. G. Tugwell, memorandum, "Functional organization of the Resettlement Administration," June 11, 1935, Folder "(AD-101. And. Gen.) Organization Charts, Etc," Box 26, General Correspondence, Washington Office, RG 96, NACP. As Brian Cannon points out, these regions did not always entirely encompass distinct geographic regions. Cannon, *Remaking the Agrarian Dream*, 14–15.

24. *Interim Report of the Resettlement Administration*, 9.

25. "Reminiscences of Robert W. Hudgens," 281–84; "Reminiscences of Will W. Alexander," 514–16.

26. Conkin, *Tomorrow a New World*, 155–56. Untitled memorandums from LeRoy Peterson to C. B. Baldwin, November 12, 1935, Folder "Department of Agriculture—Resettlement Administration. 1933–35," Box 31, Baldwin Papers. This administrative problem worked itself out not through reorganization but through the development of consensus on how to approach the problems facing the agency and a tightening of the relationships between personnel. The clunky organization persisted into the FSA, but by that point it appears that administrators and their subordinates knew each other well enough, at least in the Washington office, to work out personally any remaining divisions.

27. *Interim Report of the Resettlement Administration*, 8–13, 23–27; *First Annual Report, Resettlement Administration*, 9, 63–79; Senate, *Resettlement Administration Program*, 7–10; Gaer, *Toward Farm Security*, 54–56. A complete list of the RA's community projects as of the spring of 1936, organized in a variety of ways, is available in Senate, *Resettlement Administration Program*, 25–58.

28. Senate, *Resettlement Administration Program*, 7–8.

29. Gaer, *Toward Farm Security*, 53; Senate, *Agricultural Appropriation Bill for 1938* (1937), 429–33. On New Deal land use reform, the best recent works include Gregg, *Managing the Mountains*, 175–212, and Phillips, *This Land, This Nation*, 75–148. Land use is also an important subject in Cannon, *Remaking the Agrarian Dream*.

30. Tugwell, "Place of Government in a National Land Program," 55. See also "Reminiscences of Will W. Alexander," 384–87; C. B. Baldwin, interview by Richard K. Doud, microfilm, February 24, 1965, Smithsonian Archives of American Art, Washington, DC; Conkin, *Tomorrow a New World*, 156–60, 166; Sternsher, *Rexford Tugwell and the New Deal*, 265–66.

31. L. C. Gray's definition was "the broad objective of effecting changes in the use of any type of land where the change appears desirable from a public or social point of view, rather than merely as a program of agricultural adjustment." Gray, "Social and Economic Implications of the National Land Program," 261. Much planning was still done by the Bureau of Agricultural Economics, but it and the RA's Land Use Planning Section worked together much more efficiently than had the earlier incarnations of the land program.

32. Senate, *Resettlement Administration Program*, 2–3; House, *Payments in Lieu of Taxes on Resettlement Projects*, 12; Gray, "Social and Economic Implications of the National Land Program," 265.

33. Phillips, *This Land, This Nation*, 122–32. Phillips argues that resistance to land reform from those who presumably needed land use reform the most (such as Dust Bowl farmers) helped shape federal policy in favor of rural rehabilitation and away from land utilization. New Deal programs that improved the land did not necessarily have to be unpopular; on the popularity of the Civilian Conservation Corps, see Maher, *Nature's New Deal*, 151–65.
34. Noble Clark, "The Social and Economic Implications of the National Land Program: Discussion," *Journal of Farm Economics* 18.2 (1936): 276–77.
35. "Reminiscences of Will W. Alexander," 387. See also Gaer, *Toward Farm Security*, 53–54. A detailed, first-person look at one project, at Casa Grande, Arizona, can be found in Banfield, *Government Project*.
36. Senate, *Resettlement Administration Program*, 5.
37. Jonathan Garst to Laurence Hewes, September 20, 1935, Folder "Department of Agriculture—Resettlement Administration. 1933–35," Box 31, Baldwin Papers; Senate, *Resettlement Administration Program*, 6.
38. "Government Begins a New Town Today," *New York Times*, October 12, 1935; Senate, *Resettlement Administration Program*, 7; *Interim Report of the Resettlement Administration*, 17–21; *First Annual Report, Resettlement Administration*, 4, 48–49; Conkin, *Tomorrow a New World*, 305–25. See C. Kline Fulmer, *Greenbelt* (Washington, DC: American Council on Public Affairs, 1941); Arnold, *New Deal in the Suburbs*, and Knepper, *Greenbelt, Maryland*, for a fuller discussion of the greenbelt program. On the European influences of New Deal suburban community building, see Rodgers, *Atlantic Crossings*, 4554–61.
39. "Civic Leaders Assail Claim Baltimore Has Passed Prime," *Washington Post*, October 18, 1935.
40. John H. Bankhead Jr. to Rexford G. Tugwell, November 11, 1935, Folder 7, Box 4, Bankhead Papers.
41. Maddox, "Farm Security Administration," 40–41.
42. Biles, *The South and the New Deal*, 48–49; Kirkendall, *Social Scientists and Farm Politics in the Age of Roosevelt*, 113–20; "New Cooperative Idea," *Extension Service Review* 7.12 (December 1936): 190; John A. Overholt, et al, "Report and Recommendations on key positions in the Resettlement Administration in two parts: Part I as of July 1, 1935, Part II from July 1, 1935 to March 1, 1936," p. 2, Folder "Dept. of Agriculture—Resettlement Adm.—Report and Recommendations on Key Positions. 1935–1936." Box 31, Baldwin Papers. As Daniel Rodgers points out, in reality most New Deal agricultural policymakers had little interest in re-creating inefficient (as planners saw them) European or particularly Soviet farm programs. Rodgers, *Atlantic Crossings*, 451, 454.
43. Felix Bruner, "Utopia Unlimited," *Washington Post*, February 10–13, 1936. Bruner also wrote a three-part criticism of the Greenbelt project specifically, emphasizing its impressive quality but more its expense and experimental nature; see "Tugwelltown: Uncle Sam—Landlord Super De Luxe," *Washington Post*, August 2–4, 1936.
44. Felix Bruner, "Utopia Unlimited," *Washington Post*, February 10, 1936.
45. Felix Bruner, "Utopia Unlimited," *Washington Post*, February 13, 1936.
46. W. R. Altstaetter to W. W. Alexander, September 11, 1936, Folder "Department of Agriculture—Resettlement Administration 1936," Box 31, Baldwin Papers.
47. Roll, *Spirit of Rebellion*, 108–13, 133–39, 145–53. The FSA did make a large number of emergency grants to the demonstrators, but politics intervened: local officials lodged a complaint with the House of Representatives, and the FSA was forced to stop its grant payments to the area.

48. Schuler, *Social Status and Farm Tenure*, 82–83. This was also, incidentally, the case with many politicians. At a hearing on the RA, Minnesota congressman Harold Knutson thought the RA administered a slum-clearing program operated by the PWA, and Georgia's John W. McCormack was unaware that the RA was an independent government agency. House, *Payments in Lieu of Taxes on Resettlement Projects*, 14–15.

49. "Address by Joseph H. B. Evans, Advisor to the Director of Rural Resettlement, Resettlement Administration, before the annual meeting of the Joint Committee on National Recovery, Washington, D.C., Saturday evening, November 23, 1935," pp. 13–14, Folder "E AD-163-1 Press Releases," Box 31, General Correspondence, Washington Office, RG 96, NACP. As Sarah T. Warren and Robert E. Zabawa point out, only one land utilization project (the Tuskegee Planned Land Use Demonstration) involving southern African Americans was entirely managed by black administrators. Sarah Warren and Robert Zabawa, "The Origins of the Tuskegee National Forest: Nineteenth- and Twentieth-Century Resettlement and Land Development Programs in the Black Belt Region of Alabama," *Agricultural History* 72.2 (1998): 499. On the hiring of black employees in another large New Deal organization, the Tennessee Valley Authority, which similarly employed African Americans in primarily menial service jobs, see Nancy L. Grant, *TVA and Black Americans: Planning for the Status Quo* (Philadelphia: Temple University Press, 1990), 45–72. On the Civilian Conservation Corps, which similarly enrolled fewer blacks and put its black enrollees in less important jobs in geographically isolated camps, see Maher, *Nature's New Deal*, 109–10.

50. "Fruits of Bureaucracy," *Washington Post*, November 18, 1935.

51. Frank L. Kluckhohn, "Tugwell Has Staff of 12,089 to Create 5,012 Relief Jobs," *New York Times*, November 17, 1935; for the response, see "Tugwell Protests Published Figures," *New York Times*, November 20, 1935. If anything, the early rural rehabilitation program lacked sufficient funds to handle even a fraction of the need. See Grant, *Down and Out on the Family Farm*, 102–5.

52. "Calls 'Fire Traps' Peril to RA Loans," *New York Times*, August 30, 1936; "Farm Loan Laxity Again Is Charged," *New York Times*, August 31, 1936.

53. Louis Towley, "Gover'ment Cow," *Survey Graphic* 25.12 (1936): 647–51, 696–98.

54. The Historical/Photographic Section of the RA/FSA is by far the most well-studied area of the agencies' work. Works dealing entirely or in part with RA and FSA photography include Lili Corbus Bezner, *Photography and Politics in America: From the New Deal into the Cold War* (Baltimore: Johns Hopkins University Press, 1999); Michael Carlebach and Eugene F. Provenzo, *Farm Security Administration Photographs of Florida* (Gainesville: University of Florida Press, 1996); Birmingham Historical Society, *Digging Out of the Great Depression: Federal Programs at Work in and around Birmingham* (Birmingham, AL: Birmingham Historical Society, 2010); Stu Cohen, *The Likes of Us: America in the Eyes of the Farm Security Administration*, ed. Peter Bacon Hales (Boston: David R. Godine, 2009); Cara Finnegan, *Picturing Poverty: Print Culture and FSA Photographs* (Washington, DC: Smithsonian Books, 2003); Linda Gordon, "Dorothea Lange: The Photographer as Agricultural Sociologist," *Journal of American History* 93.3 (2006): 698–727; Gilles Mora and Beverly W. Brannan, *FSA: The American Vision* (New York: Abrams, 2006); and Charles J. Shindo, *Dust Bowl Migrants in the American Imagination* (Lawrence: University of Kansas Press, 1997). As with the resettlement projects, the aspects of the farm security program that have found the most historical and critical approval received the most contemporary criticism; Stryker's section was roundly attacked in Congress and the media

as wastefully expensive and overly political (accurately, at least in terms of its political goals; see Gordon, "Dorothea Lange: The Photographer as Agricultural Sociologist").
55. "Candid Camera Is Less than Candid, RA Critics Claim," *Washington Post*, August 30, 1936; "Explains Cow Skull Use in Drought Pictures," *New York Times*, August 30, 1936; Westbrook Pegler, "Fair Enough," *Washington Post*, September 2, 1936; "Retracts 'Fake' Charge," *New York Times*, September 6, 1936.
56. Tugwell, "The Resettlement Idea," 159.
57. Senate, *Agricultural Appropriation Bill for 1938* (1937), 414. On the 1936 fight, see numerous related letters and memos, Folder "Allocations General Fiscal Years 1935–1936, AD-410-01," Box 46, General Correspondence, Washington Office, RG 96, NACP; "Only WPA Survives in New Relief Bill," *New York Times*, May 3, 1936; "President to Keep PWA," *New York Times*, May 16, 1936; Felix Bruner, "Senate Group Delays Action on Relief Bill," *Washington Post*, May 16, 1936; "Aid Bill Signed, Work Assured for 2,300,000," *Washington Post*, June 23, 1936; and Maddox, "Farm Security Administration," 40–41.
58. Marquis W. Childs, *I Write from Washington* (New York: Harper and Brothers, 1942), 14.
59. "Tugwell Substitute in Doubt," *New York Times*, May 1, 1936.
60. "Reminiscences of Howard R. Tolley," 401.
61. Hewes, *Boxcar in the Sand*, 55.
62. Paul W. Ward, "The End of Tugwell," *The Nation*, November 28, 1936, 623.
63. Robert C. Albright, "Tugwell Appeals to President to Establish a Permanent RA," *Washington Post*, November 8, 1936; Kirkendall, *Social Scientists and Farm Politics in the Age of Roosevelt*, 118–22; Conkin, *Tomorrow a New World*, 180–82. On the development of Tugwell's bad reputation, see Sternsher, *Rexford Tugwell and the New Deal*, 223–50. Tugwell's critics got in a final shot about Tugwell's last report as RA Administrator, a relatively slick publication described as too flashy and expensive; see Felix Bruner, "Tugwell Report Sets New Highs for Gaudiness," *Washington Post*, December 25, 1936.
64. "Tugwell Urges RA Be Placed in Department," *Washington Post*, October 9, 1936; "New Storm Rises Around Tugwell," *New York Times*, November 15, 1936; Baldwin, *Poverty and Politics*, 121–22.
65. "New Storm Rises Around Tugwell," *New York Times*, November 15, 1936; Lord, *Wallaces of Iowa*, 459–62; Culver and Hyde, *American Dreamer*, 169–71; "Wallace, Aide Will Make Tour of RA Projects," *Washington Post*, November 16, 1936; Felix Belair Jr., "RA Job Marvelous, Declares Wallace," *New York Times*, November 26, 1936; "Tugwell Resigns His Post; Acceptance Today Likely; Plans Return to Columbia," *New York Times*, November 18, 1936. Wallace was not alone in being moved by the experience; Howard R. Tolley, an AAA administrator, also went, and he too was shaken by what he had seen. Tolley, never terribly impressed with Tugwell, found that seeing and experiencing the poverty had a much greater impact than that of any convincing talk. See "Reminiscences of Howard R. Tolley," 396–400. Wallace made a tour earlier in the summer of 1936 to visit twelve drought-ridden states in the West, which likely also strengthened his view that the RA had valuable work to do. See "Wallace to Tour Big Drought Area," *New York Times*, June 30, 1936; Sidney Olson, "Rain—Or Relief Millions—May Deluge the Drought-Ridden States," *Washington Post*, July 5, 1936; and "Drought May Upset New WPA Program," *New York Times*, July 6, 1936.
66. FDR, Executive Order 7530, 1936; "To Absorb Forces of Resettlement," *New York Times*, December 22, 1936; W. T. Frazier, "Farm Security Administration," vol. 1, n.d.,Box 3, FSA Historical Record, RG 96, NACP; Henry A. Wallace, "Memorandum No. 710," February 2,

1937, Folder "U.S. Dept. of Agriculture—Farm Security Administration. 1937," Box 29, Baldwin Papers.
67. Senate, *Agricultural Appropriation Bill for 1938* (1937), 413; House, *Supplemental Estimate of Appropriation, 1938* (1937).
68. Clark, "Social and Economic Implications of the National Land Program," 274; Baldwin, *Poverty and Politics*, 231–32; Gaer, *Toward Farm Security*, 57.
69. Charles B. Crow, "John H. Bankhead, 2d., United States Senator from Alabama," September 2, 1936, Folder 1, Box 1, Bankhead Papers; "Reminiscences of Howard R. Tolley," 407–8.
70. John H. Bankhead Jr. to Clinton P. Anderson, July 2, 1945, Folder 5, Box 7, Bankhead Papers.
71. Irvin M. May Jr., *Marvin Jones: The Public Life of an Agrarian Advocate* (College Station: Texas A&M University Press, 1980), 129–30, 151–53.
72. Maddox, "The Bankhead-Jones Farm Tenant Act," *Law and Contemporary Problems* 4.4 (1937): 435–36; "Reminiscences of Will W. Alexander," 390; Mertz, *New Deal Policy and Southern Rural Poverty*, 93–105. Some of the most important thinkers behind the idea were Frank Tannenbaum, Charles S. Johnson, Edwin R. Embree, and Will Alexander; USDA attorney Philip Glick wrote most of the first bill.
73. *To Create the Farm Tenant Homes Corporation, Hearings Before A Subcommittee of the Committee on Agriculture and Forestry on S. 1800*, 74th Cong., 1st sess., 1935; Maddox, "Bankhead-Jones Farm Tenant Act," 436–38; "A.F.L. For Bankhead Bill," *New York Times*, April 15, 1935; *To Create the Farm Tenant Homes Corporation, Report (to accompany S. 2367)*, 74th Cong., 1st sess., 1935, Report no. 446; Mertz, *New Deal Policy and Southern Rural Poverty*, 112–17, 127–52; Maddox, "Bankhead-Jones Farm Tenant Act," 438; Baldwin, *Poverty and Politics*, 138–39, 151–52; Maddox, "Farm Security Administration," 41–44; May, *Marvin Jones*, 128–30, 151–58; Hewes, *Boxcar in the Sand*, 75.
74. Walter L. Randolph to J. D. Pope, December 17, 1935, Folder "Cooperation A RR-070," Box 4, Rural Rehabilitation General Correspondence, RG 96, NACP. The differences between how the various state Farm Bureaus generally favored the bill and how the national AFBF showed such disinterest raises questions about how much the AFBF represented even the state bureaus, much less their membership.
75. "Stave Off Debate on Lynching Bill," *New York Times*, April 24, 1935; "Senate Sidetracks Farm Tenant Bill," *New York Times*, April 25, 1935; *Congressional Record* 79 (1935), 302; Baldwin, *Poverty and Politics*, 152–53; Maddox, "Farm Security Administration," 42–44; Mertz, *New Deal Policy and Southern Rural Poverty*, 151–61; "Reminiscences of Howard R. Tolley," 409–10; May, *Marvin Jones*, 132.
76. Hewes, *Boxcar in the Sand*, 86.
77. House, *Farm Tenancy Hearing*, 324.
78. "Share Croppers' Plight Spurs Relief Agencies Into Action," *Christian Science Monitor*, April 15, 1935; F. Raymond Daniell, "Tenant Law Clash Roils Cotton Belt," *New York Times*, April 18, 1935; F. Raymond Daniell, "Arkansas Violence Laid to Landlords," *New York Times*, April 16, 1935; Conrad, *Forgotten Farmers*, 83–93. Other tenant resistance took place, but none to the extent of the STFU's activity in Missouri and Arkansas; see, for example, Kelley, *Hammer and Hoe*. For a more personal view, see Rosengarten, *All God's Dangers*, 296–343, 390–92, 545–55. On the STFU, see Grubbs, *Cry From the Cotton*; Conrad, *Forgotten Farmers*; Kester, *Revolt Among the Sharecroppers*; Mitchell, *Mean Things Happening in This Land*; Roll, *Spirit of Rebellion*; and Holley, *Uncle Sam's Farmers*, 82–104.

79. F. Raymond Daniell, "Farm Tenant Union Hurt by Outsiders," *New York Times*, April 19, 1935. On the ideology of small southern farmers, see Roll, *Spirit of Rebellion*, and on the STFU specifically, Manthorne, "View from the Cotton."
80. On the, in retrospect, surprising slowness of agricultural experts to recognize the significance of tenancy and its relationship to erosion and soil quality, see William H. Harbaugh, "Twentieth-Century Tenancy and Soil Conservation: Some Comparisons and Questions," *Agricultural History* 66.2 (1992): 95–101. On the early history of rural sociology, see Larson and Zimmerman, *Sociology in Government*, 11–22. For more on rural poverty in popular culture (especially regarding migrant workers in the Plains states and the West), see Tindall, *Emergence of the New South*, 415–17; and T. H. Watkins, *The Hungry Years: A Narrative History of the Great Depression in America* (New York: Henry Holt, 1999), 455–59.
81. Saloutos, *American Farmer and the New Deal*, 166.
82. Tugwell, "The Resettlement Idea," 162.
83. "A Worthy Objective," *Birmingham Post*, June 22, 1936; "Eradicating a Breeder of Revolution in the South," *Birmingham News*, February 10, 1936; Wilson Gee, "Reversing the Tide toward Tenancy," *Southern Economic Journal* 2.4 (1936): 4, 9. On the actual, rather small-scale spread of communism among southern African Americans during the Depression, see Kelley, *Hammer and Hoe*.
84. National Resources Committee, *Farm Tenancy*; "Reminiscences of Will W. Alexander," 392. See also Baldwin, *Poverty and Politics*, 168–77; Maddox, "Farm Security Administration," 47–48; Saloutos, *American Farmer and the New Deal*, 167–76.
85. National Resources Committee, *Farm Tenancy*, iv, 11 (quotes), 24–28. Some critics particularly worried about the impact on black farmers; see Davis, "Survey of the Problems of the Negro," 7; Charles Houston to Will W. Alexander, February 9, 1937, Folder "AD-140," Box 26, General Correspondence, Washington Office, RG 96, NACP; and Southern Policy Association, "Recommendations regarding Tenancy Legislation submitted to the Special Committee on Farm Tenancy," December 14, 1936, Folder 2, Box 8, Bankhead Papers. See also T. W. Schultz, "A Comment on the 'Report of the President's Committee on Farm Tenancy,'" *Journal of Land and Public Utility Economics* 13.2 (1937): 207–8; and Albert H. Cotton, "Regulations of Farm Landlord-Tenant Relationships," *Law and Contemporary Problems* 4.4 (1937): 508–10.
86. Maddox, "Farm Security Administration," 45–46, 408–16; Grubbs, *Cry from the Cotton*, 143–45; Franklin D. Roosevelt to John H. Bankhead, September 17, 1936, Folder "Tenancy," Box 2439, Office of the Secretary, General Correspondence, RG 16, NACP; John H. Bankhead Jr. to Charles B. Crow, January 18, 1937, Folder 3, Box 1, Charles B. Crow Papers, LPR56, Alabama Department of Archives and History, Montgomery.
87. "Farm Tenant Bill Filed by Bankhead," *New York Times*, January 9, 1937; "Administration Farm Tenant Bill Sponsored in Congress," *Wall Street Journal*, January 9, 1937; Maddox, "Bankhead-Jones Farm Tenant Act," 441–42; Mertz, *New Deal Policy and Southern Rural Poverty*, 179–83; House, *Farm Tenancy Hearing*, 208, 218–25.
88. Leroy Simms, "Bankhead Fights for Original Farm Bill," *Birmingham News-Age-Herald*, May 23, 1937; "President Urges Tenancy Bill Cut," *Atlanta Constitution*, June 5, 1937.
89. House, *Farm Tenancy Hearing*, 276. The much-maligned RA took criticism as being too big and revolutionary, too small and ineffective, too radical, a tool of big agriculture, and everything in between.
90. House, *Farm Tenancy Hearing*, 250, 88 (quotes), 113–15, 85–86.

91. House, *Farm Tenancy Hearing,* 208. See similar criticisms from Oregon's Walter Pierce (67) and North Carolina's Harold Cooley (232).
92. House, *Farm Tenancy Hearing,* 14, 19–20, 219–20, 224–30.
93. McConnell, *Decline of Agrarian Democracy,* 99; House, *Farm Tenancy Hearing,* 315–16; "Reminiscences of Will W. Alexander," 586–87.
94. "House Group Bars Roosevelt Project for Farm Tenants," *New York Times,* April 1, 1937; "Farm Tenant Bill Voted to Senate," *New York Times,* June 11, 1937; *Bankhead-Jones Farm Tenant Act, Conference Report (to accompany H.R. 7562),* 75th Cong., 1st sess., 1937, Report no. 1198, 13–16; Maddox, "Bankhead-Jones Farm Tenant Act," 443–46. *Bankhead-Jones Farm Tenant Act,* H.R. 7562, 1937, 75th Cong., 1st sess.; Baldwin, *Poverty and Politics,* 183–85; May, *Marvin Jones,* 155–56; "Reminiscences of Will W. Alexander," 602; Paul V. Maris, *"the land is mine": From Tenancy to Farm Ownership* (Washington, DC: U.S. Government Printing Office, 1950), 5–9.
95. *Bankhead-Jones Farm Tenant Act,* HR 7562, 75th Cong., 1st sess. Valuable analysis by FSA employees is found in W. T. Frazier, "Farm Security Administration," vol. 1, n.d., Box 3, FSA Historical Record, RG 96, NACP; and Maddox, "Bankhead-Jones Farm Tenant Act," 446–50. See also Helen C. Monchow, "The Farm Tenancy Act," *Journal of Land & Public Utility Economics* 13.4 (1937): 417–18. Some part-time farmers from industrial-rural places such as West Virginia did manage to slip through loopholes and enter the program. See Cahill, "Fertilizing the Weeds," 289–92.
96. Maddox, "Farm Security Administration," 49; Banfield, "Ten Years of the Tenant Purchase Program," 471; "Reminiscences of Will W. Alexander," 601–2.
97. *Congressional Record* 81 (1937), 6438. William Bankhead (John Bankhead's brother) conceded this point but compared the proposed program with other federal activities (like rural free delivery and extension service) that had begun in small and experimental ways (6453). On small farmers and their inability to gain substantial equity in their farms, see Raymond C. Smith, "New Conditions Demand New Opportunities," in *Farmers in a Changing World,* by United States Department of Agriculture (Washington, DC: U.S. Government Printing Office, 1940), 819–20.

Chapter 4

1. Starling Jackson to Hurtis Parr, Folder "85–160 Public Relations, July 1939 thru June, 1940," Box 1, Office of the Director, General Correspondence, RG 96, NARASE.
2. Robert Shogan, *Backlash: The Killing of the New Deal* (Chicago: Ivan R. Dee, 2006), 7; Watkins, *Hungry Years,* xvi. See also Watkins's postlude, tellingly titled "Dismantling the Dream, 1939." Jess Gilbert describes this period as the third agricultural New Deal, combining self-government and social reform with a greater emphasis on coordination and planning—efforts most defined by their failures. Gilbert, "Low Modernism and the Agrarian New Deal," 133.
3. Henry A. Wallace, Memorandum No. 732, September 1, 1937, Folder "Tenancy 1 (Farmer's Home Corporation)," Box 2663, Office of the Secretary, General Correspondence, RG 16, NACP; Henry A. Wallace, "In Quest of Farm Security," *Extension Service Review* 8.3 (1937): 34.
4. Sidney Olson, "Wallace Ends Housing Plans of Rex Tugwell," *Washington Post,* September 2, 1937.

5. Brown and Baugh, *Preliminary Inventory*, 3; USDA, Untitled memorandum, April 5, 1938, Folder "Farm Security," Box 2782, Office of the Secretary, General Correspondence, RG 16, NACP; Phillips, *This Land, This Nation*, 146–47.
6. Gaer, *Toward Farm Security*, 141–47.
7. Maddox, "Farm Security Administration," 91–93.
8. Maddox, "Farm Security Administration," 81–83; 98–100.
9. Johnstone, Introduction, 2.
10. Hewes, *Boxcar in the Sand*, 114.
11. Maddox, "Farm Security Administration," 203–5.
12. *Report of the Administrator of the Farm Security Administration, 1941*, 9, 28–30.
13. "Summary of FSA Programs," Folder "Farm Security Administration, 1934–1941," Edwin G. Arnold Papers, Truman Library, Independence, MO.
14. Frederick D. Mott and Milton I. Roemer, *Rural Health and Medical Care* (New York: McGraw-Hill, 1948), 392–93. This added up to about 600,000 people.
15. Senate, *Agricultural Appropriation Bill for 1942* (1941), 467.
16. *Report of the Administrator of the Farm Security Administration, 1941*, 31–32.
17. "Farm Ownership Program," n.d. [1943?], pp. 1–2, Folder 2, Box 8, John H. Bankhead Papers, LPR 53, ADAH.
18. FSA, "Farm Security Administration Homesteads," 1941, NAL; *Report of the Administrator of the Farm Security Administration, 1941*, 33; "Summary of FSA Programs," Folder "Farm Security Administration, 1934–1941," Edwin G. Arnold Papers, Truman Library.
19. Within the USDA, the Bureau of Agricultural Economics and the Soil Conservation Service both had a strong liberal and reformist bent to their activities, but the BAE lacked the political heft of the FSA, and the SCS was less immediately concerned with rural poverty and reform. Neither could directly confront rural poverty.
20. "Reminiscences of Will W. Alexander," 401.
21. John Egerton, *Speak Now Against the Day: The Generation before the Civil Rights Movement in the South* (New York: Alfred A. Knopf, 1994), 91–98; Wilma Dykeman and James Stokely, *Seeds of Southern Change: The Life of Will Alexander* (Chicago: University of Chicago Press, 1962), 217; "Reminiscences of Will W. Alexander," 599–600.
22. "Reminiscences of Will W. Alexander," 663 (quote), 665–66; Dykeman and Stokely, *Seeds of Southern Change*, 247–48; Maddox, "Farm Security Administration," 105–6. Alexander was not the only idealist driven from government service by a dislike of the politics. It had played a part in Tugwell's resignation, and a lower-level employees like Laurence Hewes frequently commented on the personal and moral toll caused by dealing with such a situation. See Hewes, *Boxcar in the Sand*, 102.
23. Calvin Benham Baldwin, "Personnel Questionnaire," November 4, 1939, Folder "Dept. of Agriculture—Farm Security Administration—Personnel, 1933–1943," Box 31, Baldwin Papers; C. B. Baldwin, interview by Richard K. Doud, 2–6; Patricia Sullivan, *Days of Hope: Race and Democracy in the New Deal Era* (Chapel Hill: University of North Carolina Press, 1996), 124–26.
24. "Reminiscences of Will W. Alexander," 476–77.
25. Reid F. Murray to Paul H. Appleby, April 1, 1941, Folder "Farm Security Loans or Grants, Mar. 1 to Apr. 19," Box 291, Office of the Secretary, General Correspondence, RG 16, NACP.
26. Christiana McFadyen Campbell, *The Farm Bureau and the New Deal: A Study of the Making of National Farm Policy, 1933–40* (Urbana: University of Illinois Press, 1962), 173;

Sullivan, *Days of Hope*, 127–28; Marion Clawson, "Resettlement Experience on Nine Selected Resettlement Projects," *Agricultural History* 52.1 (1978): 3.
27. Conrad, *Forgotten Farmers*, 105–7; Gilbert, "Eastern Urban Liberals and Midwestern Agrarian Intellectuals," 165–70.
28. Paul H. Appleby, *Big Democracy* (New York: Russel and Russel, 1945, 1970), 12–19, 25. See also John M. Gaus and Leon O. Wolcott, *Public Administration and the United States Department of Agriculture* (Chicago: Public Administration Service, 1940), 30–81.
29. Baldwin, *Poverty and Politics*, 236–38; Gaus and Wolcott, *Public Administration and the United States Department of Agriculture*, 265–67. After those five agencies, the Rural Electrification Administration, the Farm Credit Administration, and the Commodity Credit Corporation followed in importance in 1939.
30. Gaus and Wolcott, *Public Administration and the United States Department of Agriculture*, 494.
31. For example, A. R. Blanchard, "A More Secure Farm Life," *Extension Service Review* 12.6 (1941): 88–89.
32. "Reminiscences of Robert W. Hudgens," 126. See also "Reminiscences of Howard R. Tolley," 478–79; and Saloutos, *American Farmer and the New Deal*, 240.
33. Campbell, *Farm Bureau and the New Deal*, 166–69; Block, *Separation of the Farm Bureau and the Extension Service*, 37. Not all farm bureau members or state organizations agreed with the national organization's hostility to the FSA. See *Byrd Committee Hearings, Part III*, 905–8; Senate, *Agricultural Appropriation Bill for 1944* (1943), 1107.
34. McConnell, *Decline of Agrarian Democracy*, 44–54; John Mark Hansen, *Gaining Access: Congress and the Farm Lobby, 1919–1981* (Chicago: University of Chicago Press, 1991), 83.
35. James Raley Howard, Introduction to Orville Merton Kile, *The Farm Bureau Movement* (New York: Macmillan, 1922), x–ix; Hansen, *Gaining Access*, 29.
36. DeWitt C. Wing, "Trends in National Farm Organizations," in *Farmers in a Changing World*, by United States Department of Agriculture (Washington, DC: U.S. Government Printing Office, 1940), 954, 957.
37. McConnell, *Decline of Agrarian Democracy*, 68–69; Michael W. Flamm, "The National Farmers Union and the Evolution of Agrarian Liberalism," *Agricultural History* 68.3 (1994): 61; Baldwin, *Poverty and Politics*, 238–39; Maddox, "Farm Security Administration," 496–97. The Farmers' Union did not support the entirety of the FSA program, just to the extent to which it helped promote ownership of family farms. See House, *Farmers' Home Corporation Act of 1944, Hearings before the Committee on Agriculture on H. R. 4384*, 78th Cong., 2nd sess., 1944, 17–18.
38. On the AFBF's view of the FSA, see McConnell, *Decline of Agrarian Democracy*, 97–111; Saloutos, *American Farmer and the New Deal*, 244. On farmers as a self-conscious interest group, see Richard Hofstadter, *The Age of Reform: From Bryan to FDR* (New York: Vintage Books, 1955), 122–29.
39. Maddox, "Farm Security Administration," 117–18.
40. "Petition to President Franklin D. Roosevelt and Secretary Henry A. Wallace" [summer 1939], Folder "85-160-02 Criticisms and Complaints, July, 1938 thru Dec., 1939," Box 1, Office of the Director, General Correspondence, RG 96, NARASE. See also "Prophecies of National Ruin," *Dallas Morning News*, August 11, 1935, which criticizes Tugwell and the Resettlement Administration for (in the editorial's view) overdramatic warnings about the need for the land use reform: "Strangest of all . . . another branch of our Government is doing everything to reduce crop production . . . with the clear imputation that the soil is still too productive."

41. John E. Hydrick to Homer D. Lee, March 12, 1940, and Homer D. Lee to D. H. Frazier, March 13, 1940, in Folder "85-160-02 Criticisms and Complaints, July, 1938 thru Dec., 1939," Box 1, Office of the Director, General Correspondence, RG 96, NARASE.
42. Reid F. Murray to Paul H. Appleby, April 16, 1941, Folder "Farm Security Loans or Grants, Mar. 1 to Apr. 19," Box 291, Office of the Secretary, General Correspondence, RG 16, NACP; "Reminiscences of Howard R. Tolley," 477–78; Maddox, "Farm Security Administration," 118; Saloutos, *American Farmer and the New Deal*, 240. The overlap of rural rehabilitation and similar lending aimed at the rural poor with the lending activities of the FCA had been a source of criticism since before there was an FSA; see "Mr. Tugwell's Octopus," *Washington Post*, March 27, 1936.
43. Ernest K. Lindley, "The Budget Slashes," *Washington Post*, December 11, 1939.
44. Baldwin, *Poverty and Politics*, 242–43; "Reminiscences of Will W. Alexander," 639–43.
45. Charles McKinley, "Federal Administrative Pathology and the Separation of Powers," *Public Administration Review* 11.1 (1951): 21.
46. Culver and Hyde, *American Dreamer*, 227.
47. "Reminiscences of Will W. Alexander," 643; "Reminiscences of Howard R. Tolley," 527, 551–54.
48. Michael J. McDonald and John Muldowny, *TVA and the Dispossessed: The Resettlement of Population in the Norris Dam Area* (Knoxville: University of Tennessee Press, 1999), 160–61, 173–74; Baldwin, *Poverty and Politics*, 302–5.
49. Reid F. Murray to Paul H. Appleby, April 16, 1941, Folder "Farm Security Loans or Grants, Mar. 1 to Apr. 19," Box 291, Office of the Secretary, General Correspondence, RG 16, NACP.
50. See, for example, voluminous correspondence in Boxes 6–7, Bankhead Papers.
51. Quoted in McKinley, "Federal Administrative Pathology and the Separation of Powers," 22. A position of autonomy and self-reliance could at times overshadow other political goals. For example, the AAA's displeasure at being Wickard's choice as the tool for the planned defense boards in 1941 indicates that what Albertson calls "their clannish fraternity" was more important than greater access to and influence over the secretary. Dean Albertson, *Roosevelt's Farmer: Claude R. Wickard in the New Deal* (New York: Columbia University Press, 1961), 229.
52. Albertson, *Roosevelt's Farmer*, 122.
53. House, *Agriculture Department Appropriation Bill for 1943*, (1942), 274.
54. "Reminiscences of Howard R. Tolley," 446; Baldwin, *Poverty and Politics*, 295–98, 310.
55. R. W. Hudgens to State Directors, District and County, Farm and Home Supervisors, Subject "Public Relations," and William H. Dent to County Supervisors, State Directors, State Tenant-purchase Specialists, and District Supervisors, Subject "Publicizing approval of TP Loans and Purchase of Farms" [n.d., 1939–40], both in folder "Public Relations, July 1939 thru June, 1940," Box 1, Office of the Director, General Correspondence, RG 96, NARASE. These kinds of efforts often only resulted in hostility; see *Byrd Committee Hearings, Part III*, 737.
56. Raymond A. Pearson to M. L. Wilson, June 25, 1937, Folder "Resettlement," Box 2629, Office of the Secretary, General Correspondence, RG 16, NACP.
57. Helen Kennedy to Connie J. Bonslagel, January 27, 1936, Folder "Correspondence K RR-032," Box 2, Rural Rehabilitation General Correspondence, RG 96, NACP; "Reminiscences of Robert W. Hudgens," 126.
58. "New Cooperative Idea," *Extension Service Review* 7.12 (1936): 190; H. G. Seyforth, "Low-Income Families Make Good: Agent Aids Rehabilitation," *Extension Service Review* 9.9 (1938): 131; "Cooperating with FSA," *Extension Service Review* 11.6–7 (1940): 90; Senate,

Agricultural Appropriation Bill for 1943 (1942), 957–64; "FSA Workers Plan Part in U.S. Defense," *Dallas Morning News,* July 8, 1941.
59. Holley, *Uncle Sam's Farmers,* 186–90; "Reminiscences of Robert W. Hudgens," 156–72; C. B. Baldwin interview with Richard K. Doud, 28–30.
60. Baldwin, *Poverty and Politics,* 262–63.
61. "Says New Deal Aim Is 'Soviet Industry,'" *New York Times,* August 22, 1938 (quote); "New Deal Homestead Groups Chosen for Hosiery-Making," *Christian Science Monitor,* September 1, 1938; Milo Perkins, "Memorandum for the Secretary," July 13, 1938, Folder "Farm Security 2 (Projects,),"Box 2783, Office of the Secretary, General Correspondence, RG 16, NACP.
62. Senate, *Agricultural Appropriation Bill for 1940* (1939), 820–23 (quote); "FSA Factory Loans Declared Illegal," *New York Times,* December 20, 1938. Given that these funds came from the Emergency Relief funds, not agricultural appropriations, and that construction of the mills had already been prohibited, one can only assume that Representative Moser was indulging in a bit of political grandstanding.
63. Peter Edson, "In Washington," *Monroe (LA) Morning Herald,* June 6, 1942, in Folder "U.S. Dept. of Agriculture—Farm Security Administration. 1942," Box 29, Baldwin Papers.
64. House, *Farm Security Hearings, Part I,* 192.
65. House, *Agriculture Department Appropriation Bill for 1944* (1943), 1007.
66. Baldwin, *Poverty and Politics,* 279, 250; Conkin, *Tomorrow a New World,* 200–202.
67. Adams and Gorton, "This Land Ain't My Land," 324–30. The evicted tenants were supposed to move to a new project, but mostly because of poor planning, the families scattered across the Mississippi Delta. See also Holley, *Uncle Sam's Farmers,* for more on FSA Region VI in general, and on race specifically, 179–90. On the displacement of African Americans by the Tennessee Valley Authority, see Grant, *TVA and Black Americans,* 74–86.
68. John H. Bankhead to J. L. Edwards [early December 1936?], Folder 1, Box 5, Bankhead Papers.
69. James C. Derieux to R. W. Hudgens, December 20, 1936, Folder "85-163-1 Articles & Press Releases, July, 1938 thru December, 1938," Box 2, Office of the Director, General Correspondence, RG 96, NARASE.
70. Holley, *Uncle Sam's Farmers,* 180.
71. James F. Byrnes to Mr. W. W. Alexander, October 9, 1937, Folder "U.S. Dept. of Agriculture—Farm Security Administration. 1937," Box 29, Baldwin Papers.
72. "Conference of FSA officials held at Capon Springs, W. Va., during the period May 31 1940–June 3, 1940" [n.d.], Folder "Dept. of Agriculture. Farm Security Administration. Conference at Capon Springs. 1940," Box 30, Baldwin Papers.
73. C. P. Trussell, "FSA Is Accused of Tax Loan Plan," *New York Times,* January 24, 1942.
74. J. Lewis Henderson, "In the Cotton Delta," *Survey Graphic* 36.1 (1947): 48. For a fuller discussion of the New Deal and black civil rights, see John Brueggemann, "Racial Considerations and Social Policy in the 1930s: Economic Change and Political Opportunities," *Social Science History* 26.1 (2002): 139–68.
75. Harry F. Byrd, "Note to Editors and Correspondents," July 3, 1937, Folder "U.S. Dept. of Agriculture—Farm Security Administration. 1937," Box 29, Baldwin Papers; see also "Byrd Attacks RA on 'Sinful Waste,'" *New York Times,* July 5, 1937. Byrd still used the FSA when it furthered his constituents as individually necessary, writing to ensure the FSA considered their needs; see Grover B. Hill to Harry F. Byrd, Folder "Farm Security Loans or Grants, Aug. 1 to Nov. 1," Box 292, Office of the Secretary, General Correspondence, RG 16, NACP.

76. "Governmental Waste and Extravagance (Radio address by Senator Harry F. Byrd of Virginia, over stations of the National Broadcasting Company, from Washington, D.C., Saturday, July 17, 1937.)," Folder "U.S. Dept. of Agriculture—Farm Security Administration. 1937," Box 29, Baldwin Papers. Byrd and Henry Wallace had conversed both in person and by letter regarding the Shenandoah homesteads and other such projects around Virginia, but apparently Byrd was unsatisfied by Wallace's response, as criticism of the homesteads would become one of the centerpieces of Byrd's years-long attack on the FSA. Byrd's frequent requests for information on the homesteads projects, creating work by the RA and FSA which he then used to attack them, did not improve relations, especially since Byrd put off appearances by the secretary of agriculture before the Committee to Investigate Executive Agencies of the Government in the summer of 1937, a move apparently designed simply to make things more difficult for the RA. RA/FSA administrators maintained an extensive "Byrd File" for dealing with the difficult politician. Robert Hudgens estimated that the Byrd committee alone cost the agency about three hundred man-hours. "Reminiscences of Robert W. Hudgens," 200. On Shenandoah under the RA and FSA, see Gregg, *Managing the Mountains*, 204–8, 211–12.
77. National Resources Committee, *Farm Tenancy*, 28.
78. FSA, "To the FSA Personnel," p. 4, Folder "85-161 Speeches, July thru Dec. 1940," Box 14, Office of the Director, General Correspondence, RG 96, NARASE.
79. House, *Farm Labor Program, 1943*, 96.
80. Significant studies of this change include Brinkley, *End of Reform*, and Sullivan, *Days of Hope*.
81. Larson, *Ten Years of Rural Rehabilitation*, 14.

Chapter 5

1. Rannie Anderson to Mr. Coker, August 5, 1942, Rannie Anderson to Mr. Coker, October 9, 1942, "Affidavit of Loss of Mortgaged Property," March 26, 1942, all in folder "Anderson, Rannie," Rural Rehabilitation Loan Files, RG 96, NARASE.
2. Maddox, "Farm Security Administration," 84–89. See also Grant, *Down and Out on the Family Farm*.
3. Larson, *Ten Years of Rural Rehabilitation*, 41–53.
4. Wilson, "Problem of Poverty in Agriculture," 28. For other examples related to the Rural Rehabilitation Division or its predecessors, see David R. Williams, "Rural Industrial Communities," *Rural Rehabilitation* 1.1 (1934), 4; O. E. Baker, "Rural-Urban Migration and the National Welfare," *Rural Rehabilitation* 1.2 (1935): 25; and "The Spiritual Side of Rehabilitation," *Rural Rehabilitation* 1.2 (1935), 27.
5. Clifford B. Anderson, "The Metamorphosis of American Agrarian Idealism in the 1920's and 1930's," *Agricultural History* 35.4 (1961): 184; Jess Gilbert and Steve Brown, "Alternative Land Reform Proposals in the 1930s: The Nashville Agrarians and the Southern Tenant Farmers' Union," *Agricultural History* 55.4 (1981): 352–55.
6. National Emergency Council, *Report on Economic Conditions of the South*, 21.
7. Schuler, *Social Status and Farm Tenure*, 10.
8. FSA, *The Work of the Farm Security Administration*, FSA Instruction 400.10, sheet 1–sheet 1 reverse (Washington, DC: U.S. Government Printing Office, 1942).
9. Larson, *Ten Years of Rural Rehabilitation*, 45–57.
10. In the RA, the Land Utilization Division had the largest Washington staff with two hundred, plus another twenty-five hundred in the field. "Force of 6,090 Hired in Tugwell Program," *New York Times*, July 15, 1935.

11. Senate, *Resettlement Administration Program*, 7–10; Gaer, *Toward Farm Security*, 54–56; *Interim Report of the Resettlement Administration*, 8–13; *First Annual Report, Resettlement Administration*, 9.
12. Senate, *Resettlement Administration Program*, 7–8; *First Annual Report, Resettlement Administration*, 9–10.
13. Larson, *Ten Years of Rural Rehabilitation*, 156; Maddox, "Farm Security Administration," 202–3.
14. Lawrence Lazar, "Rural Rehabilitation General Personnel Report," September 18, 1935, Folder "RR-101-3 L Reports," Box 6, Rural Rehabilitation General Correspondence, RG 96, NACP.
15. Henry A. Wallace, "In Quest of Farm Security," *Extension Service Review* 8.3 (1937): 34.
16. Senate, *Agricultural Appropriation Bill for 1938* (1937), 419–21.
17. R. W. Hudgens to R. G. Tugwell, November 3, 1936, Folder "AD-AL-17 (060) Coffee Farms," Box 52, Project Records, RG 96, NACP.
18. Pyle, *Home Country*, 368.
19. W. T. Frazier, "Farm Security Administration," vol. 2, n.d., Box 3, FSA Historical Records, RG 96, NACP; *Report of the Administrator of the Farm Security Administration, 1939*, 2; T. Swann Harding, "Farmers Home Administration," *Antioch Review* 6.4 (1946): 587.
20. FSA, *The Work of the Farm Security Administration*, FSA Instruction 400.10, sheet 1 reverse.
21. Maddox, "Farm Security Administration," 162–64, 132–34.
22. Larson, *Ten Years of Rural Rehabilitation*, 161–66. In the period between July 1, 1935, and June 30, 1944, Region IV (West Virginia, Virginia, Kentucky, Tennessee, and North Carolina) got 8.1 percent of loans; Regions V (Alabama, Georgia, South Carolina, and Florida) and VI (Arkansas, Louisiana, and Mississippi) each received 13.3 percent of the total loans; and Region VIII (most of Texas and Oklahoma) got 12.5 percent. Besides living in greater levels of poverty, part of the reason for the disparity in loan amounts is that southerners, particularly in Regions V and VI, were much more likely to require supplemental loans. Regions III (Iowa, Missouri, Illinois, Indiana, and Ohio) and VII (the Dakotas, Nebraska, and Kansas) received a similar percentage (11 percent and 10.3 percent respectively) to the southern regions.
23. Larson, *Ten Years of Rural Rehabilitation*, 103–4. On how New Deal works programs reinforced gendered and racial boundaries, see Jason Scott Smith, *Building New Deal Liberalism: The Political Economy of Public Works, 1933–1956* (Cambridge, UK: Cambridge University Press, 2006), 15, 260. In promoting the family farm, the tenant purchase and rural rehabilitation programs similarly strengthened gender and familial norms in rural America. The FSA's racial impact, particularly in terms of the rural rehabilitation program, had less to do with encouraging white-dominated industry than it did with fitting into an already-existent racial hierarchy.
24. Anonymous letter to "The United States Department of Agriculture," September 1 1939, Folder "85-160-02 Criticisms and Complaints, July, 1938 thru Dec., 1939," Box 1, Office of the Director, General Correspondence, RG 96, NARASE.
25. E. S. Morgan to R. F. Kolb, October 21, 1938, Folder "85-160-02 Criticisms and Complaints, July, 1938 thru Dec., 1939," Box 1, Office of the Director, General Correspondence, RG 96, NARASE.
26. Brown and Baugh, *Preliminary Inventory*, 16; Larson, *Ten Years of Rural Rehabilitation*, 179–81; Senate, *Resettlement Administration Program*, 9–10; *Interim Report of the*

Resettlement Administration, 11; *First Annual Report, Resettlement Administration*, 11–15. For an overview of the expansion of, and hope for, agricultural cooperatives during World War II, see also Eaton, *Exploring Tomorrow's Agriculture*, 65–191.
27. Maddox, "Farm Security Administration," 261–64.
28. Maddox, "Farm Security Administration," 274–77; Eaton, *Exploring Tomorrow's Agriculture*, 87–88; Larson, *Ten Years of Rural Rehabilitation*, 194–200; Baldwin, *Poverty and Politics*, 207–8.
29. Maddox, "Farm Security Administration," 230–36; R. W. Hudgens, "The Plantation South Tries a New Way," *Land Policy Review* 3.7 (1940): 26–29.
30. Eaton, *Exploring Tomorrow's Agriculture*, 36, 67. Thirteen of these, or about half, were in the South.
31. Hudgens, "Plantation South Tries a New Way," 25 (quote); Maddox, "Farm Security Administration," 243.
32. Reid, "Public Assistance to Low-Income Farmers of the South," 192.
33. R. M. Evans to Howard. A. Cowden, June 25, 1937, Folder "Resettlement," Box 2629, Office of the Secretary, General Correspondence, RG 16, NACP. On southern elite opposition to FSA cooperatives, see also Alston and Ferrie, *Southern Paternalism and the American Welfare State*, 85.
34. Larson, *Ten Years of Rural Rehabilitation*, 209–30.
35. Particularly valuable on the subject of the FSA's medical care program: Mott and Roemer, *Rural Health and Medical Care*, chap. 22; "Rural Health Activities of the Department of Agriculture," 389–431 (written by two doctors personally involved with the FSA medical program); and House, *Farm Security Hearings, Part IV*, 1613–32. The best secondary source is Grey, *New Deal Medicine*.
36. O. V. Wells, "Agriculture Today: An Appraisal of the Agricultural Problem," in USDA, *Farmers in a Changing World*, 388; National Emergency Council, *Report on Economic Conditions of the South*, 29–30; Johnson, *Shadow of the Plantation*, 192–96; Terry and Sims, *They Live on the Land*, 110–27; Mott and Roemer, *Rural Health and Medical Care*, 389–91; Grey, *New Deal Medicine*, 31–47.
37. Maddox, "Farm Security Administration," 307; Mott and Roemer, *Rural Health and Medical Care*, 393–94.
38. Mott and Roemer, *Rural Health and Medical Care*, 394; House, *Farm Security Hearings, Part IV*, 1616; Grey, *New Deal Medicine*, 58–60; Gaer, *Toward Farm Security*, 71–72. See also USDA, *Better Health for Rural America: Plans of Action for Farm Communities* (Washington, DC: U.S. Government Printing Office, 1945), 10–14.
39. Mott and Roemer, *Rural Health and Medical Care*, 395; Grey, *New Deal Medicine*, 53–57; A. M. Simon, "Medical Service Plans for the Farm Security Administration," *Journal of the American Medical Association* 120.16 (1942): 1315.
40. R. C. Williams to C. B. Baldwin, April 10, 1941, Folder "AD-AL-17 (060) Coffee Farms," Box 52, Project Records, RG 96, NACP.
41. Maddox, "Farm Security Administration," 306–7; House, *Farm Security Hearings, Part IV*, 1613–16.
42. Larson, *Ten Years of Rural Rehabilitation*, 253–56.
43. Pyle, *Home Country*, 366.
44. Mott and Roemer, *Rural Health and Medical Care*, 396; Grey, *New Deal Medicine*, 65.
45. Simons, "Medical Service Plans of the Farm Security Administration," 1315. On physicians' responses to the medical cooperatives, see Grey, *New Deal Medicine*, 65–71.

46. Ruth B. Register to George C. Stoney, May 17, 1941, Folder "85–160 Public Relations, July thru December, 1940," Box 14, Office of the Director, General Correspondence, RG 96, NARASE.
47. USDA, *Better Health for Rural America*, 13–14; Farm Security Administration, "Farm Security in Coffee County, Alabama," June 17, 1941, Folder "AD-AL-17 (060) Coffee Farms," Box 53, Project Records, RG 96, NACP.
48. Mott and Roemer, *Rural Health and Medical Care*, 405–6, 411, 413.
49. The story of the FSA's role in the national health care debate in the 1930s and 1940s is in Grey, *New Deal Medicine*, 152–67.
50. *First Annual Report, Resettlement Administration*, 12–14; H. C. M. Case, "Farm Debt Adjustment During the Early 1930s," *Agricultural History* 34.4 (1960): 174–82; Baldwin, *Poverty and Politics*, 209–11; Maddox, "Farm Security Administration," 142–44, 213–22. See also Ernest Feder, "Farm Debt Adjustment During the Depression—The Other Side of the Coin," *Agricultural History* 35.2 (1961): 78–81. For county supervisor policies for debt adjustment, see FSA, *Procedure Manual*, FSA Instructions 732.1 and 733.1, [Washington, 1938–40].
51. "3,661 Farm Debts Cut $6,491,866, Or 31.7%," *New York Times*, February 3, 1936.
52. Case, "Farm Debt Adjustment During the Early 1930s," 176–82; Maddox, "Farm Security Administration," 213–21.
53. FSA, "Progress of Rural Rehabilitation Region V," n.d., Folder "Rehabilitation Region V AD-560 Reports," Box 60, General Correspondence, Washington Office, RG 96, NACP. Twenty-four hundred other cases were not resolved by the committee, for a variety of reasons.
54. Larson, *Ten Years of Rural Rehabilitation*, 256; House, *Farm Security Hearings, Part I*, 160–62, 65–67.
55. Maddox, "Farm Security Administration," 224–28; Larson, *Ten Years of Rural Rehabilitation*, 263–67. See also Alston and Ferrie, *Southern Paternalism and the American Welfare State*, 78–80, 85–87. Like many farm security programs, the tenure improvement project went through a variety of bureaucratic shifts. It started out in the RA's Rural Rehabilitation Division, which moved to the FSA in September 1937. In July 1939, the section was eliminated and its work put into the general program. In the summer of 1941, a Farm Debt Adjustment and Tenure Improvement Section was reestablished, since the number of cases involving debt adjustment were declining and the two programs were closely linked.
56. Maddox, "Farm Security Administration," 223; Larson, *Ten Years of Rural Rehabilitation*, 262–64.
57. Larson, *Ten Years of Rural Rehabilitation*, 266–67.
58. Mertz, *New Deal Policy and Southern Rural Poverty*, 210. The story of this experiment is told throughout Conrad Taeuber and Rachel Rowe, *Five Hundred Families Rehabilitate Themselves* (Washington, DC: United States Department of Agriculture, Bureau of Agricultural Economics and Farm Security Administration, Cooperating, 1941); Rachel Rowe Swiger and Conrad Taeuber, *They Too Produce for Victory* (Washington, DC: United States Department of Agriculture, Bureau of Agricultural Economics and Farm Security Administration, Cooperating, 1942); Rachel Rowe Swiger and Conrad Taeuber, *Solving Problems through Cooperation* (Washington, DC: United States Department of Agriculture, Bureau of Agricultural Economics and Farm Security Administration, Cooperating, 1942);Rachel Rowe Swiger and Conrad Taeuber, *Ill Fed, Ill Clothed, Ill*

Housed—Five Hundred Farm Families in Need of Help (Washington, DC: United States Department of Agriculture, Bureau of Agricultural Economics and Farm Security Administration, Cooperating, 1942); Rachel Rowe Swiger and Olaf F. Larson, *Yesterday, Today and Tomorrow: Five Hundred Low-Income Farm Families in Wartime* (Washington, DC: United States Department of Agriculture, Bureau of Agricultural Economics and Farm Security Administration, Cooperating, 1943); and Rachel Rowe Swiger and Olaf F. Larson, *Climbing Toward Security* (Washington, DC: United States Department of Agriculture, Bureau of Agricultural Economics, 1944).

It may be valuable to compare the experience of these communities with the administrative reports on nine resettlement communities described by Marion Clawson in "Resettlement Experience on Nine Selected Resettlement Projects." Clawson, an agricultural economist working for the FSA, discovered an utter failure in resettlement, a project that the FSA began with high hopes but dropped in embarrassment in the early 1940s. Clawson recovered as much information as possible and wrote up a report. But with World War II and the concentrated congressional attack on the FSA getting underway, Clawson felt it would be better to keep the report strictly internal; it was not published until 1978 and apparently was not seen by any but the top of the FSA administration before then.

59. Taeuber and Rowe, *Five Hundred Families Rehabilitate Themselves*, 4.
60. Swiger and Larson, *Climbing Toward Security*, 2.
61. Taeuber and Rowe, *Five Hundred Families Rehabilitate Themselves*, 6–7; Swiger and Larson, *Climbing Toward Security*, 6.
62. Swiger and Taeuber, *Solving Problems through Cooperation*, 1–11.
63. Taeuber and Rowe, *Five Hundred Families Rehabilitate Themselves*, 3, 18–19, 19.
64. Swiger and Larson, *Yesterday, Today, and Tomorrow*, 1 (quote), iv.
65. Swiger and Taeuber, *They Too Produce for Victory*, 1, 3.

Chapter 6

1. "Former Tenant Farmer Proves Worth in Month on Own Acres," *Dallas Morning News*, March 3, 1940.
2. USDA, "Talking Script for program with Farm and Family Forum, WAPI, Birmingham, and the Auburn Farm Network, 12:30 to 1:00, May 27, 1941, Participants: Harwood Hyll, Jim Romine, Dorothy Wood, John H. McCullough, Mrs. McCullough, Imogene McCullough, Leroy McCullough, and Tarleton Collier for FSA," Folder 85-1-16-165-2 Radio Programs, Box 18, Project Construction Files, RG 96, NARASE.
3. Brown and Baugh, *Preliminary Inventory*, 24.
4. Paul V. Maris, "From Tenant to Owner: The First Year Shows Progress Under the Bankhead-Jones Tenant Act," *Extension Service Review* 9.10 (1938): 154. See also Baldwin, *Poverty and Politics*, 195–96.
5. Paul V. Maris, "Significant Administrative Problems Related to the Tenant Purchase Program," September 1940, p. 3, Folder "85-161 Speeches, July thru Dec. 1940," Box 14, Office of the Director, General Correspondence, RG 96, NARASE.
6. Baldwin, *Poverty and Politics*, 197–99; Maris, *"the land is mine,"* 319–20.
7. Maris, *"the land is mine,"* 17–22.
8. Will W. Alexander to Paul V. Maris, "Report of Progress for the Period January 1 to January 31, 1938, inclusive," February 1, 1938, p. 2, and Will W. Alexander to Paul V. Maris,

"Report of Progress for the Period February 1 to February 28, 1938, inclusive," March 1, 1938, p. 1, both in Folder "Tenant Purchase Division AD-183-6," Box 40, General Correspondence, Washington Office, RG 96, NACP.

9. Paul V. Maris, "From Tenant to Owner: The First Year Shows Progress Under the Bankhead-Jones Tenant Act," *Extension Service Review* 9.10 (1938): 154. Practically all of the RA/FSA programs received many more applications than could realistically be approved. See, for example, Cannon, *Remaking the Agrarian Dream*, 20–21.

10. "Bankhead-Jones Act Break for This Man," *Birmingham News*, February 17, 1940; "Bankhead Jones Invokes Bankhead-Jones Statute," *Montgomery Advertiser*, February 18, 1940.

11. Paul V. Maris, "Significant Administrative Problems Related to the Tenant Purchase Program," September 1940, pp. 2–3, Folder "85–161 Speeches, July thru Dec. 1940," Box 14, Office of the Director, General Correspondence, RG 96, NARASE; Senate, *Agricultural Appropriation Bill for 1940* (1939), 625.

12. House, *Agriculture Department Appropriation Bill for 1941* (1940), 954 (quote); House, *Agriculture Department Appropriation Bill for 1942* (1941), 87–89.

13. Banfield, "Ten Years of the Farm Tenant Purchase Program," 474–76; House, *Agriculture Department Appropriation Bill for 1944* (1943), 1646.

14. Maris, *"the land is mine,"* 299; Reid, "Public Assistant to Low-Income Farmers of the South," 193. Oklahoma Congressman Phil Ferguson claimed to have not found a single farmer against the program. Senate, *Agricultural Appropriation Bill for 1941* (1940), 667.

15. Farm Security Administration, "The FSA—Pro and Con," January 28, 1942, NAL.

16. Senate, *Agricultural Appropriation Bill for 1939* (1938), 567.

17. Paul H. Appleby, "Memorandum for Dr. Howard R. Tolley," February 16, 1939, Folder "Farm Security 4 Loans," Box 3020, Office of the Secretary, General Correspondence, RG 16, NACP.

18. Senate, *Agricultural Appropriation Bill for 1941* (1940), 429.

19. FSA, "Administrative Instruction 137," January 22, 1938, pp. 30–31, Folder "Farm Security (Adm. Orders Etc)," Box 2782, Office of the Secretary, General Correspondence, RG 16, NACP.

20. Maddox, "Farm Security Administration," 428–29.

21. Senate, *Agricultural Appropriation Bill for 1940* (1939), 638.

22. FSA, "Administrative Instruction 137," January 22, 1938, p. 4, Folder "Farm Security (Adm. Orders Etc)," Box 2782, Office of the Secretary, General Correspondence, RG 16, NACP.

23. FSA, *The Work of the Farm Security Administration*, FSA Instruction 611.1.

24. Edward C. Banfield, "Ten Years of the Farm Tenant Purchase Program," *Journal of Farm Economics* 31.3 (1949): 472.

25. FSA, "Administrative Instruction 137," January 22, 1938, pp. 6–7, 17–20, Folder "Farm Security (Adm. Orders Etc)," Box 2782, Office of the Secretary, General Correspondence, RG 16, NACP.

26. FSA, *The Work of the Farm Security Administration*, FSA Instruction 403.1; numerous letters in Folder "Tenancy 2.1 (County Committees)," Box 2663, Office of the Secretary, General Correspondence, RG 16, NACP; FSA, "Administrative Instruction 137," January 22, 1938, p. 5, Folder "Farm Security (Adm. Orders Etc)," Box 2782, Office of the Secretary, General Correspondence, RG 16, NACP; Paul V. Maris, "From Tenant to Owner: The First Year Shows Progress Under the Bankhead-Jones Tenant Act," *Extension Service Review* 9.10 (October 1938): 154.

27. John H. Bankhead to J. L. Edwards, [early December 1936?], Folder 1, Box 5, Bankhead Papers.
28. [Paul V. Maris?], "To the Members of the County FSA Committees Assembled in 1940, Tenant Purchase Schools of Instruction," pp. 2–4, 9–10, Folder "Farm Security," Box 76, Office of the Secretary, General Correspondence, RG 16, NACP.
29. "Reminiscences of Will W. Alexander," 603–5.
30. House, *Agriculture Department Appropriation Bill for 1941* (1940), 955.
31. Charles Houston to Will W. Alexander, February 9, 1937, Folder "AD-140," Box 26, General Correspondence, Washington Office, RG 96, NACP.
32. Grant, *TVA and Black Americans*, 30–34.
33. Larson, *Ten Years of Rural Rehabilitation*, 97–98.
34. William J. Roberts and Dorothy D. Fashee, "Family Narrative Robert Hixson," in Folder "Hixson, Robert" Farm Ownership Files, RG 96, NARASE. On the ideal ages for farmers in rehabilitation and resettlement programs, see Holt, *Analysis of Methods and Criteria*, 7–10.
35. James R. White and Tee H. Owings, "Narrative Statement Regarding Family of J. B. Kelly, T. P. Applicant," Folder "Kelley, J. B.," Farm Ownership Files, RG 96, NARASE.
36. Ev M. Flannigan, "Narrative for Family Information Schedule—Form TP 2," Folder "Parham, June," Farm Ownership Files, RG 96, NARASE.
37. Patty E. Kroell, and Wike C. Ikey, "Narrative of the Jack L. Harris Family," December 20, 1938, Folder "Harris, Jack L.," Farm Ownership Files, RG 96, NARASE.
38. Marjorie L. Andrews, "Narrative," Folder "Farr, Joe C.," Farm Ownership Files, RG 96, NARASE.
39. Otto Mills, County RR Supervisor, "W. B. Moss and Beulah Satterfield Moss," Folder "Moss, W. B.," Farm Ownership Files, RG 96, NARASE.
40. Holt, *Analysis of Methods and Criteria*, 6–7.
41. Carolyn A. Warnell, "Home Narrative," Folder "Ferguson, Willie F.," Farm Ownership Files, RG 96, NARASE.
42. Ibid.
43. Sallie U. Patrick, "Narrative of Home Supervisor to Accompany Josh Page's Papers," Folder "Page, Josh," Farm Ownership Files, RG 96, NARASE.
44. John C. Holmes and Eileen P. Cole, "Narrative of the Carlos A. Moore Family," Folder Moore, Carlos," Farm Ownership Files, RG 96, NARASE.
45. Mildred A. Row, and William M. DuBose, "Narrative—Billie Morring," April 5, 1940, Folder "Morring, Billie T.P. 1–45," Farm Ownership Files, RG 96, NARASE
46. William R. Mallard and Maxie H. Adkins, "Narrative, Ansel E. Harden, Lyons, RFD # 3," Folder "Harden, Ansel E.," Farm Ownership Files, RG 96, NARASE.
47. T. C. Hubbard, "Narrative Statement—Josh Dowdell," and "Remarks," Folder "Dowdell, Josh," Farm Ownership Files, RG 96, NARASE.
48. C. H. Bedgfield to Joe G. Woodruff, n.d., and E. S. Morgan to Josh Dowdell, March 20, 1942, both in Folder "Dowdell, Josh," Farm Ownership Files, RG 96, NARASE. Dowdell did in the end manage to pay off his loan.
49. Patty E. Kroell and Wike C. Ikey, "Narrative of the Jack L. Harris Family," December 20, 1938, Folder "Harris, Jack L.," Farm Ownership Files, RG 96, NARASE.
50. Ted M. Phelps, "Tenant Purchase Narrative, to Accompany Home Plan," Folder "Owens, George H.," Farm Ownership Files, RG 96, NARASE.
51. Jerome B. McMichael and Lois D. McCants, "Narrative Report B. M. Tyler," Folder "Tyler, B. M.," Farm Ownership Files, RG 96, NARASE. On the efforts to inculcate middle-class

norms of sanitation and hygiene on poor, rural southerners, see Brenda J. Taylor, "The Farm Security Administration and Rural Families in the South," in *The New Deal and Beyond: Social Welfare in the South since 1930*, ed. Elna C. Green (Athens: University of Georgia Press, 2003), 30–46.

52. Buena F. Klinhart, "German Moats and Family," Folder "Moats, German 10-81," Farm Ownership Files, RG 96, NARASE.
53. FSA, *The Work of the Farm Security Administration*, FSA Instruction 611.1. As Debra Reid points out, even when New Deal programs did provide money or aid to black farmers, acceptance of funds usually meant accepting additional white influence (and, in the case of cooperative canning centers, male control of traditionally female work). Reid, *Reaping a Greater Harvest*, 137.
54. Robert E. Perry, "Narrative Report of Eugene and Josephine Orr," May 2, 1939, and M. Elizabeth Miss, "Narrative Report, Eugene Orr and Josephine Orr," May 2, 1939, both in Folder "Orr, Eugene," Farm Ownership Files, RG 96, NARASE.
55. Paul V. Maris, "Report of Progress for the Tenant Purchase Division for the Period July 1 to 30, 1938, inclusive," September 17, 1938, p. 10, Folder "Tenant Purchase Division AD-183-6," Box 40, General Correspondence, Washington Office, RG 96, NACP.
56. Baldwin, *Poverty and Politics*, 196–7. See also Reid, *Reaping a Greater Harvest*, 133–38, and, on the New Deal and black farmers in general, 114–44.
57. M. H. Pearson to Joe G. Woodruff, May 28, 1938, Folder "Burton, Wallace," Farm Ownership Files, RG 96, NARASE.
58. Robert Boyd and Nellie D. Roberts, "Narrative to accompany FSA-RR-14 Farm and Home Plan, Dean Hardy, RFD # 2, Power Springs, Georgia," January 24, 1940, Folder "Hardy, Dean," Farm Ownership Files, RG 96, NARASE.
59. James E. Bains and Leeta C. Lowery, "William E. Moody," May 10, 1938, Folder "Moody, William E.," Farm Ownership Files, RG 96, NARASE.
60. Maddox, "Farm Security Administration," 438–40; FSA, "Administrative Instruction 137," January 22, 1938, pp. 8–13, Folder "Farm Security (Adm. Orders Etc)," Box 2782, Office of the Secretary, General Correspondence, RG 16, NACP.
61. FSA, "Rural Housing by the Farm Security Administration," May 5, 1941, p. 8, NAL. Specifically, 41.4 percent needed new homes altogether, and 55.5 percent repaired their existing homes.
62. Senate, *Agricultural Appropriation Bill for 1940* (1939), 640–41.
63. Isaac E. Davis and Elizabeth May Byrd, "Narrative on Hallie P. Martin," April 29–30, 1941, Folder "Martin, Hallie P.," Farm Ownership Files, RG 96, NARASE.
64. Buena F. Klinhart, "German Moats and Family," Folder "Moats, German 10-81," Farm Ownership Files, RG 96, NARASE.
65. Maris, *"the land is mine,"* 70, 148 (quotes), 141.
66. Maris, *"the land is mine,"* 141–47, 151.
67. J. M. Maclachlen, "Salvation for the Tenant Farmer," *Opportunity: Journal of Negro Life* 13.4 (1935): 104–5.
68. FSA, *The Work of the Farm Security Administration*, FSA Instruction 623.10.
69. Charles R. Walker, "Homesteaders-New Style," *Survey Graphic* 28.6 (1939): 378.
70. One published example of such a family and their plan can be found in Maris, *"the land is mine,"* 211–27.
71. FSA, *The Work of the Farm Security Administration*, FSA Instruction 658.10; Maris, *"the land is mine,"* 251–76; FSA, *The Work of the Farm Security Administration*, FSA Instruction 658.11.

72. Otis A. O'Dell to Joe G. Woodruff, April 24, 1942, Folder "Nations, I. B.," Farm Ownership Files, RG 96, NARASE.
73. "Tenant Purchase Family Progress Report No. 1," Folder "Clarke, Lee," Farm Ownership Files, RG 96, NARASE.
74. John C. Holmes and Eileen P. Cole, "Narrative of the Carlos A. Moore Family," and Julian Brown to Samuel T. Windham, January 18, 1945, both in Folder "Moore, Carlos," Farm Ownership Files, RG 96, NARASE.
75. Garza D. Roberts to Julian Brown, May 3, 1945, Folder "Morris, Floyd W., Entry 94," Farm Ownership Files, RG 96, NARASE.
76. C. B. Baldwin, "Rural Housing in 1942: The Work of the Farm Security Administration," in *Housing Yearbook 1943*, ed. Hugh R. Pomeroy and Edmond H. Hoben (Chicago: National Association of Housing Officials, 1943), 76.
77. "Administrative Letter 286, Exhibit C" (quote), C. B. Edwards to Julian Brown, August 14, 1944, and Julian Brown to C. B. Edwards, November 6, 1945, all in Folder "Blankenship, Waymon," Farm Ownership Files, RG 96, NARASE.
78. William E. Elsberry to E. S. Morgan, January 10, 1945, Folder "Gowan, William W.," Farm Ownership Files, RG 96, NARASE.
79. Elizie E. Wilson to Joe G. Woodruff, May 7, 1942, Folder "Crowe, John T." Farm Ownership Files, RG 96, NARASE.
80. Maris, *"the land is mine,"* 330–35.
81. Ibid., 300–305.
82. For a much greater emphasis on the conservative aspects of the farm ownership program, see Baldwin, *Poverty and Politics*, 195–99.

Chapter 7

1. Monroe Oppenheimer to W. F. Farrell, November 19, 1936, Folder "Carter, Tom," Box 12, Office of the Solicitor, Rural Rehabilitation, RG 16, NACP.
2. FSA, *Procedure Manual*, FSA Instruction 731.1. See also *Report of the Administrator of the Farm Security Administration, 1938*, 1–4; and Maddox, "Farm Security Administration," 84–89.
3. Gaer, *Toward Farm Security*, 59–62. Many postwar agricultural thinkers believed the same thing. See Murray R. Benedict, *Can We Solve the Farm Problem?* (New York: Twentieth-Century Fund, 1955), 212–13.
4. Baldwin, *Poverty and Politics*, 250. For a breakdown by region, state, and county, see the *Byrd Committee Hearings, Part II*, 370–404.
5. Baldwin, *Poverty and Politics*, 249–50; Lorena Hickock to Harry L. Hopkins, April 7, 1934, *One Third of a Nation*, 211–12; Maddox, "Farm Security Administration," 101–2; Reid, *Reaping a Greater Harvest*, 136–44.
6. "Salary Ranges for County Agents in the Extension Service" and "Salary Schedule—Farm and Home Management Supervisors," [1937], both in Folder "Dept. of Agriculture—Farm Security Administration—Personnel, 1933–1945," Box 31, Baldwin Papers.
7. Maddox, "Farm Security Administration," 100–101; FSA, *The Work of the Farm Security Administration*, FSA Instruction 403.1; Larson, *Ten Years of Rural Rehabilitation*, 75–78; FSA, *The Work of the Farm Security Administration*, FSA Instruction 400.10; Maris, *"the land is mine,"* 53–57. FERA started the practice of having an advisory committee for local county agents. These continued without much change until the Bankhead-Jones Farm Tenant Act created new committees in each county designated for the tenant purchase

program. In 1941 another committee was introduced, an eight-to-ten-member FSA Advisory Council. This provided a forum for representatives of FSA clients, public welfare and similar agencies, the extension service, and other interested parties. In late 1943 the four often-overlapping committees (including the debt adjustment committees) merged into one large county committee.

8. FSA, *The Work of the Farm Security Administration*, FSA Instruction 403.1 (quote); FSA, *The Work of the Farm Security Administration*, FSA Instruction 400.10, sheet 2 reverse. See also FSA, *A Handbook for County FSA Committeemen* (August, 1945), NAL.
9. Theodore W. Schultz, *Training and Recruiting of Personnel in the Rural Social Studies* (Washington, DC: American Council on Education, 1941), 1–8, 10, 27, 185–87. On Extension visits, see, for example, "The Extension Visit," *Extension Service Review* 1.6 (October, 1930): 88.
10. Bernard D. Joy to J. R. Allgyer, April 7, 1935, Folder "RR 101-3 I-J Reports," Box 6, Rural Rehabilitation General Correspondence, RG 96, NACP; Gaer, *Toward Farm Security*, xi.
11. FSA, *Procedure Manual*, FSA Instruction 705.1. See also "Report of the Committee on Standard Qualifications for Rehabilitation Field Supervisory Positions," *Report of Regional Personnel Advisers' Conference, 1938*, p. 3, NAL.
12. Gaer, *Toward Farm Security*, xii–xiv.
13. Swiger and Larson, *Climbing Toward Security*, 5, 18–19; Gaer, *Toward Farm Security*, 3. See also Taeuber and Rowe, *Five Hundred Families Rehabilitate Themselves*, 3
14. FSA, *Induction Training Course* (October, 1941), pp. 1–12, NAL.
15. FSA, "FSA Washington Conference, August 6–14, 1940" (quote), NAL; FSA, "Proposed Program for the District Meetings of RR and HM Supervisors," 1940, NAL.
16. Clawson, "Resettlement Experience on Nine Selected Resettlement Projects," 17.
17. "Report of the Committee on Standard Qualifications for Rehabilitation Field Supervisory Positions," *Report of Regional Personnel Advisers' Conference, 1938*, p. 3, NAL.
18. Gaer, *Toward Farm Security*, 93–94.
19. Larson, *Ten Years of Rural Rehabilitation*, 71–72; Gaer, *Toward Farm Security*, 95–97.
20. Gaer, *Toward Farm Security*, 117. Emphasis in the original.
21. Gaer, *Toward Farm Security*, 97–98; FSA, *Procedure Manual*, FSA Instruction 731.1; *Report of the Administrator of the Farm Security Administration, 1939*, 2–4; Baldwin, *Poverty and Politics*, 251–52; Blaisdell, *Government and Agriculture*, 152.
22. *Report of the Administrator of the Farm Security Administration, 1939*, 2; Clawson, "Resettlement Experience on Nine Selected Resettlement Projects," 16. See also Holt, *Analysis of Methods and Criteria*, 14–16.
23. Bernard D. Joy to J. R. Allgyer, April 7, 1935, Folder "RR 101-3 I-J Reports," Box 6, Rural Rehabilitation General Correspondence, RG 96, NACP.
24. FSA, *Procedure Manual*, FSA Instruction 731.1; FSA, *The Work of the Farm Security Administration*, FSA Instruction 731.10.
25. FSA, *Procedure Manual*, FSA Instruction 702.1.
26. Maddox, "Farm Security Administration," 161.
27. George S. Mitchell to C. B. Baldwin, "Proposal for eliminating in part the work of Loan Officers," May 11, 1936, Folder "Organization, Administration, General AD-101," Box 15, General Correspondence, Washington Office, RG 96, NACP. Similarly, in 1940 the FSA regional office's role was described as essentially "mechanical checking." "Conference of FSA officials held at Capon Springs, W. Va., during the period May 31 1940—June 3, 1940," p. 4, Folder "Dept. of Agriculture. Farm Security Administration. Conference at Capon Springs. 1940," Box 30, Baldwin Papers.

28. FSA, *The Work of the Farm Security Administration*, FSA Instruction 400.10, sheet 2; "Conference of FSA officials held at Capon Springs, W. Va., during the period May 31 1940–June 3, 1940," p. 4, Folder "Dept. of Agriculture. Farm Security Administration. Conference at Capon Springs. 1940," Box 30, Baldwin Papers; House, *Farm Security Hearings, Part I*, 125, 118. This mirrored, incidentally, a similar shift toward local control in other matters. In 1937 every FSA appointment had to be approved in Washington; by 1941 regional directors hired practically every FSA field employee, many of those having been recommended by even lower-level administrators.
29. Maddox, "Farm Security Administration," 146.
30. F. H. Robinson to P. F. Aylesworth, "Report on Visit to Coffee County Farms, RR-AL-17," October 29, 1937, p. 1, folder "Coffee County RF-AL-17," Box 67, Project Records, RG 96, NACP. Locals used the old merchant system because, while the government program might eventually end, local merchants would always be there. Local farmers feared that turning away from their traditional sources of credit would mean losing credit when the government was out of the area. See P. F. Aylesworth to F. H. Robinson, "Report on Visit to Coffee County Farms, RR-AL-17," October 29, 1937, p. 1, folder "Coffee County RF-AL-17," Box 67, Project Records, RG 96, NACP.
31. Maddox, "Farm Security Administration," 149.
32. Gaer, *Toward Farm Security*, 64–68; Maddox, "Farm Security Administration," 147–48; Grant, *Down and Out on the Family Farm*, 117–21. Some FERA officials even recommended starting clients out as laborers, not owners or managers. See S. B. Cleland, "More Jobs for Farm Labor," *Rural Rehabilitation* 1.2 (February, 1935): 7. See Baldwin, *Poverty and Politics*, 217–21, for further discussion of "skimming the cream" vs. "digging deeper."
33. Maddox, "Farm Security Administration," 149–50; "Conference of FSA officials held at Capon Springs, W. Va., during the period May 31 1940–June 3, 1940," n.d., Folder "Dept. of Agriculture. Farm Security Administration. Conference at Capon Springs. 1940," Box 30, Baldwin Papers.
34. Larson, *Ten Years of Rural Rehabilitation*, 333; Baldwin, *Poverty and Politics*, 254; Maddox, "Farm Security Administration," 153.
35. Senate, *Agricultural Appropriation Bill for 1940* (1939), 643–44; Senate, *Agricultural Appropriation Bill for 1943* (1942), 299–300; R. W. Hudgens to Paul V. Maris, July 26, 1937, Folder "Rehabilitation Region V AD-560 Reports," Box 60, General Correspondence, Washington Office, RG 96, NACP.
36. Gaer, *Toward Home Security*, 99.
37. "Leadership Called Vital in Running Farms," *Dallas Morning News*, February 16, 1939.
38. Maddox, "Farm Security Administration," 183–84 (quotes), 449.
39. Erna E. Proctor, "Home Economics in a Rural Rehabilitation Program," *Journal of Home Economics* 27.8 (1935): 504 (quote), 503–5.
40. Banfield, *Government Project*, 57–58.
41. Resettlement Administration, "Field Instruction RS 11," August 3, 1936, pp. 2–16, Folder "Organization, Administration, General AD-101," Box 15, General Correspondence, Washington Office, RG 96, NACP.
42. Paul H. Johnstone, "Memorandum for Mr. M. L. Wilson," March 8, 1939, p. 2, Folder "Farm Security 2 (Projects)," Box 3019, Office of the Secretary, General Correspondence, RG 16, NACP; Resettlement Administration, "Field Instruction RS 11," August 3, 1936, p. 101, Folder "Organization, Administration, General AD-101," Box 15, General Correspondence, Washington Office, RG 96, NACP; Gaer, *Toward Farm Security*, 101–4;

E. S. Morgan to Pat Cannon, October 27, 1939, Folder "85-160-02 Criticisms and Complaints, July, 1938 thru Dec., 1939," Box 1, Office of the Director, General Correspondence, RG 96, NARASE.
43. Larson, *Ten Years of Rural Rehabilitation*, 10.
44. Arthur F. Raper, *Tenants of the Almighty* (New York: Macmillan, 1943), 233–42.
45. Ayers, *Promise of the New South*, 188–91.
46. Larson, *Ten Years of Rural Rehabilitation*, 110.
47. Terry and Sims, *They Live on the Land*, 63.
48. Taeuber and Rowe, *Five Hundred Families Rehabilitate Themselves*, 4.
49. Gaer, *Toward Farm Security*, 314 (quote), 107; Swiger and Larson, *Yesterday, Today, and Tomorrow*, 6. Self-sufficiency was a complicated ideal in the rural South, involving notions of what it meant to be a farmer, class relations, and more. See Walker, *Southern Farmers and Their Stories*, 82–86.
50. Conkin, *Revolution Down on the Farm*, 39–42.
51. FSA, *Household Furniture and Domestic Equipment* (Washington, DC: Government Printing Office, 1940), n.p. Jess Gilbert argues that New Deal agricultural policymakers were generally "low modernists"; people such as Henry Wallace and M. L. Wilson would use modernist institutions and activities (like a powerful state engaged in scientific planning) to create a more participatory democracy, one that embraced local knowledge and the traditions of family farming. While this conception is generally accurate in terms of the intent of New Deal agricultural reforms and FSA planners, it assumes that FSA leaders and employees knew what farmers actually wanted and believed and, even more than that, that farmers entirely understood what they wanted and believed. Gilbert, "Low Modernism and the Agrarian New Deal," 129–31.
52. Gaer, *Toward Farm Security*, 108–11.
53. FSA, *Farm Family Record Book*, [1939?], NAL.
54. F. H. Robinson to P. F. Aylesworth, October 29, 1937, Folder "Coffee County, RF-AL-17," Box 67, Project Records, RG 96, NACP.
55. Swiger and Larson, *Climbing Toward Security*, 19. Rural southerners made good use of orange crates, like one newlywed Georgia couple who made a dish cabinet out of one. Bindas, *Remembering the Great Depression*, 106.
56. Gaer, *Toward Farm Security*, 111; Raper, *Tenants of the Almighty*, 280; Swiger and Larson, *Climbing Toward Security*, 20.
57. John E. Hydrick to Homer D. Lee, March 12, 1940, Folder "85-160-02 Criticisms and Complaints, July, 1938 thru Dec., 1939," Box 1, Office of the Director, General Correspondence, RG 96, NARASE (quote); Gaer, *Toward Farm Security*, 111–12.
58. "Farm Visit Report," August 27, 1940, Folder "Dickerson, Gus," Rural Rehabilitation Loan Files, RG 96, NARASE; "Farm Visit Report," [July 22, 1940?], Folder "Williams, Callie," Rural Rehabilitation Loan Files, RG 96, NARASE.
59. *Report of the Administrator of the Farm Security Administration, 1939*, 3–5; Gaer, *Toward Farm Security*, 114–15; Larson, *Ten Years of Rural Rehabilitation*, 145; Raper, *Tenants of the Almighty*, 257–58.
60. Adams and Gorton, "This Ain't My Land," 341–42; Raper and Reid, *Sharecroppers All*, 38–39; Gaer, *Toward Farm Security*, 16–17; Schuler, *Social Status and Farm Tenure*, 34–36.
61. James Edward Rice, "On a Tenant Farm," *New Republic*, April 15, 1931, 234. On the attraction of growing cotton, see also Robertson, *Red Hills and Cotton*, 7. On the cotton

market (a losing bet for most southerners), see National Emergency Council, *Report on Economic Conditions of the South* (Washington, DC: Government Printing Office, 1938), 45; Rupert B. Vance, *Human Factors in Cotton Culture: A Study in the Social Geography of the American South* (Chapel Hill: University of North Carolina Press, 1929), 108; George C. Osbourne, "The Southern Agricultural Press and Some Significant Rural Problems, 1900–1940," *Agricultural History* 29.3 (1955): 115–18; and Roger Biles, *The South and the New Deal* (Lexington: University of Kentucky Press, 1994), 36–37.

62. "Conference of FSA officials held at Capon Springs, W. Va., during the period May 31 1940—June 3, 1940," n.d., Folder "Dept. of Agriculture. Farm Security Administration. Conference at Capon Springs. 1940," Box 30, Baldwin Papers.
63. "Farm and Home Plan Summary," February 5, 1942, Folder "Anderson, Rannie," Rural Rehabilitation Loan Files, RG 96, NARASE.
64. Maddox, "Farm Security Administration," 184–86.
65. Clawson, "Resettlement Experience on Nine Selected Resettlement Projects," 68–69.
66. "Reminiscences of Robert W. Hudgens," 185–87.
67. Larson, *Ten Years of Rural Rehabilitation*, 140.
68. Maris, *"the land is mine,"* 300–302.
69. Swiger and Larson, *Climbing Toward Security*, 11–13.
70. Banfield, *Government Project*, 160, 162.
71. Swiger and Larson, *Climbing Toward Security*, 4.
72. F. E. Patch to Paul V. Maris, April 22, 1936, Folder "RR-101-3 P Reports," Box 6, Rural Rehabilitation General Correspondence, RG 96, NACP. Borrowers in other regions similarly resented government supervision; see Cannon, *Remaking the Agrarian Dream*, 82–88.
73. FSA, *The Work of the Farm Security Administration*, FSA Instruction 623.10.
74. Clawson, "Resettlement Experience on Nine Selected Resettlement Projects," 64. On joint bank accounts, see FSA, *Procedure Manual*, FSA Instruction 731.1; Maddox, "Farm Security Administration," 195–96; Cannon, *Remaking the Agrarian Dream*, 123–25.
75. James L. McCamy, "We Need More Personalized Administration," in *Public Administration Readings and Documents*, ed. Felix A. Nigro (New York: Rinehart, 1951), 467.
76. Swiger and Larson, *Climbing Toward Security*, 20; Larson, *Ten Years of Rural Rehabilitation*, 152–53.
77. "Crawford County, Georgia Farmers" to "U.S. Department of Agricultural, Farm Security Administration, Rural Farm Rehabilitation," August 9, 1939, Folder "85-160-02 Criticisms and Complaints, July, 1938 thru Dec., 1939," Box 1, Office of the Director, General Correspondence, RG 96, NARASE. A considerable number of complaints were anonymous, as farmers in the rural South often feared retribution from the targets of their complaints, who were usually connected with local elites.
78. "Narrative Report, Survey in Alabama, October 10, 1938 through October 15, 1938," in Folder 85-570 Debt Adjustment, July 1938 thru June 1940, Box 9, Office of the Director, General Correspondence, RG 96, NARASE.
79. Senate, *Agricultural Appropriation Bill for 1939* (1938), 568–69. Those drought-afflicted states were Colorado, Kansas, Montana, Nebraska, New Mexico, North Dakota, Oklahoma, South Dakota, and Wyoming.
80. FSA, *Procedure Manual*, FSA Instructions 765.1 and 765.2; Larson, *Ten Years of Rural Rehabilitation*, 175–79.

81. "Waiver and New Promise to Pay," Folder "Fongemie, Octave (Clients)," Box 23, Office of the Solicitor, Rural Rehabilitation, RG 16, NACP.
82. Larson, *Ten Years of Rural Rehabilitation*, 178–79.
83. P. G. Beck to Frank Hancock, January 22, 1944, Folder "3D-Gen Jan-April 1944," Box 1, General Administrative Records, RG 96, NACP.
84. Monroe Oppenheimer, "Memorandum for Dr. W. W. Alexander," February 17, 1940, Folder "Caplis, William L.," Box 12, Office of the Solicitor, Rural Rehabilitation, RG 16, NACP.
85. Folder "(Campbell, W. C.) Clients," Box 12, Office of the Solicitor, Rural Rehabilitation, RG 16, NACP. See esp. "Report on F.S.A. Litigation."
86. John P. Cowart to the Attorney General, January 5, 1940, Folder "Barkley, Pierce," Box 4, Office of the Solicitor, Rural Rehabilitation, RG 16, NACP.
87. Folder "Ball, S. A.," Box 4, Office of the Solicitor, Rural Rehabilitation, RG 16, NACP.
88. W. R. Fuchs to Martin G. White, September 8, 1941, Folder "Farm Security Loans or Grants Aug. 1 to Nov. 1," Box 292, Office of the Secretary, General Correspondence, RG 16, NACP.
89. Larson, *Ten Years of Rural Rehabilitation*, 333.
90. Bernard D. Joy to J. R. Allgyer, April 7, 1935, Folder "RR 101-3 I-J Reports," Box 6, Rural Rehabilitation General Correspondence, RG 96, NACP.
91. R. W. Hudgens to Paul V. Maris, July 26, 1937, Folder "Rehabilitation Region V AD-560 Reports," Box 60, General Correspondence, Washington Office, RG 96, NACP.
92. House, *Farm Security Hearings, Part I*, 190.
93. Maddox, "Farm Security Administration," 141–43.
94. Maddox, "Farm Security Administration," 170–72.
95. Murff Hawkins, "Talk Made by Murff Hawkins, TP Engineer at Tenant Purchase School, Linden, Ala. July 31, 1941," p. 2, Folder "85–161 Speeches, July thru Dec. 1940," Box 14, Office of the Director, General Correspondence, RG 96, NARASE.
96. Larson, *Ten Years of Rural Rehabilitation*, 317–18. Region IV included Virginia, West Virginia, Kentucky, Tennessee, and North Carolina. Region V included South Carolina, Alabama, Georgia, and Florida. Region VI was made up of Mississippi, Louisiana, and Arkansas.
97. Charles P. Loomis, *Social Relationships and Institutions in Seven New Rural Communities*, Social Research Report no. 18 (Washington, DC: United States Department of Agriculture, the Farm Security Administration, and the Bureau of Agricultural Economics, Cooperating, 1940), 23, 48.
98. Maddox, "Farm Security Administration," 188–89.
99. Raper and Reid, *Sharecroppers All*, 262, 263.

Chapter 8

1. Senate, *Agricultural Appropriation Bill for 1941* (1940), 430; G. T. (Jean) Walker, "G. T. (Jean) Walker, Memoir," interviewed by Joe Lilly, Oral History Research Office, University of Alabama in Birmingham (April 25, 1981), p. 8, Mervyn H. Sterne Library, Birmingham, Alabama. On the division between the (generally positive) memory of the homestead projects and the (generally negative) historical and political verdicts, see Brown, *Back to the Land*, 162–67. On the experience of resettlement projects in the Mountain West, see Cannon, *Remaking the Agrarian Dream*.

2. Douglas L. Smith, *The New Deal in the Urban South* (Baton Rouge: Louisiana State University Press, 1988), 12; Blaine A. Brownell, "Birmingham: New South City in the 1920s," *Journal of Southern History* 38.1 (1972): 25, 30–36.
3. Smith, *New Deal in the Urban South*, 17–20.
4. Rutledge, "Development of the Bankhead Farmsteads Community," 5.
5. Edward Shannon LaMonte, *Politics and Welfare in Birmingham, 1900–1975* (Tuscaloosa: University of Alabama Press, 1995), 69–70, 92–93; James H. Tuten, "Regulating the Poor in Alabama: The Jefferson County Poor Farm, 1885–1945," in *Before the New Deal: Social Welfare in the South, 1830–1930*, ed. Elna C. Green (Athens: University of Georgia Press, 1999), 40–60. Unique among major cities in the South, Birmingham had been spending less and less on social needs; its residents paid little and got less in public services than most other major American cities. See LaMonte, *Politics and Welfare in Birmingham*, 70–72.
6. Kollmorgen, "Subsistence Homesteads Near Birmingham," 65–66; Wager, *One Foot on the Soil*, 16.
7. John Beecher, "Suburban Resettlement in the Birmingham Industrial District," Folder "85-60 Projects (General Information Only) July, 1938 thru Dec., 1939," Box 2, Records of the Resettlement Division, Correspondence Relating to Proposed Projects, 1934–1941 (Entry 95), Records of the Farmers Home Administration RG 96, NARASE.
8. "Economic and Social Merits of Proposed Project," pp. 2–3, 5, 22, in "200—Plans—Project Book AL-3," Box 10, Project Records, RG 96, NACP.
9. Conkin, *Tomorrow a New World*, 111; Wager, *One Foot on the Soil*, 15–16. Other sponsors included Victor Henry Hanson, president of the Birmingham News Publishing Company; Robert Jemison Jr., president of Jemison and Co., Inc.; and Hugh Morrow, president of Sloss-Sheffield Steel and Iron. "Biographical Sketch of Sponsors of Birmingham Homesteads Project," in "Subsistence Garden Homestead Project in Birmingham, Alabama," Folder "Alabama No. 3 Birmingham 310," Box 22, Project Records, RG 96, NACP. The entire state of Mississippi was approved for six Division of Subsistence Homesteads projects. See Smith, "Tupelo Homesteads." In comparison, the Civilian Conservation Corps relatively ignored politically safe regions, instead putting the most camps in states where Franklin Roosevelt could be least certain about his political support or electoral chances. On the other hand, southerners were rewarded for their political support with an outsized investment in outdoor recreation sites. Maher, *Nature's New Deal*, 59, 69, 74.
10. E. Richardson, "Birmingham Homesteads Project #1 Summary of Studies and Reports Jefferson County, Alabama," n.p., June 28, 1935, Folder "Alabama No. 3 Birmingham 310," Box 22, Project Records, RG 96, NACP; "Homestead Body Organized Here," *Birmingham News*, January 1, 1934; "Homestead Soil Surveys Begun," *Birmingham News*, December 21, 1933; "Four Tracts Picked by Homestead Body," *Birmingham News*, December 24, 1933.
11. Wager, *One Foot on the Soil*, 18; "Homestead Body Organized Here," *Birmingham News*, January 10, 1934; "Homesteads to Be Chosen," *Birmingham Post*, January 19, 1934. Photographs of the communities can be found in Birmingham Historical Society, *Digging Out of the Great Depression*, 50–65.
12. "Procedure Manual," [1936?], Box 19, Project Records, RG 96, NACP; Wager, *One Foot on the Soil*, 20.
13. F. G. Morris to Mr. Dugger and Mr. Liles, July 30, 1934, Folder "OS-ERA-SH-AL-2 (210) Title," Box 1, Office of the Solicitor, Resettlement Cases, RG 16, NACP; Will W. Alexander,

"Memorandum for the Secretary," October 2, 1939, Folder "Farm Security 2 Projects," Box 3019, Office of the Secretary, General Correspondence, RG 16, NACP.

14. "Project History Jasper Homesteads MH-AL-12 Walker County, Alabama," pp. 1–4, Folder "Jasper Homesteads (MH-AL-12) Walker County, Alabama," Box 31, Project Records, RG 96, NACP; Milo Perkins, "Memorandum for the Secretary," July 16, 1938, Folder "Farm Security 2 (Projects)," Box 2783, Office of the Secretary, General Correspondence, RG 16, NACP.

15. Resettlement Administration Construction Division, "Greenwood, Jefferson County, Alabama, SH-AL-5 Region V," Box 26, Project Records, RG 96, NACP; Mildred Moore McCrimmon Crain, *Welcome to Greenwood: The History of a Subsistence Homesteads Project and Its People* (Montevallo, AL: Times Printing Company, [1990]), 1–4; Wager, *One Foot on the Soil*, 20.

16. Wager, *One Foot on the Soil*, 20–22; Resettlement Administration Management Division, "Semi-Monthly Report on Progress of Family Selection," April 1, 1937, through September 1, 1937, Folder "SH-AL-4," Box 23, Project Records, RG 96, NACP.

17. Wager, *One Foot on the Soil*, 22; Farm Security Administration, "Official Name: Cahaba" (untitled report), Folder "85-163-1 Articles & Press Releases, Jan. 1939 thru April, 1939," Box 2, Office of the Director, General Correspondence, RG 96, NARASE; Marion W. Ormond, "Marion W. Ormond Memoir," interviewed by Bob Haynes, Oral History Research Office, University of Alabama in Birmingham (April 17, 1981), Mervyn H. Sterne Library, Birmingham, Alabama, 2.

18. John H. Bankhead Jr. to Charles E. Pynchon, July 26, 1934, Folder 5, Box 4, Bankhead Papers.

19. "Subsistence Garden Homestead Project in Birmingham, Alabama," Folder "Alabama No. 3 Birmingham 310," Box 22, Project Records, RG 96, NACP.

20. E. Richardson, "Birmingham Homesteads Project #1 Summary of Studies and Reports Jefferson County, Alabama," June 28, 1935, Folder "Alabama No. 3 Birmingham 310," Box 22, Project Records, RG 96, NACP; Wager, *One Foot on the Soil*, 130–32.

21. Philip M. Glick, "Memorandum for Mr. Jenkins," September 1, 1934, Folder "OS-ERA-SH-AL-2 (210) Title," Box 1, Office of the Solicitor, Resettlement Cases, RG 16, NACP.

22. Wager, *One Foot on the Soil*, 27–29; FDR, Executive Order 6209, 1933. See also Crain, *Welcome to Greenwood*, 10–11.

23. Senate, *Resettlement Administration Program*, 4; R. G. Tugwell, "What About Resettlement? Sweep of Task Ahead and How Extension Can Help," *Extension Service Review* 7.3 (1936): 35–36; Wager, *One Foot on the Soil*, 30.

24. Clarence A. Wiley, "Settlement and Unsettlement in the Resettlement Administration Program," *Law and Contemporary Problems* 4.4 (1937): 466–67.

25. Wager, *One Foot on the Soil*, 32–34, 132–35.

26. Dr. Wendell Lund, "Family Selection in the Management Division," December 12, 1935, pp. 9–11, Folder 161–Speeches," Box 11, Rural Rehabilitation General Correspondence, RG 96, NACP.

27. Holt, *An Analysis of Methods and Criteria Used in Selecting Families for Colonization Projects*, 44–45; Senate, *Resettlement Administration Program*, 22–23.

28. "Ormond Memoir," 4. Residents in western projects similarly complained about keeping up with everything from haircuts to food budgets. Cannon, *Remaking the Agrarian Dream*, 83–84.

29. Dr. Wendell Lund, "Family Selection in the Management Division," pp. 4–6, December 12, 1935, Folder 161–Speeches," Box 11, Rural Rehabilitation General Correspondence, RG 96, NACP.
30. "Walker Memoir," 2.
31. Baldwin, *Poverty and Politics*, 214–16.
32. Wager, *One Foot on the Soil*, 136–40; Harding, "The Record of the Subsistence Homesteads," 500; Jack House, "547 Homesteaders in District Now Enjoy More Abundant Life," *Birmingham News–Age-Herald*, May 9, 1943.
33. J. O. Walker, "Memorandum for Dr. W. W. Alexander," May 31, 1938, Folder "A," Box 2, Farmers Home Administration, General Correspondence, 1935–42, RG 96, NACP.
34. Kollmorgen, "Subsistence Homesteads Near Birmingham," 78; Clarence I. Blau, "Memorandum for Mr. C. B. Baldwin," March 17, 1941, Folder "Farm Security 3 Projects," Box 292, Office of the Secretary, General Correspondence, RG 16, NACP; Wager, *One Foot on the Soil*, 140–41; "Government Offers County Homestead Projects for Sale," *Birmingham News*, January 30, 1940; Finance Division, FSA, "Report on Examination, SH-AL 4, Mount Olive, Jefferson County, Alabama, Rental Accounts for the Period May 15, 1935 to April 30, 1940," Folder OS-ERA-SH-AL-4 (460) Mount Olive Homesteads," Box 2, Office of the Solicitor, Resettlement Cases, RG 16, NACP; Finance Division, FSA, "Report on Examination, Greenwood Homesteads, SH-AL 5, Bessemer, Jefferson County, Alabama, Rental and Utility Accounts for the Period May 15, 1935 to April 30, 1940," Folder OS-ERA-SH-AL-5 (460) Greenwood Homesteads Audit," Box 2, Office of the Solicitor, Resettlement Cases, RG 16, NACP; Finance Division, FSA, "Report on Examination, Bankhead Farms, SH-AL 13, Jasper, Walker County, Alabama, Rental Accounts for the Period May 15, 1935 to September 30, 1938," Folder "OS-ERA-SH-AL-13 (460) Bankhead Farms," Box 3, Office of the Solicitor, Resettlement Cases, RG 16, NACP.
35. J. O. Walker, "Memorandum for the Solicitor," June 20, 1938, Folder "S-ERA-SH-AL-13 (922-5) Transfer of Assets," Box 3, Office of the Solicitor, Resettlement Cases, RG 16, NACP; Mastin G. White, "Memorandum for Dr. Will W. Alexander," June 29, 1938, Folder "S-ERA-SH-AL-13 (922-5) Transfer of Assets," Box 3, Office of the Solicitor, Resettlement Cases, RG 16, NACP.
36. Kollmorgen, "Subsistence Homesteads Near Birmingham," 80; House, *Agriculture Department Appropriation Bill for 1944* (1943), 1616–30, 1653–57, 1691–93.
37. June Gilliland Andrews, *A Palmerdale Tale* [Palmerdale, AL: self-published, 1995], 116.
38. Wager, *One Foot on the Soil*, 93, 97.
39. "Ormond Memoir," 11–12.
40. "Annual Report of the Administrator, Farm Security Administration, 1941–42," 18; Wager, *One Foot on the Soil*, 142–46.
41. Francis E. Brill to Kurt M. Loewy, Subject "Status of Accounting Records for Industrial Co-operative Associations," January 16, 1948, Folder "Subsistence Homesteads," Box 1, Records Relating to Subsistence Homesteads and the PHA's Rural Housing Programs (NC-196-80-01) (Entry 54), in Records of the Public Housing Administration, RG 196, NACP.
42. House, *Activities of the FSA*, 2.
43. W. T. Frazier, "Farm Security Administration," vol. 1, n.d., Box 3, FSA Historical Record, RG 96, NACP.

44. Conkin, *Tomorrow a New World*, 201–2. See also Holley, *Uncle Sam's Farmers*, 112–13, 183–86. On New Deal displacement of resident farmers, see McDonald and Muldowny, *TVA and the Dispossessed*.
45. "Procedure Manual," [1936?], Box 19, Project Records, RG 96, NACP.
46. Oscar M. Dugger to M. L. Wilson, March 12, 1934, Folder "030 Birmingham—General," Box 17, Project Records, RG 96, NACP.
47. "Excerpt from Minutes of Meeting of Birmingham Homesteads, Inc.," March 8, 1934, Folder "030 Birmingham—General," Box 17, Project Records, RG 96, NACP.
48. Donald Holley, "The Negro in the New Deal Resettlement Program," *Agricultural History* 45.3 (1971): 179–93; Baldwin, *Poverty and Politics*, 279–86.
49. "Homestead Plans Near Completion," *Birmingham News*, January 18, 1934; Bob Kincey, "Palmer Station Project to Offer Opportunity to Obtain Homes," *Birmingham News*, August 12, 1934.
50. Kollmorgen, "Subsistence Homesteads Near Birmingham," 67–70.
51. John Beecher, "Suburban Resettlement in the Birmingham Industrial District," Folder "85-60 Projects (General Information Only) July, 1938 thru Dec., 1939," Box 2, Records of the Resettlement Division, Correspondence Relating to Proposed Projects, 1934–1941 (Entry 95), Records of the Farmers Home Administration, RG 96, NARASE.
52. Crain, *Welcome to Greenwood*, 14.
53. "Ormond Memoir," 11.
54. Kollmorgen, "The Subsistence Homesteads Near Birmingham," 71; George Nagel, "Cahaba Village—a Dwelling for Neither the Rich nor Poor," *Birmingham News*, March 29, 1940. The Tupelo Homesteads had a similar dynamic; low rent meant few vacancies, but residents had little interest in subsistence homesteading. Smith, "Tupelo Homesteads," 105–6.
55. "Ormond Memoir," 1–2.
56. Andrews, *A Palmerdale Tale*, 41; Crain, *Welcome to Greenwood*, 14.
57. "Walker Memoir," 5.
58. Wager, *One Foot on the Soil*, 61.
59. Kollmorgen, "Subsistence Homesteads Near Birmingham," 72. For comparison, farmers relocating to western resettlement projects had to learn entirely new ways of farming; see Cannon, *Remaking the Agrarian Dream*, 57–72. As Cannon points out, the historiographic trend of emphasizing the limitations of resettled families ignores the fact that most responded rationally or at least understandably to the conditions they found in the new communities. See Cannon, *Remaking the Agrarian Dream*, 89–95, 149–55.
60. J. L. Liles to Oscar M. Dugger, February 22, 1934, Folder "Oscar M. Dugger 181.18," Box 17, Project Records, RG 96, NACP.
61. Kollmorgen, "Subsistence Homesteads Near Birmingham," 72.
62. Wager, *One Foot on the Soil*, 79, 91; "Walker Memoir," 9.
63. Jack House, "547 Homesteaders in District Now Enjoy More Abundant Life." *Birmingham News–Age-Herald*, May 9, 1943.
64. Kollmorgen, "Subsistence Homesteads Near Birmingham," 73–74; Crain, *Welcome to Greenwood*, 18.
65. Wager, *One Foot on the Soil*, 115–16; Audit Division, "Audit Report, Bankhead Cooperative Association, Jasper, Walker County, Alabama, for the Period June 12, 1937 to October 15, 1938," pp. 3–5, 13, 17, Folder "OS-ERA-SH-AL-13 (460) Bankhead Coop. Assn. Audit," Box 3, Office of the Solicitor, Resettlement Cases, RG 16, NACP; Rutledge, "Development of the Bankhead Farmsteads Community," 32.

66. Kollmorgen, "Subsistence Homesteads Near Birmingham," 74.
67. Wager, *One Foot on the Soil*, 108; Kollmorgen, "Subsistence Homesteads Near Birmingham," 74.
68. Kollmorgen, "Subsistence Homesteads Near Birmingham," 76.
69. Audit Division, FSA, "Audit Report, Palmerdale Cooperative Association, Palmerdale, Jefferson County, Alabama, for the Period September 1, 1938 to December 31, 1939," pp. 6–8, 19, Folder "OS-ERA-SH-AL-3 Audit (460) Palmerdale Coop. Assn.," Box 1, Office of the Solicitor, Resettlement Cases, RG 16, NACP.
70. Wager, *One Foot on the Soil*, 109–11.
71. "Memorandum for Major Walker," June 22, 1937, Folder "OS-ERA-SH-AL-5 (922–1) (Greenwood Coop. Assn.) Loan Application," Box 2, Office of the Solicitor, Resettlement Cases, RG 16, NACP.
72. Wager, *One Foot on the Soil*, 116–18; Crain, *Welcome to Greenwood*, 18.
73. Finance Division, FSA, "Audit Report, Greenwood Cooperative Association, Bessemer, Jefferson County, Alabama, for the Period September 1, 1938 to December 31, 1939," pp. 6, 18, 21, Folder OS-ERA-SH-AL-5 (460) Greenwood Coop. Assn. Audit," Box 2, Office of the Solicitor, Resettlement Cases, RG 16, NACP.
74. Wager, *One Foot on the Soil*, 112.
75. Audit Division, FSA, "Audit Report, Mount Olive (Formerly Gardendale) Cooperative Association, Mount Olive, Jefferson County, Alabama, for the Period October 20, 1938 to December 31, 1939," pp. 15–16, 6, Folder OS-ERA-SH-AL-4 (460) Garden Coop. Assn. Audit," Box 2, Office of the Solicitor, Resettlement Cases, RG 16, NACP.
76. "By-Laws of Cahaba Cooperative Association," Folder "OS-ERA-SH-AL-1 (921) Organization (Cahaba Coop. Ass'n," Box 1, Office of the Solicitor, Resettlement Cases, RG 16, NACP.
77. Lewis B. Woodson to Mastin G. White, June 19, 1937, Folder "OS-ERA-SH-AL-1 (921) Organization (Cahaba Mutual Assn.)," Box 1, Office of the Solicitor, Resettlement Cases, RG 16, NACP; J. O. Walker, "Memorandum for Mr. Mastin G. White," August 10, 1939, and Donald B. MacGuineas, "Memorandum for Dr. W. W. Alexander," August 18, 1939, both in Folder "OS-ERA-SH-AL-1 (921) Organization (Cahaba Coop. Ass'n," Box 1, Entry 113, RG 96, NACP.
78. Wager, *One Foot on the Soil*, 120.
79. Kollmorgen, "Subsistence Homesteads Near Birmingham," 74–75; "Ormond Memoir," 8.
80. Wager, *One Foot on the Soil*, 119–20.
81. Roy M. Little to H. L. Wooten, Subject "Mount Olive Cooperative Association, Gardendale, Alabama," April 1, 1948, Irving N. Goodman to Kurt M. Loewy, Subject "Mt. Olive Cooperative Association, Mt. Olive, Alabama," June 9, 1948, both in Folder "Subsistence Homesteads," Box 1, Records Relating to Subsistence Homesteads and the PHA's Rural Housing Programs (NC-196-80-01) (Entry 54), in Records of the Public Housing Administration, RG 196, NACP.
82. "First Project in Jefferson Busy," *Birmingham News*, June 7, 1936.
83. Wager, *One Foot on the Soil*, 70–71; Andrews, *A Palmerdale Tale*, 123.
84. Wager, *One Foot on the Soil*, 68; House, *Farm Security Hearings, Part III*, 1030–33. A list of the FSA's resettlement projects in 1943, with description, can be found on pages 1034–117; transfers, costs, and liquidation information is on pages 1118–37.
85. George Nagel, "City Dwellers Turn Farmer on Resettlement Projects," *Birmingham News—Age-Herald*, March 31, 1940.
86. Crain, *Welcome to Greenwood*, 27–30.

87. Wager, *One Foot on the Soil*, 76, 81–83.
88. Jack House, "547 Homesteaders in District Now Enjoy More Abundant Life," *Birmingham News—Age-Herald*, May 9, 1943. Similarly, on community sentiment in the western resettlement projects, see Cannon, *Remaking the Agrarian Dream*, 97–113.
89. "Bankhead Farmsteads Near Jasper Are Proving Successful Project," *Birmingham News*, December 13, 1939; Rutledge, "Development of the Bankhead Farmsteads Community," 24.
90. "Walker Memoir," 22; "Ormond Memoir," 17. On rural memory of Franklin Roosevelt and the New Deal in general, see Bindas, *Remembering the Great Depression*, esp. 37–63.
91. "Palmerdale," *Birmingham Age-Herald*, May 6, 1936.
92. "The Spirit of Right," *Birmingham News*, March 2, 1946.
93. Jerry Thrailkill, "All He Knew about Farming in 1937 Was the Gentle Art of How to Milk a Cow," *Birmingham News*, January 8, 1939.

Chapter 9

1. Raper and Reid, *Sharecroppers All*, 3–17.
2. Kennedy, *Freedom from Fear*, 782–88; Phillips, *This Land, This Nation*, 198–200, 215–23.
3. Baldwin, *Poverty and Politics*, 326–28; "Reminiscences of Howard R. Tolley." The FSA also had to move many of its offices from Washington, D.C., to make room for expanding defense work. Several hundred employees ended up in Cincinnati, Ohio. On wartime agricultural production and planning, see Bela Gold, *Wartime Economic Planning in Agriculture: A Study in the Allocation of Resources* (New York: Columbia University Press, 1949).
4. Kennedy, *Freedom from Fear*, 465–79.
5. Brinkley, *End of Reform*, 143–46, 154–64.
6. *Report of the Administrator of the Farm Security Administration, 1941*, 1–12; "Annual Report of the Administrator, Farm Security Administration, 1941–42," 2–9, 20–25; Maddox, "Farm Security Administration," 105–8; Brown and Baugh, *Preliminary Inventory*, 4; House, *Farm Labor Program, 1943*, esp. 72–74, 85–86, 113–18; E. R. Stettinius Jr. and John A. Vieg, "Progress Report on the Rehabilitation Division," December 3, 1942, Folder "U.S. Dept. of Agriculture—Farm Security Administration. 1942," Box 29, Baldwin Papers.
7. Phillips, *This Land, This Nation*, 212–18. On the FSA's reorientation to wartime footing, see Baldwin, *Poverty and Politics*, 326–32. Japanese and Japanese-American farmers dominated California's production of strawberries and grew a majority of the state's snap beans, celery, and cauliflower. Protecting Japanese property and equipment proved impossible, as did maintaining the farms at full production. Laurence I. Hewes Jr., "Final Report of the Participation of the Farm Security Administration in the Evacuation Program of the Wartime Civil Control Administration, Civil Affairs Division, Western Defense Command and Fourth Army," June 5, 1942, Box 1, Final Report of the Farm Security Administration for the Period March 15, 1942–May 31, 1942 (Entry 21), in Records of the Farmers Home Administration, RG 96, NACP; U.S. Army, Western Defense Command and Fourth Army, *Final Report, Japanese Evacuation from the West Coast, 1942* (Washington, DC: Government Printing Office, 1943), 136–43. See also Greg Robinson, *A Tragedy of Democracy: Japanese Confinement in North America* (New York: Columbia University Press, 2009). On social workers and the evacuation, see Yoosun Park, "Facilitating Injustice: Tracing the Role of Social Workers in the World War II Internment of the Japanese," *Social Service Review* 82.3 (2008): 447–83; on Japanese farms

and evacuation, see Page Smith, *Democracy on Trial: The Japanese American Evacuation and Relocation in World War II* (New York: Simon & Schuster, 1995), 163–69.
8. Maddox, "Farm Security Administration," 106–8. The Civilian Conservation Corps similarly tried, and failed, to make an argument for its necessity in the war effort, including an argument about the strengthening of the country's young men; see Maher, *Nature's New Deal*, 212–14.
9. Lawrence E. Davies, "Women 'Okies' Win California Niche," *New York Times*, January 28, 1942.
10. C. B. Baldwin, correspondence to "Dear Fellow-workers," July 22, 1940, Folder "85–153–1 Articles & Press Releases (Oct-Dec, 1940)," Box 15, Office of the Director, General Correspondence, RG 96, NARASE; "Physical Status of Farm Security Borrowers as Indicated by Preliminary Studies of Examinations Conducted in Typical Counties," May 17, 1941, p. 1, NAL. This argument had its roots in what Sarah Phillips calls the "New Conservationists" of the 1920s and 1930s, who argued for balance between rural and urban life and for the government's role in making that happen. Phillips, *This Land, This Nation*, 21–25.
11. FSA, "Relocation of Farm Families Displaced by the Defense Program," April 9, 1941, NAL; FSA, "Rural Housing by the Farm Security Administration," May 5, 1941, NAL; "FSA Seeks Housing For Defense Help," *Christian Science Monitor*, May 22, 1941; "Uncle Sam 'Whips Up Houses' For Families Moved En Masse From Defense Areas," *Washington Post*, July 6, 1941; "Prefabricated Steel Houses Made for Occupancy by Defense Workers," *Christian Science Monitor*, January 29, 1941.
12. FSA, "Rural Housing by the Farm Security Administration," May 5, 1941, pp. 1–3, NAL; FSA, "Relocation of Farm Families Displaced by the Defense Program," April 9, 1941, NAL.
13. Albertson, *Roosevelt's Farmer*, 310–13; "Reminiscences of Howard R. Tolley," 493; Senate, *Agricultural Appropriation Bill for 1943* (1942), 285.
14. "FSA Head Gives Three Talks at Thursday Rallies," *St. Cloud (Minnesota) Times*, August 7, 1942, in Folder "U.S. Dept. of Agriculture—Farm Security Administration. 1942," Box 29, Baldwin Papers; "Toward a Full Mobilization of Manpower: Speech for delivery by C. B. Baldwin, Administrator, Farm Security Administration, before the National Farmers Union Convention, Oklahoma City, Tuesday, November 17 [1942], at 8: pm," Folder "U.S. Dept. of Agriculture—Farm Security Administration. 1942," Box 29, Baldwin Papers.
15. C. B. Baldwin, "Memorandum for Mr. H. W. Parisius," January 6, 1943, Folder "Dept. of Agriculture—Farm Security Administration. 1943. II," Box 30, Baldwin Papers. This may have also been a criticism aimed at the democratic planning committees put forward by leaders in the sympathetic Bureau of Agricultural Economics.
16. W. L. McArthur, April 18, 1941, attached to letter from Will Alexander to Henry B. Steagall, June 17, 1941, Folder "AD-AL-17 Coffee Farms (060)," Box 53, Project Records, RG 96, NACP.
17. Swiger and Taeuber, *They Too Produce for Victory*, 22.
18. Eleanor Roosevelt, "My Day: Definite Service for Every One," *Atlanta Constitution*, March 11, 1942. See also "Farm, Labor, Religious Leaders Ask President to Take to the People Issue of Adequate FSA Appropriations, Sale of Grain for Livestock Feed," June 21, and James G. Patton, Murray Lincoln, William Green, Philip Murray, J. G. Luhrsen, L. G. Ligutti, and Benson Y. Landis to Franklin D. Roosevelt, June 20, 1942, both in Folder "U.S. Dept. of Agriculture—Farm Security Administration. 1942," Box 29, Baldwin Papers.

19. Senate, *Agricultural Appropriation Bill for 1944* (1943), 617–22.
20. *The Annual Report of the Farm Security Administration, 1942–43*, 12.
21. Gold, *Wartime Economic Planning in Agriculture*, 224–40.
22. McConnell, *The Decline of Agrarian Democracy*, 100; Senate, *Agricultural Appropriation Bill for 1943* (1942), 751–52.
23. Baldwin, *Poverty and Politics*, 331; Maddox, "Farm Security Administration," 108. See also Philip Murray to John H. McCormack, March 7, 1942, Folder "U.S. Dept. of Agriculture—Farm Security Administration. 1942," Box 29, Baldwin Papers.
24. Senate, *Agricultural Appropriation Bill for 1944* (1943), 957. On Johnston and the FSA, see Lawrence J. Nelson, *King Cotton's Advocate: Oscar G. Johnston and the New Deal* (Knoxville: University of Tennessee Press, 1999), 211–26. Johnston's allegations about the FSA's alleged mistreatment of clients and Baldwin's reply can be found in House, *Agriculture Department Appropriation Bill for 1944* (1943), 1616–30, 1653–57, 1691–93.
25. For example, see House, *Farm Labor Program, 1943*, 23–24. On the FSA's role in moving laborers and its relationship to American immigration policy, see Cindy Hahamovitch, "In America Life is Given Away," in *The Countryside in the Age of the Modern State*, ed. Catherine McNicol Stock and Roberts D. Johnston (Ithaca: Cornell University Press, 2001), 134–60.
26. George P. West, "Growers Condemn Tugwell's Camps," *New York Times*, August 2, 1936.
27. House, *Farm Labor Program, 1943*, 40–42, 46, 105–13. This carnivorous obligation turned out not to be a requirement of the FSA, but of a commissary company and fruit company that had trouble with missed work because of food-related illness. The FSA was often the target of criticisms for things beyond its control, like complaints that it required owners to sign contracts to meet certain standards for housing, pay, and so on—standards set by the State Department in agreement with the Mexican government (49–53).
28. House, *Farm Labor Program, 1943*, 118–21.
29. House, *Agriculture Department Appropriation Bill for 1942* (1941), 407. The Chamber of Commerce made a similar argument; see Chamber of Commerce of the United States, *Rural Relief and Rehabilitation under the Farm Security Administration* (Washington, D.C.: May 1942), p. 4, in Folder "U.S. Dept. of Agriculture—Farm Security Administration. 1942," Box 29, Baldwin Papers.
30. House, *Agriculture Department Appropriation Bill for 1942* (1941), 409–13, 517–18 (quote).
31. House, *Agriculture Department Appropriation Bill for 1942* (1941), 518.
32. *Byrd Committee Hearings, Part III*, 792–838; for southern discussion, see 801–12, 818–38; quotes on 802–3. During Kirkpatrick's testimony, he recounted telling his investigators to "get the best you can. Let us get a true picture to the committee" (815), which resulted in open laughter from FSA employees watching the proceedings.
33. *Byrd Committee Hearings, Part III*, 839–46, 855–58, 873–75; see also Holley, *Uncle Sam's Farmers*, 246. Baldwin introduced a lengthy response to many of the specific charges anyway, conceding in a few cases that local FSA employees had shown bad judgment but mostly showing that the AFBF's specific allegations were misleading at best. *Byrd Committee Hearings, Part III*, 893–902.
34. House, *Agriculture Department Appropriation Bill for 1943* (1942), 607–13.
35. *Byrd Committee Hearings, Part III*, 743–46; "Attacks FSA Stand on Poll Tax Loans," *New York Times*, February 7, 1942; House, *Agriculture Department Appropriation Bill for 1943* (1942), 734.
36. *Byrd Committee Hearings, Part II*, 368–70. Critics argued this about Hudgens's trip despite the fact that he paid for about half of it as a personal vacation because he did not

consider it to be FSA business. Richard Russell called the treatment of Hudgens a "grave injustice" (1059). Senate, *Agricultural Appropriation Bill for 1943* (1942), 1049–59. See also "Reminiscences of Robert W. Hudgens."
37. Senate, *Agricultural Appropriation Bill for 1942* (1941), 355.
38. House, *Farm Security Hearings, Part I*, 177–81, 179 (quote). This was part of a related turf war between the Appropriations and Agriculture Committees.
39. House, *Agriculture Department Appropriation Bill for 1943* (1942), 248–49, 252, 237. See also Baldwin, *Poverty and Politics*, 354.
40. *Congressional Record* 88 (1942), 4283, 4286.
41. Westbrook Pegler, "Phone Records," *Atlanta Constitution*, June 21, 1944. Pegler claimed that Baldwin was "on leave serving with the political leader of the New York Communist faction of the union movement," the CIO's Sidney Hillman. See "Reds Control PAC, Dies Aide Charges," *New York Times*, September 28, 1944. Pegler described the FSA as working closely with communists to get Roosevelt reelected, with Baldwin "on loan" to the CIO. Critics were slightly correct: Nathan Gregory Silvermaster, who worked with the FSA and the Board of Economic Warfare, was in fact a communist spy.
42. *Byrd Committee Hearings, Part III*, 699–704; House, *Agriculture Department Appropriation Bill for 1943* (1942), 323; C. P. Trussell, "FSA Is Accused of Tax Loan Plan," *New York Times*, January 24, 1942; "FSA Defends 'Payment' Of Farmers' Poll Taxes," *Christian Science Monitor*, January 23, 1942.
43. *Byrd Committee Hearings, Part III*, 706–9, 711–42. See also Senate, *Agricultural Appropriation Bill for 1943* (1942), 304; "Attacks FSA Stand on Poll Tax Loans," *New York Times*, February 7, 1942.
44. "Attacks FSA Stand on Poll Tax Loans," *New York Times*, February 7, 1942; *Byrd Committee Hearings, Part III*, 742.
45. C. P. Trussell, "FSA Is Accused of Tax Loan Plan," *New York Times*, January 24, 1942.
46. Quoted in FSA, "Region V Bulletin," February 7, 1942, Folder "85-160 Public Relations, July thru December, 1940," Box 14, Office of the Director, General Correspondence, RG 96, NARASE.
47. Robert De Vore, "Roosevelt Rebukes FSA for Paying Poll Tax; Also Hits at Farm Bureau Dues Deductions," *Washington Post*, February 11, 1942; C. P. Trussell, "Roosevelt Scores Levying Poll Tax," *New York Times*, February 14, 1942; FDR, *PPA*, 11:95–96.
48. House, *Agriculture Department Appropriation Bill for 1944* (1943), 1036.
49. Maddox, "Farm Security Administration," 491–93.
50. House, *Agriculture Department Appropriation Bill for 1944* (1943), 1006.
51. "Statement Made by James S. Heizer, Acting Director, Management Division, before French Lick Conference—July 15, 1943," pp. 1 (quote), 6, Folder "F2," Box 30, Baldwin Papers.
52. Sullivan, *Days of Hope*, 127; House, *Agriculture Department Appropriation Bill for 1942* (1941), 87–89.
53. House, *Agriculture Department Appropriation Bill for 1942* (1941), 99–107.
54. Senate, *Agricultural Appropriation Bill for 1942* (1941), 456–77.
55. Richard N. Chapman, *Contours of Public Policy, 1939–1945* (New York: Garland Publishing, 1981), 139. On the Byrd Committee in general, see 129–42, 152–66.
56. Alston and Ferrie, *Southern Paternalism and the American Welfare State*, 97.
57. House, *Agriculture Department Appropriation Bill for 1943* (1942), 300–303. The specific document in question was the loan proposal for Lord Scully Estate in Bates County, Missouri. Tarver criticized the FSA's actions in organizing an association made largely of FSA employees to purchase the land and convey it to future settlers.

58. House, *Agriculture Department Appropriation Bill for 1943* (1942), 607–13, 618 (quote).
59. C. P. Trussell, "Farm Tenant Aid Slashed By House," *New York Times*, March 13, 1942.
60. Robert C. Albright, "Agriculture Dept. Fundless as New Fiscal Year Starts," *Washington Post*, July 1, 1942; C. P. Trussell, "Deadlock Holds on Wickard Funds," *New York Times*, July 2, 1942; "Text of President's Letter on Farm Problem," *Washington Post*, July 4, 1942; "House Lines Hold Against Any Move in Farm Deadlock," *New York Times*, July 7, 1942; "Congress Tying Up War Food Supplies, President Asserts," *New York Times*, July 10, 1942; "Senate, House Conferees Compromise on Farm Bill; Vote May Come Today," *Wall Street Journal*, July 15, 1942; "Agreement on Appropriation," *New York Times*, July 15, 1942; "House Yields to Senate in Parity Issue," *Christian Science Monitor*, July 15, 1942; Grain Sales Votes at Below Parity as House Yields," *New York Times*, July 16, 1942; "Signs Sub-Parity Bill," *New York Times*, July 24, 1942.
61. Larson, *Ten Years of Rural Rehabilitation*, 80. Between budget reductions, congressional interference, and decreasing cooperation with other agencies, the spring and summer of 1942 was a despairing time for FSA administrators. One regional director even recommended to Baldwin that he demand the FSA be withdrawn from the USDA entirely. Laurence I. Hewes Jr. telegram to Mr. C. B. Baldwin, April 22, 1942, Folder "U.S. Dept. of Agriculture—Farm Security Administration. 1942," Box 29, Baldwin Papers.
62. Maher, *Nature's New Deal*, 214.
63. John MacCormac, "Wickard Named Food Administrator," *New York Times*, December 7, 1942; Brown and Baugh, *Preliminary Inventory*, 4; Baldwin, *Poverty and Politics*, 367–74; Albertson, *Roosevelt's Farmer*, 276, 327–44; "Reminiscences of Howard R. Tolley," 575; "FSA Left Out of New Farm Loan Setup," *Washington Post*, January 24, 1943. Along with Parisius, Donald Montgomery, USDA consumers counsel, resigned, and Gardner Jackson, special assistant to Undersecretary Paul Appleby, was fired; all three were sympathetic to the FSA. "Wickard Ousts Jackson, FSA Proponent," *Washington Post*, January 28, 1943.
64. "Reminiscences of Howard R. Tolley," 487; Ben W. Gilbert, "Food System Delay Laid to Wickard," *Washington Post*, January 8, 1943.
65. House, *Farm Labor Program, 1943*, 188.
66. Franklin D. Roosevelt, "Centralizing and Delegating Authority with Respect to the Production and Distribution of Food," March 25, 1943, Folder "Dept. of Agriculture—Farm Security Administration. 1943. II," Box 30, Baldwin Papers; Edward T. Folliard, "Resignation Caused by Lack of Authority; Doubts Subsidy Plan Would Be Effective; Marvin Jones New Administrator," *Washington Post*, June 26, 1943; Baldwin, *Poverty and Politics*, 374–76; Maddox, "Farm Security Administration," 489–91.
67. Robert M. Hallett, "FSA Streamlined Farm Camps Play Big Part in Saving Crops," *Christian Science Monitor*, October 31, 1942; "Huge Crop Losses Faced in Jersey," *New York Times*, May 8, 1942; "Crops Spoiling as Farm Labor Grows Scarce; Wickard Warns of Shortages," *Washington Post*, May 8, 1942. On the FSA's role in bringing Mexican laborers to the United States, see the first-person account in Hewes, *Boxcar in the Sand*, 176–201.
68. Ben W. Gilbert, "Government Defied on 2 Labor Fronts," *Washington Post*, November 5, 1942.
69. "FSA, Farmers' Union Attacked by Cotton Head," *Atlanta Constitution*, November 25, 1942.
70. House, *Farm Labor Program, 1943*, 131–36.
71. Senate, *Agricultural Appropriation Bill for 1943* (1942), 309–11.
72. Ben W. Gilbert, "Roosevelt Puts Nonwar Economy Up to Congress," *Washington Post*, December 30, 1942; Drew Pearson, "White House Yielding; FSA Losing Out," *Washington Post*, March 21, 1943; Henry N. Dorriss, "Farm Labor Bill Under House Fire," *New*

York Times, March 17, 1943; Mason Barr to Jesse B. Gilmer, July 20, 1943, Folder "30 Organization & Administration July 16—31, 1943," Box 2, General Administrative Records, RG 96, NACP; Jack H. Bryan to Regional information Specialists, July 30, 1943, Folder "30 Organization & Administration July 16—31, 1943," Box 2, General Administrative Records, RG 96, NACP; Baldwin, *Poverty and Politics,* 382–83.

73. "FSA Funds Restored," *Washington Post,* June 4, 1943; "Compromise on FSA Scores Senate Victory," *Washington Post,* June 29, 1943. The reduction in funding was from $40 million in administration and $97.5 million for new loans to $20 million and $60 million, respectively.

74. C. P. Trussell, "House Group Cuts Wickard's Funds," *New York Times,* April 14, 1943; Christine Sadler, "Agriculture Bill Arouses Congress," *Washington Post,* April 15, 1943; "FSA Fight Looms on Floor of House," *New York Times,* April 15, 1943.

75. *Congressional Record* 89 (1943), 3411. South Carolina Congressman Hampton Fulmer had similar complaints (3416).

76. "House Farm Battle Ends in Rules Second," *Washington Post,* April 16, 1943; "FSA Fund Raised by Senate Group," *New York Times,* June 6, 1943. Similarly, the AFBF's success against the FSA ended up backfiring to a degree, as it helped solidify the movement to split the Farm Bureau from the Extension Services. By the 1950s, formal relations between the two had ceased almost entirely. See Block, *Separation of the Farm Bureau and the Extension Service,* esp. 243–77.

77. Senate, *Agricultural Appropriation Bill for 1944* (1943), 615. Specifically, for the fiscal year ending June 30, 1943, the resettlement projects got 0.29 percent of FSA funds. *The Annual Report of the Farm Security Administration, 1942–43,* 9.

78. Christine Sadler, "FSA Selling 500 Family Units a Month," *Washington Post,* May 12, 1943; Senate, *Agricultural Appropriation Bill for 1944* (1943), 630–31; *The Annual Report of the Farm Security Administration, 1942–43,* 4–5.

79. Maddox, "Farm Security Administration," 112–15.

80. Senate, *Agricultural Appropriation Bill for 1944* (1943), 850.

81. C. B. Baldwin to E. R. Stettinus Jr., May 13, 1943, Folder "Dept. of Agriculture—Farm Security Administration. 1943. I," Box 30, Baldwin Papers.

82. C. B. Baldwin to Eleanor Roosevelt, October 12, 1943, Folder "Dept. of Agriculture—Farm Security Administration. 1943. I," Box 30, Baldwin Papers; Baldwin, *Poverty and Politics,* 393. See also the numerous (apparently unsent) versions of his resignation letter, reflecting different levels of Baldwin's sense of professional, ideological, and personal insult in each, in Folder "Dept. of Agriculture—Farm Security Administration. 1943. I," Box 30, Baldwin Papers.

83. Lee Mannify, "On the Farm Front," September 6, 1943, Folder "Dept. of Agriculture—Farm Security Administration. 1943. I," Box 30, Baldwin Papers.

84. Samuel B. Bledsoe, "Baldwin May Quit as Farm Unit Chief," *New York Times,* August 31, 1943; "Agriculture's Loss," *Washington Post,* September 6, 1943. Baldwin was one of many New Dealers pushed out by the stresses of war; those who remained, like Harry Hopkins, took on jobs far removed from the liberal reforms of the New Deal. See Brinkley, *End of Reform,* 144–46. The job in Italy may never have existed, at least according to Will Alexander; he saw the whole thing as a sop to Baldwin. Baldwin began to put together a staff and plans, but the State Department and military had no interest in giving him a job and FDR was not willing to push for it. This humiliation apparently contributed to his postwar bitterness and shift to the left. "Reminiscences of Will W. Alexander," 482–85.

85. Baldwin, *Poverty and Politics,* 395–97; May, *Marvin Jones,* 213.

86. Marvin Jones, "Administrator's Memorandum No. 33," November 25, 1943, Folder "War Food Administrator's Memoranda," Box 1, War Food Administrators Memoranda (Entry I-54) in Records of the Office of the Secretary of Agriculture, RG 16, NACP.
87. House, *Farmers' Home Corporation Act of 1944, Hearings before the Committee on Agriculture on H. R. 4384*, 225–26, 239; Maddox, "Farm Security Administration," 209–11.
88. Baldwin also lamented that his appointment resulted in "the whole atmosphere of the agency [changing] pretty rapidly." C. B. Baldwin interview by Richard K. Doud, 30.
89. FSA, *Farm Security*; FSA, *Services to Rural Families* (Washington, DC: Government Printing Office, 1945); House, *Agriculture Department Appropriation Bill for 1946* (1945), 508.
90. Quoted by E. Lee Ozbirn, "To All Employees," December 2, 1943, Folder "30-Gen Jan-April 1944," Box 1, General Administrative Records, RG 96, NACP; similar language is in several other letters to other regional directors. This letter was apparently sent at the urging of J. H. Wood, director of Region I, who worried that FSA morale was in bad shape because of concerns about the future of the organization. J. H. Wood to R. W. Hudgens, November 10, 1943, Folder "30-Gen Jan-April 1944," Box 1, General Administrative Records, RG 96, NACP.
91. House, *Agriculture Department Appropriation Bill for 1946* (1945), 508.
92. House, *Farmers' Home Corporation Act of 1944, Hearings before the Committee on Agriculture on H. R. 4384*, 78th Cong., 2nd sess., 1944, 223–24.
93. House, *Farm Security Hearings, Part IV*, 1475
94. FSA, "Making Supervised Credit Mean What It Implies," 1945, NAL.
95. FSA, *A Handbook for County FSA Committeemen*, August 1945, pp. 45 (quote), 46–47, NAL.
96. House, *Farm Security Hearings, Part I*, 1. The Committee finally released its report a year later, timed to improve the chances of Cooley's FHA bill (H.R.4384, included in the report), House, *Activities of the FSA*.
97. House, *Activities of the FSA*, 2.
98. "FSA Leadership," *Washington Post*, December 7, 1945.
99. C. B. Baldwin to Pete Hudgens, December 17, 1945, Folder "Dept. of Agriculture—Farm Security Administration. 1944–1945," Box 30, Baldwin Papers; "Dillard Lasseter Appointed FSA Administrator," *Atlanta Constitution*, December 11, 1945.
100. Jean Wunderlich to Clinton P. Anderson, January 21, 1946, Folder "Organization 1 FSA AGL Dept Work," Box 1325, Office of the Secretary, General Correspondence, RG 16, NACP.
101. Imogene Rousseau to Clinton P. Anderson, [January 1946], Folder "Organization 1 FSA AGL Dept Work," Box 1325, Office of the Secretary, General Correspondence, RG 16, NACP.
102. See Anderson's responses to critics like Wunderlich and Rousseau in Folder "Organization 1 FSA AGL Dept Work," Box 1325, Office of the Secretary, General Correspondence, RG 16, NACP.
103. Maddox, "Farm Security Administration," 111–13; Baldwin, *Poverty and Politics*, 400–401; Conkin, *Tomorrow a New World*, 228–33. The FSA continued to be criticized after it had given up the fight against rural poverty and even after it had ceased to exist. See, for example, "Minutiae of Life in U.S. Art, $750,000 Lot, Irks Capehart," *Washington Post*, May 21, 1948.
104. "Wallace Would Give Farms to Servicemen," *New York Times*, October 27, 1943.

105. Senate, *Agricultural Appropriation Bill for 1947* (1946), 384, 387–91.
106. Senate, *Agricultural Appropriation Bill for 1944* (1943), 744–47, 949–50; House, *Farmer's Home Corporation Act of 1944*, 78th Cong., 2nd sess., H.R. 4384; *House Farmers' Home Corporation Act of 1944, Hearings before the Committee on Agriculture on H.R. 4384*, 1–17; "Bill Would Merge FSA In Home Agency," *New York Times*, March 14, 1944; Baldwin, *Poverty and Politics*, 401.
107. Harding, "Farmers Home Administration," 585–92; United States Department of Agriculture, Farmers Home Administration, *Strengthening the Family Farm: A Report on Activities of the Farmers Home Administration in the 1946–47 Fiscal Year* (Washington, D.C.: United States Department of Agriculture, 1947), 1–4.
108. USDA-FmHA, *Strengthening the Family Farm*, 5–13; Benedict, *Can We Solve the Farm Problem?* 199–205. On the FmHA as a reactionary force in postwar southern life, see Daniel, *Dispossession*.
109. Shover, *First Majority—Last Minority*, 237; "Reminiscences of Howard R. Tolley," 508–11; Phillips, *This Land, This Nation*, 212–23. On the destruction of the Bureau of Agricultural Economics, see Kirkendall, *Social Scientists and Farm Politics in the Age of Roosevelt*, 195–254.

Conclusion

1. United States Department of Agriculture, Economic Research Service, "Geography of Poverty," http://www.ers.usda.gov/topics/rural-economy-population/rural-poverty-well-being/geography-of-poverty.aspx (last modified February 28, 2014). The rest are largely rural western counties, usually near Native American reservations, and counties along the Mexican border.
2. Harding, "Farmers Home Administration," 587.
3. Larson, *Ten Years of Rural Rehabilitation*, 84, 93–95. FERA made about 397,000 rehabilitation loans, but most of these would have been classified as "nonstandard" or "grants only" clients by the RA/FSA. On the reach and impact of the FSA, see also Mertz, *New Deal Policy and Southern Rural Poverty*, 190–91. As discussed above, determining exactly how many families the FSA and its predecessors served can be tricky.
4. Gilbert and O'Connor, "Leaving the Land Behind," 115.
5. Banfield, "Ten Years of the Farm Tenant Purchase Program," 469. As Brenda J. Taylor points out, it can be difficult to quantify much of the FSA's impact because of the program's "abrupt end." Taylor, "The Farm Security Administration and Rural Families in the South," 43.
6. Grant, *Down and Out on the Family Farm*, 155–60; Mertz, *New Deal Policy and Southern Rural Poverty*, 261–62; Lester M. Salamon, "The Time Dimension in Policy Evaluation: The Case of the New Deal Land-Reform Experiments," *Public Policy* 27.2 (1979): 163; Cannon, *Remaking the Agrarian Dream*, 133–47.
7. Gilbert and O'Connor, "Leaving the Land Behind," 114.
8. See Brinkley, *End of Reform*, for a larger discussion of this process.
9. Phillips, *This Land, This Nation*, 227–31.
10. Brinkley, *End of Reform*, 170–74.
11. O. E. Baker and Conrad Taeuber, "The Rural People," in *Farmers in a Changing World*, by United States Department of Agriculture (Washington, DC: Government Printing Office, 1940), 841–44.

12. Terry and Sims, *They Live on the Land*, 37.
13. Conkin, *A Revolution Down on the Farm*, 104.
14. Wright, *Old South, New South*, 242. On the postwar southern economy in general, see 239–74. See also Holley, *Second Great Emancipation*, 35–54, 93–118; Charles S. Aiken, *The Cotton Plantation South since the Civil War* (Baltimore: Johns Hopkins University Press, 1998), 97–132.
15. Ham, "Status of Agricultural Labor," 562. See also Herman Clarence Nixon, *Forty Acres and Steel Mules* (Chapel Hill: University of North Carolina Press, 1938). By no means did all agree. "By the end of the decade," Gavin Wright says of the South in the 1930s, "many observers thought that little had changed." Wright, *Old South, New South*, 236
16. Rupert B. Vance, *All These People: The Nation's Human Resources in the South* (Chapel Hill: University of North Carolina Press, 1945), 156.
17. Morton Rubin, *Plantation County* (Chapel Hill: University of North Carolina Press, 1951), 12–27.
18. See Conkin, *Revolution Down on the Farm*, esp. chap. 5, "Dimensions of an Agricultural Revolution," 97–122; Daniel, *Breaking the Land*, 153–298; and Fite, *Cotton Fields No More*, 180–225.
19. Senate, *Agricultural Appropriation Bill for 1947* (1946), 389–91. On rural memory of this transformation, see Walker, *Southern Farmers and Their Stories*, 146–52. Tellingly, Walker finds that tenants by and large saw mechanization pushing them off the farm, while large landowners believed mechanization to be a response to decreasing labor availability.
20. Holley, *Second Great Emancipation*, xiv, 72.
21. Paul V. Maris to Frank Hancock, March 2, 1944, Folder "30 Organization & Administration March—1944," Box 2, General Administrative Records, RG 96, NACP. Maris was making a relatively conservative case—that the FSA must loan to higher-income clients and stick more closely to congressional will while recognizing that some poor farmers were stuck with a lower standard of living—but with a Tugwellian ideal of shifting population and land resources into "reasonable balance."
22. Biles, *South and the New Deal*, 56.
23. Schulman, *From Cotton Belt to Sunbelt*, xii.
24. See, for example, the analysis of a religiously infused notion of agrarian independence in Roll, *Spirit of Rebellion*. On the failures of such efforts in postwar America, see Daniel, *Dispossession*.
25. Conkin, *Revolution Down on the Farm*, 164.
26. Benedict, *Can We Solve the Farm Problem?* 173–74.

Bibliography

Archival and Manuscript Sources

Alabama Relief Administration. State publications, SG014348. Alabama Department of Archives and History, Montgomery.
Arnold, Edwin G. Papers. Truman Library. Independence, MO.
Baldwin, C. B. Papers. MsC 343. University of Iowa Libraries, Iowa City.
———. Interview by Richard K. Doud, February 24, 1965. Microfilm. Smithsonian Archives of American Art, Washington, DC.
Bankhead, John H., Jr. Papers. LPR53. Alabama Department of Archives and History, Montgomery.
Crow, Charles B. Papers. LPR56. Alabama Department of Archives and History, Montgomery.
Farm Security Administration. "Rural Housing by the Farm Security Administration." 1941. Francis Loeb Library, Harvard University Graduate School of Design, Cambridge, MA.
Farmers Home Administration. Records. Record Group 96. National Archives at College Park, MD.
———. Records. Record Group 96. National Archives Southeast Region, Morrow, GA.
Jackson, Gardner. Papers. Franklin D. Roosevelt Library, Hyde Park, NY.
Maddox, James G. "The Farm Security Administration." PhD diss., Harvard University, 1950.
Office of the Secretary of Agriculture. Records. Record Group 16. National Archives at College Park, MD.
Ormond, Marion W. Memoir. Interviewed by Bob Haynes, April 17, 1981. Oral History Research Office, Mervyn H. Sterne Library, University of Alabama in Birmingham.
"The Reminiscences of Will W. Alexander." 1972. Microfiche. Oral History Research Office, Columbia University, New York, NY.
"The Reminiscences of Robert W. Hudgens." 1957. Microfiche. Oral History Research Office, Columbia University, New York, NY.
"The Reminiscences of Howard R. Tolley." 1975. Microfiche. Oral History Research Office, Columbia University, New York, NY.
Tugwell, Rexford. Papers. Franklin D. Roosevelt Library, Hyde Park, NY.
Walker, G. T. (Jean). Memoir. Interviewed by Joe Lilly, April 25, 1981. Oral History Research Office, Mervyn H. Sterne Library, University of Alabama in Birmingham.

Government Documents

Federal Emergency Relief Administration. *Purposes and Activities of the Federal Emergency Relief Administration*. Washington, DC: U.S. Government Printing Office, 1935.
———. *Rural Rehabilitation Forms. Section I, Forms FRR 1, 2, and 3*. Washington, DC, [1934?].
———. *Rural Rehabilitation Forms. Section II, Forms FRR 20–27*. Washington, DC, [1934?].
National Emergency Council. *Report on Economic Conditions of the South*. Washington, DC: U.S. Government Printing Office, 1938.

National Resources Committee. *Farm Tenancy: The Report of the Special Committee on Farm Tenancy.* Washington, DC: U.S. Government Printing Office, 1937.

Resettlement Administration. *First Annual Report, Resettlement Administration.* Washington, DC: U.S. Government Printing Office, 1936.

———. *Interim Report of the Resettlement Administration.* Washington, DC: U.S. Government Printing Office, 1936.

———. *Report of the Administrator of the Resettlement Administration, 1937.* Washington, DC: U.S. Government Printing Office, 1937.

U.S. Army. Western Defense Command and Fourth Army. *Final Report, Japanese Evacuation from the West Coast, 1942.* Washington, DC: U.S. Government Printing Office, 1943.

U.S. Census Bureau. *Fifteenth Census of the United States: 1930, Agriculture.* Washington, DC: U.S. Government Printing Office, 1932.

U.S. Congress. House. *Activities of the Farm Security Administration, Report of Select Committee of the House Committee on Agriculture to Investigate the Activities of the Farm Security Administration.* 78th Cong., 2d sess., 1944. H.R. 1430.

———. House. *Agriculture Department Appropriation Bill for 1942, Hearings before the Subcommittee of the Committee on Appropriation, Part II.* 77th Cong., 1st sess., 1941.

———. House. *Agriculture Department Appropriation Bill for 1943, Hearings before the Subcommittee of the Committee on Appropriations, Part II.* 77th Cong., 2d sess., 1942.

———. House. *Agriculture Department Appropriation Bill for 1944, Hearings before the Subcommittee of the Committee on Appropriation, Part II.* 78th Cong., 1st sess., 1943.

———. House. *Agriculture Department Appropriation Bill for 1946, Hearings before the Subcommittee of the Committee on Appropriation, Part II.* 79th Cong., 1st sess., 1945.

———. House. *Bankhead-Jones Farm Tenant Act.* 75th Cong., 1st sess., 1937. H.R. 7562.

———. House. *Bankhead-Jones Farm Tenant Act, Conference Report (to accompany H.R. 7562).* 75th Cong., 1st sess., 1937. Report no. 1198.

———. House. *Hearings Before the Select Committee of the House Committee on Agriculture, to Investigate the Activities of the Farm Security Administration.* 78th Cong., 1st sess., 1943–1944.

———. House. *Farm Labor Program, 1943, Hearings before the Subcommittee of the Committee on Appropriations, on the Appropriation for the Farm Labor Program, Calendar Year 1943.* 78th Cong., 1st sess., 1943.

———. House. *Farm Security Act of 1937.* 75th Cong., 1st sess., 1937. Report no. 1065.

———. House. *Farm Security Act of 1937.* 75th Cong., 1st sess., 1937. H.R. 6240.

———. House. *Farm Security Act of 1937, Report (to accompany H.R. 6240).* 75th Cong., 1st sess., 1937. Report no. 586.

———. House. *Farm Tenancy, Hearing before the Committee on Agriculture on H.R. 8.* 75th Cong., 1st sess., 1937.

———. House. *Farmers' Home Corporation Act of 1944, Hearings before the Committee on Agriculture on H. R. 4384.* 78th Cong., 2nd sess., 1944.

———. House. *National Industrial Recovery Act.* 73rd Cong., 1st sess., 1933. H.R. 5755.

———. House. *Payments in Lieu of Taxes on Resettlement Projects, Hearings before a Subcommittee of the Committee on Ways and Means.* 74th Cong., 2nd sess., 1936.

———. House. *Supplemental Estimate of Appropriation, 1938, Department of Agriculture.* 75th Cong., 1st sess., 1937. Doc. 330.

———. Joint Committee on Reduction of Nonessential Federal Expenditures. *Reduction of Nonessential Federal Expenditures, Part II.* 77th Cong., 1st–2nd sess., December 1, 1941, January 23, 1942, and February 3, 1942.

———. Joint Committee on Reduction of Nonessential Federal Expenditures. *Reduction of Nonessential Federal Expenditures, Part III.* 77th Cong., 2nd sess., February 6, 10, 13, and 27, 1942.
———. Senate. *Agricultural Appropriation Bill for 1938, Hearings before the Subcommittee of the Committee on Appropriations on H.R. 6523.* 75th Cong., 1st sess., 1937.
———. Senate. *Agricultural Appropriation Bill for 1939, Hearings before the Subcommittee of the Committee on Appropriations on H.R. 10238.* 75th Cong., 3rd sess., 1938.
———. Senate. *Agricultural Appropriation Bill for 1940, Hearings before the Subcommittee of the Committee on Appropriations on H.R. 5269.* 76th Cong., 1st sess., 1939.
———. Senate. *Agricultural Appropriation Bill for 1941, Hearings before the Subcommittee of the Committee on Appropriations on H.R. 8202.* 76th Cong., 3rd sess., 1940.
———. Senate. *Agricultural Appropriation Bill for 1942, Hearings before the Subcommittee of the Committee on Appropriations on H.R. 3735.* 77th Cong., 1st sess., 1941.
———. Senate. *Agricultural Appropriation Bill for 1943, Hearings before the Subcommittee of the Committee on Appropriations on H.R. 6709.* 77th Cong., 2nd sess., 1942.
———. Senate. *Agricultural Appropriation Bill for 1944, Hearings before the Subcommittee of the Committee on Appropriations on H.R. 2481.* 78th Cong., 1st sess., 1943.
———. Senate. *Agricultural Appropriation Bill for 1945, Hearings before the Subcommittee of the Committee on Appropriations on H.R. 4443.* 78th Cong., 2nd sess., 1944.
———. Senate. *Agricultural Appropriation Bill for 1946, Hearings before the Subcommittee of the Committee on Appropriations on H.R. 2689.* 79th Cong., 1st sess., 1945.
———. Senate. *Agricultural Appropriation Bill for 1947, Hearings before the Subcommittee of the Committee on Appropriations on H.R. 5605.* 79th Cong., 2nd sess., 1946.
———. Senate. *Resettlement Administration Program.* 74th Cong., 2nd sess., 1936. Doc. 213.
———. Senate. *To Create the Farm Tenant Homes Corporation, Report (to accompany S. 2367).* 74th Cong., 1st sess., 1935. Report no. 446.
———. Senate. *To Create the Farm Tenant Homes Corporation, Hearings Before A Subcommittee of the Committee on Agriculture and Forestry on S. 1800.* 74th Cong., 1st sess., 1935.
U.S. Department of Agriculture. *Better Health for Rural America, Plans of Action for Farm Communities.* Washington, DC: U.S. Government Printing Office, 1945.
———. *Farmers in a Changing World.* Washington, DC: U.S. Government Printing Office, 1940.
———. Economic Research Service. "Geography of Poverty." http://www.ers.usda.gov/topics/rural-economy-population/rural-poverty-well-being/geography-of-poverty.aspx. Last modified February 28, 2014.
———. Farmers Home Administration. *Strengthening the Family Farm: A Report on Activities of the Farmers Home Administration in the 1946–47 Fiscal Year.* Washington, DC: United States Department of Agriculture, 1947.
———. Farm Security Administration. *Annual Report of the Administrator, Farm Security Administration, 1941–42.* Washington, DC: United States Department of Agriculture, 1943.
———. *Farm Security: The Work of the Farm Security Administration.* Washington, DC: U.S. Government Printing Office, [1939?].
———. *Homestead Projects.* [Washington, DC]: 1939.
———. *Household Furniture and Domestic Equipment.* Washington, DC: U.S. Government Printing Office, 1940.
———. *Postwar Development in Farm Security: The Annual Report of the Farm Security Administration for 1945–46.* Washington, DC: U.S. Government Printing Office, 1946.
———. *Procedure Manual.* [Washington, 1938–40].
———. *Report of the Administrator of the Farm Security Administration, 1938.* Washington, DC: U.S. Government Printing Office, 1938.

———. *Report of the Administrator of the Farm Security Administration, 1939.* Washington, DC: U.S. Government Printing Office, 1939.
———. *Report of the Administrator of the Farm Security Administration, 1940.* Washington, DC: U.S. Government Printing Office, 1940.
———. *Report of the Administrator of the Farm Security Administration, 1941.* Washington, DC: U.S. Government Printing Office, 1942.
———. *Services to Rural Families.* Washington, DC: U.S. Government Printing Office, 1945.
———. *The Work of the Farm Security Administration.* Washington, DC: U.S. Government Printing Office, 1941.
———. Interbureau Committee on Post-War Programs. *Better Health for Rural America: Plans of Action for Farm Communities.* Washington, DC: U.S. Government Printing Office, 1945.
———. Soil Conservation Service. *The Preparation of the Standard State Soil Conservation Districts Law: An Interview with Philip M. Glick.* Washington, DC: U.S. Government Printing Office, 1990.
———. War Food Administration. *The Annual Report of the Farm Security Administration, 1942–43.* Washington, DC: United States Department of Agriculture, 1943.
———. *The Annual Report of the Farm Security Administration, 1943–44.* Washington, DC: United States Department of Agriculture, 1944.
———. *The Annual Report of the Farm Security Administration, 1944–45.* Washington, DC: United States Department of Agriculture, 1945.
Whiting, Theodore E., dir. *Final Statistical Report of the Federal Emergency Relief Administration.* Washington, DC: U.S. Government Printing Office, 1942.
Works Progress Administration. *Farm Tenants in the Southeast, 1920–1935.* [Birmingham, AL: Works Progress Administration, 1938].
———. *Farm Tenure Statistics in Southeastern States, 1900–1935.* [Birmingham, AL: Works Progress Administration, 1938].

Magazines and Periodicals

Atlanta Constitution
Atlanta Daily World
Birmingham Age-Herald
Birmingham News
Birmingham Post
Christian Science Monitor
Congressional Record
Extension Service Review
Montgomery Advertiser
The Nation
The New Republic
New York Times
Rural Rehabilitation
Survey Graphic
Wall Street Journal
Washington Post

Published Primary Sources

Agee, James, and Walker Evans. *Let Us Now Praise Famous Men.* Boston: Houghton Mifflin, 1939, 1960.
Alexander, Will W. "A Review of the Farm Security Administration's Housing Activities." In *Housing Yearbook, 1939,* edited by Coleman Woodbury, 139–50. Chicago: National Association of Housing Officials, 1939.
Andrews, June Gilliland. *A Palmerdale Tale.* [Palmerdale, AL: Self-Published, 1995].
Appleby, Paul H. *Big Democracy.* New York: Russel and Russel, 1945, 1970.
Armstrong, Louise V. *We Too Are the People.* Boston: Little, Brown, 1938; reprint, New York: Arno Press, 1971.
Asch, Berta, and A. R. Mangus. *Farmers on Relief and Rehabilitation.* Washington, DC: U.S. Government Printing Office, 1937.
Baker, Gladys. *The County Agent.* Chicago: University of Chicago Press, 1939.
Baldwin, C. B. "Rural Housing in 1942: The Work of the Farm Security Administration." In *Housing Yearbook, 1943,* edited by Hugh R. Pomeroy and Edmond H. Hoben, 70–76. Chicago: National Association of Housing Officials, 1943.
Banfield, Edward C. *Government Project.* Glencoe, IL: Free Press, 1951.
———. "Ten Years of the Farm Tenant Purchase Program." *Journal of Farm Economics* 31.3 (1949): 469–86.
Bean, Louis H. "Planning Our 1935 Farm Program." *Annals of the American Academy of Political and Social Science* 176 (1934): 111–20.
Benedict, Murray R. *Can We Solve the Farm Problem?* New York: Twentieth Century Fund, 1955.
Black, John D. "Fundamental Elements in the Current Agricultural Situation." *Journal of Farm Economics* 23.4 (1941): 712–25.
Black, John D., and R. H. Allen. "The Growth of Farm Tenancy in the United States." *Quarterly Journal of Economics* 51.3 (1937): 393–425.
Blackwell, Gordon W. "The Displaced Tenant Farm Family in North Carolina." *Social Forces* 13.1 (1934): 65–73.
Blaisdell, Donald C. *Government and Agriculture: The Growth of Federal Farm Aid.* New York: Farrar and Rinehart, 1940.
Book, A. B. "A Note on the Legal Status of Share-Tenants and Share-Croppers in the South." *Law and Contemporary Problems* 4.4 (1937): 539–45.
Branson, E. C. "Farm Tenancy in the Cotton Belt: How Farm Tenants Live." *Journal of Social Forces* 1.3 (1923): 213–21.
———. "Farm Tenancy in the South: II. The Social Estate of White Farm Tenants." *Journal of Social Forces* 1.4 (1923): 450–57.
Brown, Douglass V., Edward Chamberlin, Seymour E. Harris, Wissily W. Leontief, Edward S. Mason, Joseph A. Schumpeter, and Overton H. Taylor. *The Economics of the Recovery Program.* New York: Whittlesey House, 1934.
Brown, Josephine C. "Rural Families on Relief." *Annals of the Academy of Political and Social Science* 176, Social Welfare in the National Recovery Program (1934): 90–94.
Brown, Sean, Jr., ed. *Up before Daylight: Life Histories from the Alabama Writers' Project, 1938–1939.* Tuscaloosa: University of Alabama Press, 1982, 1997.
Brown, Stanley W., and Virgil E. Baugh. *Preliminary Inventory of the Records of the Farmers Home Administration (Record Group 96).* Vol. 118. Washington, DC: National Archives, 1959.
Caldwell, Erskine, and Margaret Bourke-White. *You Have Seen Their Faces.* New York: Viking Press, 1937.

Campbell, J. Phil. "The Government's Farm Policies and the Negro Farmer." *Journal of Negro Education* 5.1 (1936): 32–39.

Cash, W. J. *The Mind of the South*. New York: Alfred A. Knopf, 1941; reprint, New York: Vintage Books, 1991.

Childs, Marquis W. *I Write from Washington*. New York: Harper and Brothers, 1942.

Clark, Noble. "The Social and Economic Implications of the National Land Program: Discussion." *Journal of Farm Economics* 18.2 (1936): 274–80.

Clawson, Marion. "Resettlement Experience on Nine Selected Resettlement Projects." *Agricultural History* 52.1 (1978): 1–92.

Cotton, Albert H. "Regulations of Farm Landlord-Tenant Relationships." *Law and Contemporary Problems* 4.4 (October, 1937): 508–38.

Couch, W. T., ed. *Culture in the South*. Chapel Hill: University of North Carolina Press, 1935.

Crain, Mildred Moore McCrimmon. *Welcome to Greenwood: The History of a Subsistence Homestead Project and Its People*. Montevallo, AL: Times Printing Company, [1990].

Davis, John P. "A Survey of the Problems of the Negro Under the New Deal." *Journal of Negro Education* 5.1 (1936): 3–12.

Eaton, Joseph F. *Exploring Tomorrow's Agriculture: Co-Operative Group Farming—A Practical Program of Rural Rehabilitation*. New York: Harper and Brothers, 1943.

Farm Tenancy Committee. *Farm Tenure in Alabama*. Wetumpka, AL: Wetumpka Printing Co., 1944.

Farnham, Rebecca, and Irene Link. *Effects of the Works Program on Rural Relief: A Survey of Rural Relief Cases Closed in Seven States, July Through November 1935*. Washington, DC: U.S. Government Printing Office, 1938.

Forster, G. W. "Progress and Problems in Agricultural Planning in the Southern States." *Journal of Farm Economics* 18.1 (1936): 86–94.

Fulmer, C. Kline. *Greenbelt*. Washington, DC: American Council on Public Affairs, 1941.

Gaer, Joseph. *Toward Farm Security: The Problem of Rural Poverty and the Work of the Farm Security Administration, Prepared Under the Direction of the FSA Personnel Training Committee, for FSA Employees*. Washington, DC: U.S. Government Printing Office, 1941.

Gaus, John M., and Leon O. Wolcott. *Public Administration and the United States Department of Agriculture*. Chicago: Public Administration Service, 1940.

Gee, Wilson. "Reversing the Tide toward Tenancy." *Southern Economic Journal* 2.4 (1936): 1–11.

Glick, Philip M. "The Federal Subsistence Homesteads Program." *Yale Law Journal* 44.8 (1935): 1324–79.

Gray, L. C. "Our Land Policy Today." *Land Policy Review* 1.1 (1938): 3–8.

———. "The Social and Economic Implications of the National Land Program." *Journal of Farm Economics* 18.2 (1936): 257–73.

Gwin, J. Blaine. "Subsistence Homesteads." *Social Forces* 12.4 (1934): 522–25.

Hagood, Margaret Jarman. *Mothers of the South: Portraiture of the White Tenant Farm Woman*. Chapel Hill: University of North Carolina Press, 1939; reprint, New York: Greenwood Press, 1969.

Ham, William T. "The Status of Agricultural Labor." *Law and Contemporary Problems* 4.4 (1937): 559–72.

Harding, T. Swann. "Farmers Home Administration." *Antioch Review* 6.4 (1946): 585–92.

———. "The Record of the Subsistence Homesteads." *American Journal of Economics and Sociology* 4.4 (1945): 499–504.

Harper, Roland M. "Rural Standards of Living in the South." *Journal of Social Forces* 2.1 (1923): 13–17.

Heer, Clarence. *Income and Wages in the South*. Chapel Hill: University of North Carolina Press, 1930.
Hewes, Laurence. *Boxcar in the Sand*. New York: Alfred A, Knopf, 1957.
Hickock, Lorena. *One Third of a Nation: Lorena Hickock Reports on the Great Depression*. Edited by Richard Lowitt and Maurine Beasley. Urbana: University of Illinois Press, 1981.
Hinckley, Russell J., and John J. Haggerty. "Taxation in Aid of Farm Security." *Law and Contemporary Problems* 4.4 (1937): 546–58.
Hoffsommer, Harold. "The AAA and the Cropper." *Social Forces* 13.4 (1935): 494–502.
———. *Land-Lord Tenant Relations and Relief in Alabama*. Research Bulletin, Series II, No. 9. Washington, DC: Division of Research, Statistics and Finance, Federal Emergency Relief Administration, 1935.
———. "Social Aspects of Farm Labor in the South." *Rural Sociology* 3.4 (1938): 434–45
Holt, John B. *An Analysis of Methods and Criteria Used in Selecting Families for Colonization Projects*. Social Research Report, No. 1. Washington, DC: United States Department of Farm Agriculture, the Farm Security Administration, and the Bureau of Agricultural Economics, Cooperating, 1937.
Hopkins, Harry L. *Spending to Save: The Complete Story of Relief*. New York: W. W. Norton, 1936.
Hudgens, R. W. "The Plantation South Tries a New Way." *Land Policy Review* 3.7 (1940): 26–29.
Ickes, Harold. *The Secret Diaries of Harold L. Ickes*. Vol. 1. *The First Thousand Days, 1933–1936*. New York: Simon and Schuster, 1953.
Johnson, Charles S. *Shadow of the Plantation*. Chicago: University of Chicago Press, 1934.
Johnson, Charles S., Edwin R. Embree, and W. W. Alexander. *The Collapse of Cotton Tenancy: Summary of Field Studies and Statistical Surveys, 1933–1935*. Chapel Hill: University of North Carolina Press, 1935.
Johnstone, Paul H. Introduction. Lord and Johnstone, *A Place on Earth*, 1–3.
Johnstone, Paul H., and Dorothy C. Goodwin. "The Administration of the Subsistence-Homesteads Program." In Lord and Johnstone, *A Place on Earth*, 38–55.
———. "The Back-to-the-Land Movement." In Lord and Johnstone, *A Place on Earth*, 11–22.
———. "The Drive for Legislation." In Lord and Johnstone, *A Place on Earth*, 23–37.
Kester, Howard. *Revolt among the Sharecroppers*. New York: Covici, Friede, 1936; reprint, Knoxville: University of Tennessee Press, 1997.
Kile, Orville Merton. *The Farm Bureau Movement*. New York: Macmillan, 1922.
———. *The Farm Bureau through Three Decades*. Baltimore: Waverly Press, 1948.
Kirkpatrick, E. L. *Analysis of 70,000 Rural Rehabilitation Families*. Social Research Report, No. 9. Washington, DC: United States Department of Farm Agriculture, the Farm Security Administration, and the Bureau of Agricultural Economics, Cooperating, 1938.
Kollmorgen, Walter M. "The Subsistence Homesteads Near Birmingham." In Lord and Johnstone, *A Place on Earth*, 65–81.
Landis, Paul H. "The New Deal and Rural Life." *American Sociological Review* 1.4 (1936): 592–603.
Larson, Olaf F. *Ten Years of Rural Rehabilitation in the United States*. Washington, DC: Bureau of Agricultural Economics, United States Department of Agriculture, 1947.
Lewis, E. E. "Black Cotton Farmers and the AAA." *Opportunity: Journal of Negro Life* 13.3 (1935): 72–74.
Loomis, Charles P. "The Development of Planned Rural Communities." *Rural Sociology* 3.4 (1938): 385–409.
———. *Social Relationships and Institutions in Seven New Rural Communities*. Social Research Report, No. 18. Washington, DC: United States Department of Farm Agriculture, the Farm Security Administration, and the Bureau of Agricultural Economics, Cooperating, 1940.

Lord, Russell. *The Wallaces of Iowa*. Boston: Houghton Mifflin, 1947.
Lord, Russell, and Paul H. Johnstone, eds. *A Place on Earth: A Critical Appraisal of Subsistence Homesteads*. Washington, DC: U.S. Department of Agriculture, Bureau of Agricultural Economics, 1942.
Louchheim, Katie, ed. *The Making of the New Deal: The Insiders Speak*. Cambridge, MA: Harvard University Press, 1983.
Maclachlen, J. M. "Salvation for the Tenant Farmer." *Opportunity: Journal of Negro Life* 13.4 (1935): 104–8.
Maddox, James G. "The Bankhead-Jones Farm Tenant Act." *Law and Contemporary Problems* 4.4 (1937): 434–55.
———. "An Historical Review of the Nation's Efforts to Cope with Rural Poverty." *American Journal of Agricultural Economics* 50.5 (1968): 1351–61.
Maris, Paul V. *"the land is mine": From Tenancy to Family Farm Ownership*. Washington, DC: U.S. Government Printing Office, 1950.
McCamy, James L. "We Need More Personalized Administration." In *Public Administration Readings and Documents*, edited by Felix A. Nigro, 460–81. New York: Rinehart, 1951.
McDowell, M. S. "What the Agricultural Extension Service Has Done for Agriculture." *Annals of the American Academy of Political and Social Science* 142 (March 1929): 250–56.
McKinley, Charles. "Federal Administrative Pathology and the Separation of Powers." *Public Administration Review* 11.1 (1951): 17–25.
Melvin, Bruce L. "Emergency and Permanent Legislation with Special Reference to the History of Subsistence Homesteads." *American Sociological Review* 1.4 (1936): 622–31.
Meriam, Lewis. *Relief and Social Security*. Washington, DC: Brookings Institute, 1946.
Mitchell, H. L. *Mean Things Happening in This Land: The Life and Times of H. L. Mitchell, Co-founder of the Southern Tenant Farmers Union*. Montclair, NJ: Allanheld, Osmun, 1979; reprint, Norman: University of Oklahoma Press, 2008.
Monchow, Helen C. "The Farm Tenancy Act." *Journal of Land and Public Utility Economics* 13.4 (1937): 417–18.
Mott, Frederick D., and Milton I. Roemer. *Rural Health and Medical Care*. New York: McGraw-Hill, 1948.
Morgan, E. L. "National Policy and Rural Public Welfare." *Rural Sociology* 1.1 (1936): 8–19.
Murray, William G. "Governmental Farm Credit and Tenancy." *Law and Contemporary Problems* 4.4 (1937): 489–507.
Myers, Howard B. "Relief in the Rural South." *Southern Economic Journal* 3.3 (1937): 281–91.
Nixon, Herman Clarence. *Forty Acres and Steel Mules*. Chapel Hill: University of North Carolina Press, 1938.
Nourse, Edwin G., Joseph S. Davis, and John D. Black. *Three Years of the Agricultural Adjustment Administration*. Washington, DC: Brookings Institution, 1937.
Odum, Howard W. *Southern Regions of the United States*. Chapel Hill: University of North Carolina Press, 1936.
Oppenheimer, Monroe. "The Development of the Rural Rehabilitation Loan Program." *Law and Contemporary Problems* 4.4 (1937): 473–88.
Percy, William Alexander. *Lanterns on the Levee: Recollections of a Planter's Son*. New York: Alfred A. Knopf, 1941.
Pickett, Clarence E. *For More Than Bread: An Autobiographical Account of Twenty-Two Year's Works with the American Friends Service Committee*. Boston: Little, Brown, 1953.
Proctor, Erna E. "Home Economics in a Rural Rehabilitation Program." *Journal of Home Economics* 27.8 (1935): 501–5.

Pyle, Ernie. *Home Country*. New York: William Sloane Associates, 1947.
Pynchon, Charles E. "Security for Low-Income Families." *Journal of Farm Economics* 27.6 (1935): 337–41.
Raper. Arthur F. "Gullies and What They Mean." *Social Forces* 16.2 (1937): 201–7.
———. *Preface to Peasantry: A Tale of Two Black Belt Counties*. Chapel Hill: University of North Carolina Press, 1936.
———. *Tenants of the Almighty*. New York: Macmillan, 1943.
Raper, Arthur F., and Ira De Reid. *Sharecroppers All*. Chapel Hill: University of North Carolina Press, 1941.
Reid, T. Roy. "Public Assistance to Low-Income Farmers of the South." *Journal of Farm Economics* 21.1 (1939): 188–94.
Richards, Henry I. *Cotton and the AAA*. Washington, DC: Brookings Institute, 1936.
Robertson, Ben. *Red Hills and Cotton: An Upcountry Memory*. New York: Alfred A. Knopf, 1942.
Roosevelt, Franklin D. *The Public Papers and Addresses of Franklin D. Roosevelt*. 13 vols. Edited by Samuel I. Rosenman. New York: Random House, 1938–1950.
Rubin, Morton. *Plantation County*. Chapel Hill: University of North Carolina Press, 1951.
Salter, Leonard A., Jr., "Research and Subsistence Homesteads." *Rural Sociology* 2.2 (1937): 206–10.
Schultz, T. W. "A Comment on the 'Report of the President's Committee on Farm Tenancy.'" *Journal of Land and Public Utility Economics* 13.2 (1937): 207–8.
Schultz, Theodore W. *Training and Recruiting of Personnel in the Rural Social Studies*. Washington, DC: American Council on Education, 1941.
Schuler, E. A. *Social Status and Farm Tenure—Attitudes and Social Conditions of Corn Belt and Cotton Belt Farmers*. Social Research Report, No. 4. Washington, DC: United States Department of Agriculture, the Farm Security Administration, and the Bureau of Agricultural Economics, Cooperating, 1938.
Schuyler, Daniel M. "Constitutional Problems Confronting the Resettlement Administration." *Journal of Land and Public Utility Economics* 12.3 (1936): 304–6.
Shafer, Karl. *A Basis for Social Planning in Coffee County, Alabama*. Social Research Report, No. 6. Washington, DC: United States Department of Farm Agriculture, the Farm Security Administration, and the Bureau of Agricultural Economics, Cooperating, 1937.
Simon, A. M. "Medical Service Plans of the Farm Security Administration." *Journal of the American Medical Association* 120.16 (1942): 1315–17.
Smith, C. B. "The Origin of Farm Economics Extension." *Journal of Farm Economics* 14.1 (1932): 17–22.
Southern Regional Committee of the Social Science Research Council. *Problems of the Cotton Economy: Proceedings of the Southern Social Science Research Conference, New Orleans, March 8 and 9, 1935*. Dallas: Arnold Foundation, 1936.
Spillman, W. J. "The Agricultural Ladder." *American Economic Review* 9.1 Supplement (1919): 170–79.
Stanfield, Mattie Cole. *Palmerdale, Alabama: 50-year History, 1934-1984*. [Palmerdale, AL: self-published, 1984].
Stigler, George J. "The Cost of Subsistence." *Journal of Farm Economics* 27.2 (1945): 303–14.
Swiger, Rachel Rowe, and Olaf F. Larson. *Climbing Toward Security*. Washington, DC: United States Department of Agriculture, Bureau of Agricultural Economics, 1944.
———. *Yesterday, Today, and Tomorrow: Five Hundred Low-Income Farm Families in Wartime*. Washington, DC: United States Department of Agriculture, Bureau of Agricultural Economics and Farm Security Administration Cooperating, 1943.

Swiger, Rachel Rowe, and Conrad Taeuber. *Ill Fed, Ill Clothed, Ill Housed—Five Hundred Families in Need of Help.* Washington, DC: United States Department of Agriculture, Bureau of Agricultural Economics and Farm Security Administration Cooperating, 1942.

———. *Solving Problems through Cooperation.* Washington, DC: United States Department of Agriculture, Bureau of Agricultural Economics and Farm Security Administration Cooperating, 1942.

———. *They Too Produce for Victory.* Washington, DC: United States Department of Agriculture, Bureau of Agricultural Economics and Farm Security Administration Cooperating, 1942.

Taueber, Conrad, and Rachel Rowe. *Five Hundred Families Rehabilitate Themselves.* Washington, DC: United States Department of Agriculture, Bureau of Agricultural Economics and Farm Security Administration Cooperating, 1941.

Taylor, Carl C. "Research Needed as Guidance to the Subsistence Homesteads Program." *Journal of Farm Economics* 16.2 (1934): 310–14.

———. "Social and Economic Significance of the National Subsistence Homesteads Program: From the Viewpoint of a Sociologist." *Journal of Farm Economics* 17.4 (1935): 720–31.

Taylor, Carl C., Helen W. Wheeler, and E. L. Kirkpatrick. *Disadvantaged Classes in American Agriculture.* Social Research Report, No. 8. Washington, DC: United States Department of Farm Agriculture, the Farm Security Administration, and the Bureau of Agricultural Economics, Cooperating, 1938.

Terry, Paul W., and Verner M. Sims. *They Live on the Land: Life in an Open-Country Southern Community.* Tuscaloosa: University of Alabama Press, 1940, 1993.

Tolley, H. R. "The Program Planning Division of the Agricultural Adjustment Administration." *Journal of Farm Economics* 16.4 (1934): 582–90.

Tugwell, Rexford G. *The Brains Trust.* New York: Viking Press, 1968.

———. *In Search of Roosevelt.* Cambridge, MA: Harvard University Press, 1972.

———. "National Significance of Recent Trends in Farm Population." *Social Forces* 14.1 (1935): 1–7.

———. "The Place of Government in a National Land Program." *Journal of Farm Economics* 16.1 (1934): 55–69.

———. "The Resettlement Idea." *Agricultural History* 33.4 (1959): 159–64.

Turner, Howard A. "Farm Tenancy Distribution and Trends in the United States." *Law and Contemporary Problems* 4.4 (1937): 424–33.

Vance, Rupert B. *All These People: The Nation's Human Resources in the South.* Chapel Hill: University of North Carolina Press, 1945.

———. *How the Other Half Is Housed: A Pictorial Record of Subminimum Farm Housing in the South.* Chapel Hill: University of North Carolina Press, 1936.

———. *Human Factors in Cotton Culture: A Study in the Social Geography of the American South.* Chapel Hill: University of North Carolina Press, 1929.

———. *Human Geography of the South: A Study in Regional Resources and Human Adequacy.* Chapel Hill: University of North Carolina Press, 1935.

Vance, Rupert, and Gordon W. Blackwell. *New Farm Homes for Old: A Study of Rural Public Housing in the South.* Tuscaloosa: University of Alabama Press, 1946.

Verne, Lewis B. *Budgetary Administration in the United States Department of Agriculture.* Chicago: Public Administration Service, 1941.

Wager, Paul W. *One Foot on the Soil: A Study of Subsistence Homesteads in Alabama.* Tuscaloosa: Bureau of Public Administration, University of Alabama, 1945.

Wehrwein, G. S. "An Appraisal of Resettlement." *Journal of Farm Economics* 19.1 (1937): 190–202.
Westbrook, Lawrence. "The Program of Rural Rehabilitation of the FERA." *Journal of Farm Economics* 17.1 (1935): 89–100.
———. *Rural-Industrial Communities for Stranded Families: An Outline of Suggested Procedure for the Guidance of State and County Public Relief Administration and Cooperating Public Agencies*. Washington, DC: Federal Emergency Relief Administration, Division of Rural Rehabilitation and Stranded Populations, [1934?].
Wiley, Clarence A. "Settlement and Unsettlement in the Resettlement Administration Program." *Law and Contemporary Problems* 4.4 (1937): 456–72.
Wilson, M. L. "The Place of Subsistence Homesteads in Our National Economy." *Journal of Farm Economics* 16.1 (1934): 73–84.
———. "Problem of Poverty in Agriculture." *Journal of Farm Economics* 22.1 (1940): 10–29.
———. "The Subsistence Homestead Program." In *Proceedings of the Institute of Public Affairs, Eighth Annual Session, Athens, Georgia, May 8–15, 1934*, 158–75. Athens: University of Georgia Press, 1934.
Woodson, Cartern Godwin. *The Rural Negro*. Washington, DC: Association for the Study of Negro Life and History, 1930.
Woofter, T. J., Jr. *Landlord and Tenant on the Cotton Plantation*. Washington, DC: Works Progress Administration Division of Social Research, 1936.
———. "Southern Population and Social Planning." *Social Forces* 14.1 (1935): 16–22.
Woofter, T. J., Jr., and Ellen Winston. *Seven Lean Years*. Chapel Hill: University of North Carolina Press, 1939.

Secondary Sources

Adams, Jane, and D. Gorton. "This Land Ain't My Land: The Eviction of Sharecroppers by the Farm Security Administration." *Agricultural History* 83.3 (2009): 323–51.
Aiken, Charles S. *The Cotton Plantation South since the Civil War*. Baltimore: Johns Hopkins University Press, 1998.
Albertson, Dean. *Roosevelt's Farmer: Claude R. Wickard in the New Deal*. New York: Columbia University Press, 1961.
Alston, Lee J., and Joseph P. Ferrie. *Southern Paternalism and the American Welfare State: Economics, Politics, and Institutions in the South, 1865–1965*. Cambridge, UK: Cambridge University Press, 1999.
Anderson, Clifford B. "The Metamorphosis of American Agrarian Idealism in the 1920's and 1930's." *Agricultural History* 35.4 (1961): 182–88.
Arnold, Joseph L. *The New Deal in the Suburbs: A History of the Greenbelt Town Program, 1935–1954*. Columbus: Ohio State University Press, 1971.
Ayers, Edward L. *The Promise of the New South: Life after Reconstruction*. New York: Oxford University Press, 1992, 2007.
Badger, Anthony J. *The New Deal: The Depression Years, 1933–1940*. New York: Hill and Wang, 1989; Chicago: Ivan R. Dee, 2002.
———. *New Deal/New South: An Anthony J. Badger Reader*. Fayetteville: University of Arkansas Press, 2007.
Baldwin, Sidney. *Poverty and Politics: The Rise and Decline of the Farm Security Administration*. Chapel Hill: University of North Carolina Press, 1968.

Bartley, Numan V. *The New South, 1945–1980*. Vol. 11 of *A History of the South*. Baton Rouge: Louisiana State University Press and the Littlefield Fund for Southern History, the University of Texas, 1995.
Bezner, Lili Corbus. *Photography and Politics in America: From the New Deal into the Cold War*. Baltimore: Johns Hopkins University Press, 1999.
Biles, Roger. *The South and the New Deal*. Lexington: University Press of Kentucky, 1994.
Bindas, Kenneth J. *Remembering the Great Depression in the Rural South*. Gainesville: University Press of Florida, 2007.
Birmingham Historical Society. *Digging Out of the Great Depression: Federal Programs at Work in and around Birmingham*. Birmingham, AL: Birmingham Historical Society, 2010.
Block, William J. *The Separation of the Farm Bureau and the Extension Service: Political Issue in a Federal System*. Urbana: University of Illinois Press, 1960.
Brinkley, Alan. *The End of Reform: New Deal Liberalism in Recession and War*. New York: Random House, 1995.
Brown, Dona. *Back to the Land: The Enduring Dream of Self-Sufficiency in Modern America*. Madison: University of Wisconsin Press, 2011.
Browne, William P. *Cultivating Congress: Constituents, Issues, and Interests in Agricultural Policymaking*. Lawrence: University Press of Kansas, 1995.
Brownell, Blaine A. "Birmingham, Alabama: New South City in the 1920s." *Journal of Southern History* 38.1 (1972): 21–48.
Brueggemann, John. "Racial Considerations and Social Policy in the 1930s: Economic Change and Political Opportunities." *Social Science History* 26.1 (2002): 139–77.
Burns, James MacGregor. *Roosevelt: The Lion and the Fox*. Vol. 1, *1882–1940*. San Diego: Harcourt Brace, 1956.
Cahill, Kevin J. "Fertilizing the Weeds: The Rural Rehabilitation Program in West Virginia." *Journal of Appalachian Studies* 4.2 (1998): 285–97.
Campbell, Christiana McFadyen. *The Farm Bureau and the New Deal: A Study of Making National Policy*. Urbana: University of Illinois Press, 1962.
Campbell, David, and David Coombs. "Skyline Farms: A Case Study of Community Development and Rural Rehabilitation." *Appalachian Journal* 10.3 (1983): 244–54.
Cannon, Brian Q. *Remaking the Agrarian Dream: New Deal Rural Resettlement in the Mountain West*. Albuquerque: University of New Mexico Press, 1996.
Carlebach, Michael, and Eugene F. Provenzo Jr. *Farm Security Administration Photographs of Florida*. Gainesville: University Press of Florida, 1993.
Case, H. C. M. "Farm Debt Adjustment during the Early 1930s." *Agricultural History* 34.4 (1960): 173–82.
Chapman, Richard N. *Contours of Public Policy, 1939–1945*. New York: Garland Publishing, 1981.
Cobb, James C., and Michael V. Namorato, eds. *The New Deal and the South*. Jackson: University Press of Mississippi, 1984.
Cohen, Adam. *Nothing to Fear: FDR's Inner Circle and the Hundred Days That Created Modern America*. New York: Penguin Books, 2009.
Cohen, Stu. *The Likes of Us: America in the Eyes of the Farm Security Administration*. Edited by Peter Bacon Hales. Boston: David R. Godine, 2009.
Conkin, Paul K. *A Revolution Down on the Farm: The Transformation of American Agriculture since 1929*. Lexington: University Press of Kentucky, 2008.
———. *Tomorrow a New World: The New Deal Community Program*. Ithaca, NY: Cornell University Press, 1959.

Conrad, David Eugene. *The Forgotten Farmers: The Story of Sharecroppers in the New Deal.* Urbana: University of Illinois Press, 1965.
Culver, John C., and John Hyde. *American Dreamer: The Life and Times of Henry A. Wallace.* New York: W. W. Norton, 2000.
Danbom, David B. *The Resisted Revolution: Urban America and the Industrialization of Agriculture, 1900–1930.* Ames: Iowa State University Press, 1979.
Daniel, Pete. *Breaking the Land: The Transformation of Cotton, Tobacco, and Rice Cultures since 1880.* Urbana: University of Illinois Press, 1985.
———. *Dispossession: Discrimination against African American Farmers in the Age of Civil Rights.* Chapel Hill: University of North Carolina Press, 2013.
———. *The Shadow of Slavery: Peonage in the South, 1901–1969.* Urbana: University of Illinois Press, 1972.
Dykeman, Wilma, and James Stokely. *Seeds of Southern Change: The Life of Will Alexander.* Chicago: University of Chicago Press, 1962.
Egerton, John. *Speak Now against the Day: The Generation before the Civil Rights Movement in the South.* New York: Alfred K. Knopf, 1994.
Feder, Ernest. "Farm Debt Adjustment during the Depression—The Other Side of the Coin." *Agricultural History* 35.2 (1961): 78–81.
Finegold, Kenneth, and Theda Skocpol. *State and Party in America's New Deal.* Madison: University of Wisconsin Press, 1995.
Finnegan, Cara. *Picturing Poverty: Print Culture and FSA Photographs.* Washington, DC: Smithsonian Books, 2003.
Fite, Gilbert C. *Cotton Fields No More: Southern Agriculture, 1865–1980.* Lexington: University Press of Kentucky, 1984.
———. *George N. Peek and the Fight for Farm Parity.* Norman: University of Oklahoma Press, 1954.
Flamm, Michael W. "The National Farmers Union and the Evolution of Agrarian Liberalism, 1937–1946." *Agricultural History* 68.3 (1994): 54–80.
Flynt, Wayne. *Alabama in the Twentieth Century.* Tuscaloosa: University of Alabama Press, 2004.
Garrett, Martin A., Jr., and Zhenhui Xu. "The Efficiency of Sharecropping: Evidence from the Postbellum South." *Southern Economic Journal* 69.3 (2003): 578–95.
Giesen, James C. *Boll Weevil Blues: Cotton, Myth, and Power in the American South.* Chicago: University of Chicago Press, 2011.
Gilbert, Jess. "Eastern Urban Liberals and Midwestern Agrarian Intellectuals: Two Group Portraits of Progressives in the New Deal Department of Agriculture." *Agricultural History* 74.2 (2000): 162–80.
———. "A Usable Past: New Dealers Henry A. Wallace and M. L. Wilson Reclaim the American Agrarian Tradition." In *Rationality and the Liberal Spirit: A* Festschrift *Honoring Ira Lee Morgan*, edited by the Centenary College Department of English, 134–42. [Shreveport, La.]: Centenary College of Louisiana, 1997.
———. "Low Modernism and the Agrarian New Deal: A Different Kind of State." In *Fighting for the Farm: Rural America Transformed,* edited by Jane Adams, 129–46. Philadelphia: University of Pennsylvania Press (2003).
Gilbert, Jess, and Steve Brown. "Alternative Land Reform Proposals in the 1930s: The Nashville Agrarians and the Southern Tenant Farmers' Union." *Agricultural History* 55.4 (1981): 351–69.

Gilbert, Jess, and Carolyn Howe. "Beyond 'State vs. Society': Theories of the State and New Deal Agricultural Policies." *American Sociological Review* 56.2 (1991): 204–20.

Gilbert, Jess, and Alice O'Connor. "Leaving the Land Behind: Struggles for Land Reform in U.S. Federal Policy, 1933–1965." In *Who Owns America? Social Conflict over Property Rights*, edited by Harvey M. Jacobs, 114–30. Madison: University of Wisconsin Press, 1998.

Gisolfi, Monica Richmond. "From Crop Lien to Contract Farming: The Roots of Agribusiness in the American South, 1929–1939." *Agricultural History* 80.2 (2006): 167–89.

Gold, Bela. *Wartime Economic Planning in Agriculture: A Study in the Allocation of Resources*. New York: Columbia University Press, 1949.

Gordon, Linda. "Dorothea Lange: The Photographer as Agricultural Sociologist." *Journal of American History* 93.3 (2006): 698–727.

Grant, Michael Johnston. *Down and Out on the Family Farm: Rural Rehabilitation in the Great Plains, 1929–1945*. Lincoln: University of Nebraska Press, 2002.

Grant, Nancy L. *TVA and Black Americans: Planning for the Status Quo*. Philadelphia: Temple University Press, 1990.

Green, Elna C., ed. *Before the New Deal: Social Welfare in the South, 1830–1930*. Athens: University of Georgia Press, 1999.

Gregg, Sara M. *Managing the Mountains: Land Use Planning, the New Deal, and the Creation of a Federal Landscape in Appalachia*. New Haven, CT: Yale University Press, 2010.

Grey, Michael R. *New Deal Medicine: The Rural Health Programs of the Farm Security Administration*. Baltimore: Johns Hopkins University Press, 1996.

Grubbs, Donald H. *Cry from the Cotton: The Southern Tenant Farmers' Union and the New Deal*. Chapel Hill: University of North Carolina Press, 1971.

Hahamovitch, Cindy. "In America Life Is Given Away." In *The Countryside in the Age of the Modern State*, edited by Catherine McNicol Stock and Robert D. Johnston, 134–60. Ithaca, NY: Cornell University Press, 2001.

Hamilton, David E. *From New Day to New Deal: American Farm Policy from Hoover to Roosevelt, 1928–1933*. Chapel Hill: University of North Carolina Press, 1991.

Hansen, John Mark. *Gaining Access: Congress and the Farm Lobby, 1919–1981*. Chicago: University of Chicago Press, 1991.

Hayes, Jack Irby, Jr. *South Carolina and the New Deal*. Columbia: University of South Carolina Press, 2001.

Hersey, Mark D. *My Work Is That of Conservation: An Environmental Biography of George Washington Carver*. Athens: University of Georgia Press, 2011.

———. "'What We Need Is a Crop Ecologist': Ecology and Agricultural Science in Progressive-Era America." *Agricultural History* 85.3 (2011): 297–321.

Hofstadter, Richard. *The Age of Reform: From Bryan to FDR*. New York: Vintage Books, 1955.

Holley, Donald. "The Negro in the New Deal Resettlement Program." *Agricultural History* 45.3 (1971): 179–93.

———. *The Second Great Emancipation: The Mechanical Cotton Picker, Black Migration, and How They Shaped the Modern South*. Fayetteville: University of Arkansas Press, 2000.

———. *Uncle Sam's Farmers: The New Deal Communities in the Lower Mississippi Valley*. Urbana: University of Illinois Press, 1975.

Horton, Carol A. *Race and the Making of American Liberalism*. New York: Oxford University Press, 2005.

Johnson, William R. "National Farm Organizations and the Reshaping of Agricultural Policy in 1932." *Agricultural History* 37.1 (1963): 35–42.

———. "Rural Rehabilitation in the New Deal: The Ropesville Project." *Southwestern Historical Quarterly* 79.3 (1976): 279–95.
Jones, Lu Ann. *Mama Learned Us to Work: Farm Women in the New South.* Chapel Hill: University of North Carolina Press, 2002.
Kelley, Robin D. G. *Hammer and Hoe: Alabama Communists During the Great Depression.* Chapel Hill: University of North Carolina Press, 1990.
Kennedy, David M. *Freedom from Fear: The American People in Depression and War, 1929–1945.* New York: Oxford University Press, 2005.
Key, V. O., Jr. *Southern Politics in State and Nation.* New York: Alfred A. Knopf, 1949; reprint, Knoxville: University of Tennessee Press, 1984.
Kirby, Jack Temple. *Rural Worlds Lost: The American South, 1920–1960.* Baton Rouge: Louisiana State University Press, 1987.
Kirkendall, Richard S. *Social Scientists and Farm Politics in the Age of Roosevelt.* Columbia: University of Missouri Press, 1966.
Knepper, Cathy D. *Greenbelt, Maryland: A Living Legacy of the New Deal.* Baltimore: Johns Hopkins University Press, 2001.
Kyriakoudes, Louis M. "'Lookin' for Better All the Time': Rural Migration and Urbanization in the South, 1900–1950." In *African American Life in the Rural South, 1900–1950,* edited by R. Douglas Hurt, 10–26. Columbia: University of Missouri Press, 2003.
LaMonte, Edward Shannon. *Politics and Welfare in Birmingham, 1900–1975.* Tuscaloosa: University of Alabama Press, 1995.
Larson, Olaf F., and Julie N. Zimmerman. *Sociology in Government: The Galpin-Taylor Years in the U.S. Department of Agriculture, 1919–1953.* University Park: Pennsylvania State University Press, 2003.
Layton, Stanford J. *To No Privileged Class: The Rationalization of Homesteading and Rural Life in the Early Twentieth-Century American West.* Salt Lake City: Charles Redd Center for Western Studies, Brigham Young University, 1988.
Leighninger, Robert D., Jr. *Long-Range Public Investment: The Forgotten Legacy of the New Deal.* Columbia: University of South Carolina Press, 2007.
Leuchtenburg, William E. *Franklin D. Roosevelt and the New Deal, 1932–1940.* New York: Harper and Row, 1963.
Maher, Neil M. *Nature's New Deal: The Civilian Conservation Corps and the Roots of the American Environmental Movement.* New York: Oxford University Press, 2008.
Maloney, C. J. *Back to the Land: Arthurdale, FDR's New Deal, and the Cost of Economic Planning.* Hoboken, NJ: John Wiley and Sons, 2011.
Manthorne, Jason. "The View from the Cotton: Reconsidering the Southern Tenant Farmers' Union." *Agricultural History* 84.1 (2010): 20–45.
May, Irvin M., Jr. *Marvin Jones: The Public Life of an Agrarian Advocate.* College Station: Texas A&M University Press, 1980.
McConnell, Grant. *The Decline of Agrarian Democracy.* Berkeley: University of California Press, 1953.
McDonald, Michael J., and John Muldowny. *TVA and the Dispossessed: The Resettlement of Population in the Norris Dam Area.* Knoxville: University of Tennessee Press, 1982.
Mertz, Paul E. *New Deal Policy and Southern Rural Poverty.* Baton Rouge: Louisiana State University Press, 1978.
Mora, Gilles, and Beverly W. Brannan. *FSA: The American Vision.* New York: Abrams, 2006.

Nelson, Lawrence J. *King Cotton's Advocate: Oscar G. Johnston and the New Deal.* Knoxville: University of Tennessee Press, 1999.
Park, Yoosun. "Facilitating Injustice: Tracing the Role of Social Workers in the World War II Internment of Japanese Americans." *Social Service Review* 82.3 (2008): 447–83.
Pasquill, Robert G., Jr. *Planting Hope on Worn-Out Land: The History of the Tuskegee Land Utilization Project, Macon County, Alabama, 1935–1959.* Montgomery, AL: NewSouth Books, 2008.
Phillips, Sarah T. *This Land, This Nation: Conservation, Rural America, and the New Deal.* Cambridge, UK: Cambridge University Press, 2007.
Porter, Jane M. "Experiment Stations in the South, 1877–1940." *Agricultural History* 53.1 (1979): 84–101.
Radford, Gail. *Modern Housing for America: Policy Struggles in the New Deal Era.* Chicago: University of Chicago Press, 1996.
Rasmussen, Chris. "'Never a Landlord for the Good of the Land': Farm Tenancy, Soil Conservation, and the New Deal in Iowa." *Agricultural History* 73.1 (1999): 70–95.
Rasmussen, Wayne D. "The New Deal Farm Programs: What They Were and Why They Survived." *American Journal of Agricultural Economics* 65.5 (1983): 1158–62.
Reid, Debra A. *Reaping a Greater Harvest: African Americans, the Extension Service, and Rural Reform in Jim Crow Texas.* College Station: Texas A&M University Press, 2007.
Robinson, Greg. *By Order of the President: FDR and the Internment of Japanese Americans.* Cambridge, MA: Harvard University Press, 2001.
——. *A Tragedy of Democracy: Japanese Confinement in North America.* New York: Columbia University Press, 2009.
Rodgers, Daniel T. *Atlantic Crossings: Social Politics in a Progressive Age.* Cambridge: Belknap Press/Harvard University Press, 1998.
Roll, Jarod. *Spirit of Rebellion: Labor and Religion in the New Cotton South.* Urbana: University of Illinois Press, 2010.
Romine, Scott. *The Narrative Forms of Southern Community.* Baton Rouge: Louisiana State University Press, 1999.
Rosengarten, Theodore. *All God's Dangers: The Life of Nate Shaw.* New York: Alfred A. Knopf, 1974.
Rutledge, Ray. "The Development of the Bankhead Farmsteads Community." M.A. thesis, Auburn University, 1951.
Salamon, Lester M. "The Time Dimension in Policy Evaluation: The Case of the New Deal Land-Reform Experiments." *Public Policy* 27.2 (1979): 129–83.
Saloutos, Theodore. "New Deal Agricultural Policy: An Evaluation." *Journal of American History* 61.2 (1974): 394–416.
——. *The American Farmer and the New Deal.* Ames: Iowa State University Press, 1982.
Schapsmeier, Edward L., and Frederick H. Schapsmeier. *Henry A. Wallace of Iowa: The Agrarian Years, 1910–1940.* Ames: Iowa State University Press, 1968.
——. *Prophet in Politics: Henry A. Wallace and the War Years, 1940–1965.* Ames: Iowa State University Press, 1970.
Schulman, Bruce J. *From Cotton Belt to Sunbelt: Federal Policy, Economic Development, and the Transformation of the South, 1938–1980.* Durham, NC: Duke University Press, 1994.
Schwartz, Bonnie Fox. *The Civil Works Administration, 1933–1934: The Business of Emergency Employment in the New Deal.* Princeton, NJ: Princeton University Press, 1984.
Scribner, Christopher MacGregor. *Renewing Birmingham: Federal Funding and the Promise of Change, 1929–1979.* Athens: University of Georgia Press, 2002.

Selznick, Philip. *TVA and the Grass Roots: A Study of Politics and Organization.* New York: Harper and Row, 1966.
Shindo, Charles J. *Dust Bowl Migrants in the American Imagination.* Lawrence: University Press of Kansas, 1997.
Shogan, Robert. *Backlash: The Killing of the New Deal.* Chicago: Ivan R. Dee, 2006.
Shover, John L. *First Majority—Last Minority: The Transforming of Rural Life in America.* De Kalb: Northern Illinois University Press, 1976.
Smith, Douglas L. *The New Deal in the Urban South.* Baton Rouge: Louisiana State University Press, 1988.
Smith, Fred C. "The Tupelo Homesteads: A New Deal Agrarian Experimentation." *Journal of Mississippi History* 68.2 (2006): 85–112.
Smith, Jason Scott. *Building New Deal Liberalism: The Political Economy of Public Works, 1933–1956.* Cambridge, UK: Cambridge University Press, 2006.
Smith, Page. *Democracy on Trial: The Japanese American Evacuation and Relocation in World War II.* New York: Simon and Schuster, 1995.
Sternsher, Bernard. *Rexford Tugwell and the New Deal.* New Brunswick, NJ: Rutgers University Press, 1964.
Sullivan, Patricia. *Days of Hope: Race and Democracy in the New Deal Era.* Chapel Hill: University of North Carolina Press, 1996.
Summers, Mary. "The New Deal Farm Programs: Looking for Reconstruction in American Agriculture." *Agricultural History* 74.2 (2000): 241–57.
———. "Putting Populism Back In: Rethinking Agricultural Politics and Policy." *Agricultural History* 70.2 (1996): 395–414.
Sutter, Paul S. "What Gullies Mean: Georgia's 'Little Grand Canyon' and Southern Environmental History." *Journal of Southern History* 76.3 (2010): 579–616.
Taylor, Brenda J. "The Farm Security Administration and Rural Families in the South: Home Economists, Nurses, and Farmers, 1933–1946." In *The New Deal and Beyond: Social Welfare in the South since 1930*, edited by Elna C. Green, 30–46. Athens: University of Georgia Press, 2003.
Tindall, George B. *The Emergence of the New South, 1913–1945.* Vol. 10 of *A History of the South*. Baton Rouge: Louisiana State University Press, 1967.
Volanto, Keith J. *Texas, Cotton, and the New Deal.* College Station: Texas A&M University Press, 2005.
Walker, Melissa. *All We Knew Was to Farm: Rural Women in the Upcountry South, 1919–1941.* Baltimore: Johns Hopkins University Press, 2000.
———. *Southern Farmers and Their Stories: Memory and Meaning in Oral History.* Lexington: University Press of Kentucky, 2006.
Warren, Sarah T., and Robert E. Zabawa. "From Company to Community: Agricultural Community Development in Macon County, Alabama, 1881 to the New Deal." *Agricultural History* 72.2 (1998): 459–86.
———."The Origins of the Tuskegee National Forest: Nineteenth- and Twentieth-Century Resettlement and Land Development Programs in the Black Belt Region of Alabama." *Agricultural History* 72.2 (1998): 487–508.
Watkins, T. H. *The Great Depression: America in the 1930s.* New York: Little, Brown, 1993.
———. *The Hungry Years: A Narrative History of the Great Depression in America.* New York: Henry Holt, 1999.
Wilkinson, Kenneth P. *The Community in Rural America.* New York: Greenwood Press, 1991.

Williams, Arthur R., and Karl F. Johnson. "Race, Social Welfare, and the Decline of Postwar Liberalism: A New or Old Key?" *Public Administration Review* 60.6 (2000): 560–72.

Wright, Gavin. *Old South, New South: Revolutions in the Southern Economy since the Civil War.* New York: Basic Books, 1986.

Wunderlich, Gene. *American Country Life: A Legacy.* Landham, MD: University Press of America, 2003.

Index

Page numbers in **boldface** refer to illustrations.

agents. *See* Extension Service; supervisors
Agricultural Adjustment Administration (AAA): community building, 43; disproportionate impact, ix-x, xxiii, 1, 6, 29, 65, 205, 217n35; and FSA, 77–81; ideology of, 10, 49–50, 217n30; land use, 25; operation of, 11–13, 216n25; and RA, 51, 54, 97; in World War II, 181, 237n51
Agricultural Adjustment Act of 1933, ix, 10, 12
Agricultural Adjustment Act of 1938, 12
Alabama Relief Administration, 5, 15–16, 21
Alexander, Will W.: and race, 167; in RA/FSA, xxiii, 53, 3, 61–62, 74–76, 85, 105, 157, 166; and subsistence homesteads, 44; and supervised credit, 139, 146, 153; and tenant purchase program, 113, 115–16, 232n97
American Farm Bureau Federation (AFBF): and BJFTA, 67; Extension Service ties, 10, 77, 81; opposition to FSA before World War II, 76–78, 82; opposition to FSA during World War II, xxiii, 90, 185–88, 190–94, 196, 200
Anderson, Clinton, 199
Appleby, Paul H., 63, 76, 80–81
Arthurdale (WV), 46, 83

back-to-the-land movement, xviii, 1–2, 30–33, 45–46, 78, 128
Baldwin, Calvin B. (C. B.), criticism of, xxiii, 76, 83–84, 112, 166, 189; as FSA administrator, 61, 75, 76, 80; and Franklin W. Hancock, 197; and Dillard P. Lasseter, 199; and Herbert W. Parisius, 193; and race, 85; in RA, 53, 60, 63, 227n21; resignation from FSA, 196; responds to attacks on FSA, 187–92, 195; and rural rehabilitation, 137, 139, 146; and World War II, 182–84

Bankhead (homesteads): clients, **39**, **143**, 167, **168**, 176; community center school, 34, **35**; construction of, 161, 163; cooperative association, 166, 171; origins of, 40, 159–60; sale to residents, 164
Bankhead, John H., Jr.: and BJFTA, 63–64, 66–68, 81, 127; and Birmingham-area subsistence homesteads, 111, 159, 161; disappointment with farm security programs, 42, 57, 166; and subsistence homesteads, 1, 29, 32–35, 37, 45; support for FSA, 189, 191; and race, xxv, 84
Bankhead-Jones Farm Tenant Act (BJFTA): and legitimacy of FSA, xxiii, 110, 197–98; and operations of FSA, 124, 126, 189; and origins of FSA, 63, 69, 71–72; passage of, 64–68, 80, 182
Bankhead, William, 234n97
Birmingham (AL): cooperative associations, 171–74; and Great Depression, ix, 158; subsistence homesteads, 30, 37–41, 44, 57, 90, 135, 157, 159–70; success of homesteads, 174–77. See also Bankhead (homesteads); Cahaba; Greenwood; Mt. Olive; Palmerdale
Bureau of Agricultural Economics: abolished, 200; and Economic Research Service, 201; and land use reform, 72, 228n31; and rural poverty, x, xx, 49; support for FSA, 80, 235n19; and tenant purchase program, 114; and war effort, 259n15
Byrd Committee, 188–90, 192, 260n32
Byrd, Harry, 83, 86, 188–89, 192, 196, 238–39nn75–76
Byrnes, James, 85

Cahaba: cooperative association, 166, 171, 173; origins of, 40, 159–61, 163–64; and race, 167; resident satisfaction, 176; subsistence farming, 168–69

INDEX

Cannon, Clarence, 191
Carr, William C., 187
Casa Grande project, 44, 148
Civilian Conservation Corps, 80, 180, 193, 230n49, 253n9, 259n8
Civil Works Administration (CWA), 13, 16, 44
Coffee County (AL), **xxviii**, 23–24, 94, 101–2, 144
committees (local): in AAA, 10–11, 92; and debt adjustment, 74, 103–4; and rural rehabilitation, 20, 89, 131–33, 135–36, 196, 198; and subsistence homesteads, 159; and tenant purchase program, 86, 111, 114–17, 122
Commodity Credit Corporation, 12, 236n29
Congress: and AFBF, 77, 185, 187; and appropriations for FSA, 75; and BJFTA, 63–64, 66–68, 72; and cooperatives, 99, 184; and effort to abolish FSA, 186–88, 191–93, 195–200; and FERA, 13, 28; and Great Depression, xix; and knowledge of FSA programs, 81–82, 188–89; and New Deal, 47; and RA, 3, 58, 61; and resettlement, 165, 190–91, 196; and rural rehabilitation, 137, 196, 198; and tenant purchase program, 110–13, 127, 196
Connally, Tom, 67
Construction Division (Section), 44, 55, 162–63
Cooley, Harold, 67, 195, 198–200, 234n91
cooperative associations: around Birmingham (AL), 157, 165–66, 171–74, 177; cooperative farming, 97–99; criticisms of, 111, 187, 189, 192; and Cumberland Homesteads, 45; and Farmers' Union, 78; and rural rehabilitation, xi, 52, 55, 91, 93, 106, 131, 145–46, 184, 196–97
Country Life Movement, 8, 30, 78
county agents. *See* Extension Service; supervisors
Cumberland Project. *See* Skyline Farms

Davis, Chester, 12, 194, 196
debt adjustment: in FCA, 2, 103; in FSA, 86, 131, 196, 200; in RA, 55, 93; and rural rehabilitation, xi, 17, 26, 74, 91, 105, 135, 145; Voluntary Debt Adjustment Committees, 103–104, 136
Department of the Interior, xx, xxiii, 1, 46, 56
Dirksen, Everett, 192
Division of Subsistence Homesteads: client selection, 38–39, 116; criticisms of, 33, 45–46; in Department of the Interior, xix, 1, 6–7; in FERA, 21–22; goals of, 1–2, 29–30, 38–40, 87, 92; and Great Depression, 159; and idealized rural life, 14; and land use reform, 25, 47; NIRA origins, 34–38, 40–44, 159, 161; and race, 167; and subsistence homesteads corporations, 36–37, 39, 41–42, 52, 159–61; support for, 31–33, 220n83, 223n27. *See also* Bankhead (homesteads); Cahaba; Greenwood; Mt. Olive; Palmerdale; Resettlement Division; subsistence homesteads
Dugger, Oscar M., 32–33, 159, 161, 167

Emergency Relief Appropriation Acts, 13, 50–52, 61
Extension Services: and opposition to RA/FSA, 58, 82, 186–87, 192, 194; and poor farmers, 8–9; and rural rehabilitation, 20, 25, 98–99, 130–33; in USDA, 10, 27, 65, 76–78

Farley, James A., 62
farm and home management plans: creation, 135–139–42, 144–45; criticisms of, 187; in FSA, 74, 196; in FERA, 1, 17–18, 23–24; and landlords, 104; problems with, 146–49; as rehabilitation, 55, 90, 93, 95, 106–7, 130, 153–54, 201; in RA, 94; in tenant purchase program, 111, 114, 121–22, 124–25, 127–28
Farm Credit Administration (FCA): debt adjustment, 2, 103; conflict with FSA, 79, 184; and FmHA, 200; proposals to merge with FSA, 186, 237n42; and RA, 53; in USDA, 236n29
Farm Security Administration (FSA): and AFBF, 77–78, 82, 90, 185–88, 190–94, 196, 200; Will W. Alexander as administrator of, 74–76; C. B. Baldwin appointed

administrator, 76; and BJFTA, 63, 71–72, 110; and cooperative farming, 97–99; criticisms of prior to World War II, 79–87, 98–99, 112, 129, 157; criticisms of during World War II, 166, 184–92, 194; and debt adjustment, 103–4; employee training, 132–34; and five hundred families, 105–7; and FmHA, 200–1; historical analysis of, xxii, 202, 204–5; and Japanese evacuation during World War II, 182; and land use reform, 72, 129; and loan collections, 150–53; and medical care program, 99–102, 197; opposition during World War II, 90, 179–180, 184–88, 191–200; organization of, xix-xx, xxiii, 2, 72, 228n26; political support for, 80–83, 189–92; and poll tax, 85, 189–90, 192; and RA, 3, 51, 53, 58, 66, 163–66; and race, 84–85, 96–97, 115–16, 120–21, 167; and resettlement, 73–74, 90, 129, 157–58, 167, 191, 197, 202; and rural poverty, x, xii-xiii, xviii-xxii, xxiv, 49, 87, 128, 202–6; and rural rehabilitation after 1943, 196, 198; and rural rehabilitation goals and results, 92–93, 153–55, 190, 201; and rural rehabilitation operations, 89–91, 93, 95–97, 136–37; and rural rehabilitation as solution to rural poverty, xi, 72–74, 129; and subdivision of land, 123; and subsistence homesteads, 34, 163–64; and supervised credit, 95, 134–36, 146, 149; and supervision, 74, 129–32, 137–50, 154–55; and tenant purchase program, 73–74, 109–118, 120–28, 192, 197, 200; and tenure improvement, 104–5, 197; in USDA, 76–77, 79–80, 179, 185, 192–97; and veterans, 199–200; and war effort, 90, 179–84; and women employees, 85, 130–31
Farmers Home Administration (FmHA), xiii, xx, xxiv, 90, 123, 166, 200–201
Farmers' Union, 10, 78, 196, 199
Federal Emergency Relief Administration (FERA): and Birmingham (AL) homesteads, 159, 162; and client selection, 20–21, 24, 26, 95–96, 138; and land use reform, 25; RA move, 27, 52–56, 93; and rural rehabilitation, xx, xxiii, 1, 5–8, 29, 32, 91–92, 107; and rural rehabilitation origins, 13–17, 78; and tenure improvement, 105
Federal Public Housing Authority (FPHA), 165–66
Federal Subsistence Homesteads Corporation, 37, 41
five hundred families program, 105–7
Flannagan, John W., 189
Forest Service, 25, 54, 77
Food Production Administration (FPA), 193–94
Frank, Jerome, 10, 12
Fulmer, Hampton, 67

Glass, Carter, 190
Glick, Philip M., 36, 161, 226n84, 232n72
Gray, Chester, 67
Gray, L. C., 25, 221n86, 228n31
Great Depression: and agriculture, xi-xii, xx, 22, 203; and Birmingham (AL), 158–60; and FSA employees, 132; memory of, ix-x, 60; and New Deal, xviii-xix, 6, 9, 26, 47; and rural poor, xvii, xxi, 1, 15, 22, 101; and subsistence homesteads, 29–30, 32–33, 176; and tenancy, xxv, xxvii, xxx, 19, 65 and tenant purchase program, 109–110
greenbelt projects, 57
Greenwood: construction of, **162**, 163, 169; cooperative associations, 166, 172–73; origins of, 40, 159–61; resident satisfaction, 175–77; sale to residents, 164; subsistence farming, 168–69

Hancock, Franklin W., 197–99
Hatch Act of 1887, 8
Hewes, Laurence, 52, 61, 73, 227n21, 235n22
Hiss, Alger, 12
Historical Section. *See* Information Division
Holt, Thad, 15
home management agents, and Extension Service, 8, 82; in FSA, 130–31, 134–35; in FERA, 18; and health, 101; and home management plans, 140–41; and loan applications, 136; problems, 148; and race, 84, 120; and tenant purchase program, 117–19, 123. *See also* supervisors

Hopkins, Harry, 14, 16–17, 45, 61, 263n84
Hudgens, Robert W.: and Coffee County (AL), 24, 94; and cooperative farming, 99; criticisms of, 188; and Extension Service, 77, 82; FSA appointments, 83, 196–97, 199; and rural rehabilitation, xxi, 20, 84–85, 139, 147, 153, 218n46; and subsistence homesteads, 42

Ickes, Harold, 37, 41–42, 45–46, 48
Information Division, x–xi, 60–61, 81–82

Jasper (AL), 24, 37, 40, 42, 57, 111, 159–60. *See also* Bankhead (homesteads)
Johnston, Oscar, 185, 194
Joint Committee on Reduction of Nonessential Federal Expenditures. *See* Byrd Committee
Jones, J. Marvin, 63–64, 66–68, 196–97

Kirkpatrick, Donald, 187

La Follette, Robert, 190
land use reform: and BJFTA, 68; in FSA, 72, 83, 129; in FERA, 25–26, 54; historical analysis of, xxii, 229n33; and race, 230n49; in RA, 51–52, 55–56, 236n40; and rural poverty, x, xiii, xviii, xix; and subsistence homesteads, 30–31
Land Utilization Division, 52, 55–56, 72, 228n31, 239n10. *See also* land use reform
Lasseter, Dillard P., 199–200
Lemke, William, 68
Long, Huey, 82–83
Lucas, Scott, 67

Maddox, James, 66
Management Division, 55, 162–63, 191
Maris, Paul V.: in RA, 53–54; and rural rehabilitation, 17–18, 205; and tenant purchase program, 76, 110–11, 123, 127
McKellar, Kenneth, 189
medical care program, 23, 74, 91, 99–103, 145, 183–84, 197
migrant farm labor, xiii, xviii, 14, 66, 91–92, 100, 182, 195
migratory labor camps, 56, 74, 99, 182, 186, 191, 194–95

Mitchell, H. L., xxvii, 190
Mitchell, John, 67
Morgan, E. S., 189–90
Moser, Guy L., 83
Mt. Olive: construction of, 161, 163; cooperative associations, 166, 172–74; origins of, 40, 159–60; sale to residents, 164; subsistence farming, 168–69

National Cotton Council, 185, 194, 200
National Industrial Recovery Act (NIRA), 34–35, 38, 43–44, 46
National Park Service, 25
National Public Housing Authority, 196
National Youth Administration, 80, 180, 195
New Deal: and agricultural policy, xx–xxii, 9–10, 12, 205; evolution of, 180, 184, 193, 196–97, 203; and FSA, 71–72, 74, 181–82; and Great Depression, 47–48, 87, 90; memory of, ix; and race, 85, 96; and rural poor, ix–xiii, xvii–xix, xxii, 1–3, 6, 13, 202; and RA, 49, 51; and rural rehabilitation, 5, 14, 25–27; and subsistence homesteads, 29, 33–36, 46, 157; and tenant purchase program, 110, 127–28; and USDA, 76–81, 130

O'Neal, Edward, 67, 185–88, 191–92

Pace, Stephen, 84
Palmerdale: construction of, 161–63; cooperative associations, 166, 172, 174; origins of, 38, 40, 159–60; and race, 167; resident satisfaction, ix-**x**, 157, 170, 174; sale to residents, 164–65; subsistence farming, 168–69, **175**, 176
Parisius, Herbert W., 193–94
parity (agricultural policy), 9–10, 81, 186, 192–93, 215n16
Peek, George N., 9–10, 12
Pepper, Claude, 189
Photographic Section. *See* Information Division
Pierce, Walter, 64
poll tax, xxx, 85, 189–90, 192, 214n50
Pope-Jones Water Facilities Act of 1937, 72
President's Committee on Farm Tenancy, xxvii, 62, 66, 68

Public Housing Administration (PHA), 166
Public Works Administration (PWA), 34, 230n48
Pynchon, Charles E., 42–43

race: and Extension Service, 9; and FSA, 84–85, 134, 188, 190; and liberalism, xxi; and RA, 59; and relief, 6, 15; and rural rehabilitation, 96–97; and subsistence homesteads, 24, 160, 165, 167, 170; and supervision, 149; and tenancy, xvii, xxv-xxviii, xxx, 12, 65, 92, 204; and tenant purchase program, 115–118, **119**, 120–21
Ramsay, Erskine, 39, 159
Reconstruction Finance Corporation, 12, 33, 110
resettlement: applications for, **135**, 136, 140; and cooperatives, 99, 171–73; criticisms of, 113, 149, 157–58, 166–67, 190–92, 198–99; decline of, 195–97; and Division of Subsistence Homesteads, 14; in FERA, 21–25; in FSA, xviii, 3, 72–73, 154, 164, 197; and greenbelt projects, 57; memory of, ix, xi, xxii; organization, xix, 45; origins of, xxiii, 26, 43, 97; in RA, 51–56, 105; and race, 85, 167; as solution to rural poverty, x, 2, 17, 63, 71–72, 74, 129; success of, xii, 90, 157, 174–77, 202; and supervision, 89–91, 111, 128, 143–44, 147; and TVA, 80–81. *See also* Division of Subsistence Homesteads; Resettlement Division; subsistence homesteads
Resettlement Administration (RA): and Will W. Alexander, 74–75; and BJFTA, 63–64, 124; and Coffee County (AL), 24; and cooperatives, 99; creation of, 50–52; criticism of, 49, 53, 57–62, 77, 81–83, 186, 226–27n79, 236n40; and debt adjustment, 103; and Robert W. Hudgens, 196; and land use reform, 51–52, 56, 72; and medical care, 99–100; organization of, xx, xxiii, 2–3, 53–55, 129, 227n13; politics of, 82; and race, 59, 83–84; and resettlement, 56–57; and rural rehabilitation, 18–19, 23, 55, 91–96, 110, 152–54, 201; and subsistence homesteads, 30, 34, 158, 161–63, 174; and supervision, 129, 132, 137–38; support for, 49, 174; and tenant purchase program, 128; and tenure improvement, 105; and transition to FSA, 71–72; and USDA, xxiii, 62–63, 74
Resettlement Division: and cooperatives, 173; in FSA, 72–74, 89, 164; in RA, 52–57, 93, 162; and subsistence living, 143–44; and World War II, 181–82, 191. *See also* Division of Subsistence Homesteads; resettlement; subsistence homesteads
Robinson, Joseph, 75, 82
Roosevelt, Eleanor, 46, 184
Roosevelt, Franklin D.: and agricultural policy, xvii, 6; and BJFTA, 64, 66; criticisms of, ix; and FERA, 13; and FSA, 78, 80, 90, 190, 192; and New Deal, xix, 47, 71, 87, 180; and RA, 48–51, 62; and rural life, xviii; and rural rehabilitation, 25; and subsistence homesteads, xvii, 26, 31–34, 36, 41, 45, 162, 176; and World War II, 90, 180–81, 193–95
Rural Electrification Administration, 192, 236n29
rural rehabilitation: applications for, 19–20, 136–39; and BJFTA, 68, 72; criticisms of, 59–60, 189; and debt-adjustment, 103–4; decline of, 196–98; definitions of, 7, 92–93; in FERA, 1–2, 5–6, 8, 48; and five hundred families, 105–7; in FSA, 73–74, 81, 196; goals, 92–93, 130, 153–54; historical analysis of, xi, xxii–xxiii; and Paul V. Maris, 110; and land use reform, 25–26, 56; and loan collections, 150–53; and medical care, xxviii, 99–101; organization of, xix-xx, 3, 227n13; origins of, 13–17; and poll tax, 190; popularity of, 20–21; problems, xii–xiii, xxi, 79, 146–47, 149–50; in RA, 52–53, 55; and race, 96, 149–50; and resettlement, 22–24, 29; as solution to rural poverty, x–xi, xviii, xx, 7, 27, 63, 87, 89–90, 107; success of, 71, 91, 139, 154–55, 201–2; and supervised credit, 18–19, 94–96; and supervisors, 89, 129–36, 140–42, 148; and tenant purchase program, 111, 113–14, 116, 118, 125–28; and tenure improvement, 104–5; and World War II, 107, 180, 182–83, 196, 199. *See also* Rural Rehabilitation Division

Rural Rehabilitation Corporations, 16–17, 23–24, 52–53, 152, 154
Rural Rehabilitation Division: creation of, 16; in FERA, xx, 1, 6–8, 14, 17–10, 45, 87, 91; in FSA, 72, 89, 91; in RA, 52–56, 93; and race, 116; and subsistence homesteads, 32; and World War II, 182. *See also* rural rehabilitation
Rural Resettlement Division, 54–55, 93. *See also* Rural Rehabilitation Division; Resettlement Division
Russell, Richard, 113, 157, 188, 191, 195

Shafer, Paul, 83
sharecropping. *See* tenancy
Skyline Farms, 21, 24
Smith, Ellison "Cotton Ed," 115–16
Smith-Lever Act of 1914, 8
Social Security Administration, 53, 102, 180
Soil Conservation and Domestic Allotment Act of 1936, 12
Soil Conservation Service (SCS), 77, 80–81, 235n19
Soil Erosion Service, 25, 54
Southern Tenant Farmers Union (STFU), xxvii, 12, 65–66, 92, 217n30
Stanfield, Mattie Cole, ix
State Emergency Relief Administrations, 13, 17, 60. *See also* Alabama Relief Administration; Texas Relief Administration
Suburban Resettlement Division, 54–57
subsistence homesteads: and John H. Bankhead Jr., 29, 166; cooperatives in, 171–74; criticisms of, 45–47, 192, 202; decline of, 128; and farming, 169–70; in FSA, 73, 78, 80; goals of, 31, 41; and Great Depression, 159; and idealized rural life, 14, 29–30; in New Deal, xix; in RA, 52–56, 162–63; resident satisfaction, 174–77; and Franklin D. Roosevelt, xvii, 31–32; sale to residents of, 164; and M. L. Wilson, 32, 36. *See also* Bankhead (homesteads); Cahaba; Division of Subsistence Homesteads; Greenwood; Mt. Olive; Palmerdale; resettlement
supervised credit: criticisms of, 67, 186, 199; effectiveness of, xii, 3, 26, 91, 153–55, 201; in FSA, 72–74; in FERA, 7, 16–18, 21; and five hundred families, 105–7; problems with, 146–48, 150; purpose of, 95, 129–30, 138–40; in RA, 55, 93–94; as solution to rural poverty, xi, xiii, xviii, xxii, 89–90, 179; and tenant purchase program, 109–10, 117–118, 121, 124–25, 127–28; in World War II, 180, 191, 196–98
supervisors: caseload, 74, 134–36; and clients, xi, xiii, **23**, 81, 90, 98, 122, 130, 145; and debt adjustment, 104; and Extension Service, 82, 131–32; and farm and home management plans, 139–44; in FSA, 73, 130; in FERA, 16, 18; and five hundred families, 105–7; and health and dental programs, 101–2; and loan applications, 94–95; and loan approval, 73, 89, 96, 137–39; and loan collections, 151–53; and poll tax, 190; problems, 146–50; and race, 84; in RA, 93; and tenant purchase program applications, 114–23, 125–28; training and support, 131–34

Tarver, Malcolm C., 84, 112, 116, 123, 186, 189–92, 199
Taylor, Carl C., 35, 54
tenancy: and AAA, ix, xxiii, 1, 11–13, 29, 51; and BJFTA, 63–68; and Extension Service, 9; and FSA, 77–79, 84, 146, 202; and land use reform, 25–26; perception of, 8, 65, 185; and RA, 59–60; and relief, 14–15, 47, 50; and rural rehabilitation, 5–6, 19–21, 55, 93–94, 97–99, 142; sharecropper demonstrations, 59, 65; in South, xvii, xviii, xxiv-xxx, 7, 203–205, 211n17, 212n27; and subdivision of land, 123; and subsistence homesteads, 34, 176; and tenant purchaser program, xi-xiii; and tenure improvement, 105
tenant purchase program: applications for, 113–122; client failures in, 126–27; and county committees, 68, 89, 111, 114–17, 122, 131; criticisms of, 86, 112, 187, 200, 205; in FSA, xi-xiii, xviii-xix, xxiii, 3, 71–74, 76, 154–55, 196–97, 202; and Paul V. Maris, 110–111; in New Deal, 127–28; perceptions

of, 82–83, 109, 113, 192; problems of, 146–48, **149**; and race, 84–85, 96, 116, 120–21; and rural rehabilitation, 91–92, 95, 107, 136; and subdivision of land, 123; and Tarver Amendment, 112; and World War II, 81

Tennessee Valley Authority (TVA), ix, 24, 36, 51, 80, 116, 230n49, 238n67

tenure improvement, xi, 91, 104–5 131, 135, 196–97, 200

Texas Relief Administration, 16, 22

Tolley, Howard, 183, 190, 231n65

Tugwell, Rexford G.: and agricultural price supports, 215n16; and Cahaba project, 40; criticisms of, xxiii, 57, 59–61, 76, 86, 166, 187, 189, 197; and land use reform, 25–26, 56; as RA administrator, 2–3, 45, 50–55, 58, 61, 74–75, 81; resignation from RA, xxiii, 62, 80, 235n22; and subsistence homesteads, 33

United States Department of Agriculture (USDA): and AAA, 6, 11; and BJFTA, 64–65, 67; and Extension Service, 8; and loan collections, 129, 151–53, 165; political divisions in, 3, 50–51, 76–82, 84, 129, 184–86, 191–97; and poor farmers, 6, 49, 78, 130; and RA/FSA organization, xx, xxiii, 3, 56, 58, 61–63, 69, 74, 99, 113, 163; and rural rehabilitation, 7, 89, 132, 134; and subsistence homesteads, 45, 165; and World War II, 181, 183

Voluntary Debt Adjustment Committees, 103–104, 136. *See also* debt adjustment

Wallace, Henry: and Paul Appleby, 76; criticism of, 83, 149; and FSA, 72, 80, 185, 239n76, 250n51; and RA move to USDA, 45, 62; and supervised credit, 94; and tenancy, 66, 71–72, 78; and Claude Wickard, 193; during World War II, 181, 199

Walter, Francis E., 76

War Food Administration (WFA), 194–96

War Production Board, 126

Westbrook, Lawrence, 8, 16–17, 22

Wickard, Claude, 80–81, 181, 183, 193–94, 196, 237n51

Wilson, M. L.: and BJFTA, 64; as director of Division of Subsistence Homesteads, 36–38, 41–43, 45–46, 92, 159; and FSA, 80; and subsistence homesteads, 31–32, 250n51, 222n12

Works Progress Administration (WPA), 6, 44, 50, 188

World War II: and antireform sentiment, xi, xiii, 180–81, 198, 201–2; and farmers leaving land, xx, 20, 204–5; and FSA, xxiii, 3, 63, 90, 179, 184–85, 199–200; and liberalism, xxi, 87, 203; and rural rehabilitation, 93, 102, 104, 107, 134, 138–39; and subsistence homesteads, 157, 165–66, 170–71, 174–76; and tenant purchase program, 125–26; and USDA, 80, 81

The Farm Security Administration and Rural Rehabilitation in the South was designed and typset by Kelly Gray in 9.75/13 Minion Pro with Clarendon LT Std display.

Minion is an Adobe Originals typeface designed by Robert Slimbach. It was inspired by classical, old style typefaces of the late Renaissance, a period of elegant, beautiful, and highly readable type designs. Minion Pro exhibits the aesthetic and functional qualities that make text type highly readable, yet is also suitable for display settings. Clarendon LT is a modern era iteration of a 19th century publishing classic slab serif by Hermann Eidenbenz.